Linear Control Theory

Linear Control Theory
The State Space Approach

Frederick Walker Fairman
Queen's University,
Kingston, Ontario, Canada

John Wiley & Sons
Chichester • New York • Weinheim • Brisbane • Singapore • Toronto

Baffins Lane, Chichester,
West Sussex PO19 1UD, England

National 01243 779777
International (+44) 1243 779777

e-mail (for orders and customer service enquiries): cs-books@wiley.co.uk

Visit our Home Page on http://www.wiley.co.uk
or
http://www.wiley.com

Other Wiley Editorial Offices

John Wiley & Sons, Inc., 605 Third Avenue,
New York, NY 10158-0012, USA

Wiley-VCH Verlag GmbH, Pappelallee 3,
D-69469 Weinheim, Germany

Jacaranda Wiley Ltd, 33 Park Road, Milton,
Queensland 4064, Australia

John Wiley & Sons (Asia) Pte Ltd, 2 Clementi Loop #02-01,
Jin Xing Distripark, Singapore 129809

John Wiley & Sons (Canada) Ltd, 22 Worcester Road,
Rexdale, Ontario M9W 1L1, Canada

Library of Congress Cataloguing-in-Publication Data

Fairman, Frederick Walker.
Linear control theory : The state space approach / Frederick Walker Fairman.
p. cm.
Includes bibliographical references and index.
ISBN 0-471-97489-7 (cased : alk. paper)
1. Linear systems. 2. Control theory. I. Title.
QA402.3.F3 1998
629.8'312–dc21 97–41830
CIP

British Library Cataloguing in Publication Data

A catalogue record for this book is available from the British Library

ISBN 0 471 97489 7

To Nancy for her untiring support

Contents

Preface

This book was written with the intent of providing students and practicing control engineers with the basic background in control theory needed to use control system design software more productively. The book begins with a detailed treatment of those aspects of the state space analysis of linear systems that are needed in the remainder of the text. The book is organized in the following manner:

- The first four chapters develop linear system theory including model reduction via balanced realization.
- Chapters 5 and 6 deal with classical optimal control theory.
- The final four chapters are devoted to the development of suboptimal H_∞ control theory.

The mathematical ideas required in the development are introduced as they are needed using a "just-in-time" approach. This is done to motivate the reader to venture beyond the usual topics appearing in introductory undergraduate books on "automatic control", to more advanced topics which have so far been restricted to postgraduate level books having the terms "mathematical control theory" and "robust control" in their titles.

This book can be used as the text for either a one or two-semester course at the final year undergraduate level or as a one semester course at the beginning postgraduate level. Students are assumed to have taken a basic course in either "signals and systems" or "automatic control". Although not assumed, an introductory knowledge of the state space analysis of systems together with a good understanding of linear algebra would benefit the reader's progress in acquiring the ideas presented in this book.

Ideas presented in this book which provide the reader with a slightly different view of control and system theory than would be obtained by reading other textbooks are as follows:

- The so-called PBH test which is usually presented as a test for controllability and/or observability is used throughout the present book to characterize eigenvalues in control problems involving eigenvalue assignment by state feedback and/or output injection.
- An easy to understand matrix variational technique is used to simplify the development of the design equations for the time invariant, steady state, quadratic and LQG controllers.
- The relatively simple idea of the L_2 gain is used as a basis for the development of the H_∞ controller.

Concerning the style of the book, the beginning section, "Introduction", for each chapter contains motivational material and an overview of the ideas to be introduced in subsequent sections in that chapter. Each chapter finishes with a section called "Notes and References", which indicates a selection of other sources for the material treated in the chapter, as well as an indication of recent advances with references.

I would like to thank the following colleagues in the Department of Electrical and Computer Engineering at Queen's University for proof-reading parts of the manuscript: Norm Beaulieu, Steve Blostein, Mingyu Liu, Dan Secrieu and Chris Zarowski. Special thanks go to my former research student Lacra Pavel for proof-reading and advice on Chapters 6, 9 and 10 as well as to Jamie Mingo in the Department of Mathematics and Statistics at Queen's University for his help with some of the ideas in Chapter 7. Thanks go also to Patty Jordan for doing the figures. Finally, I wish to acknowledge the contribution to this book made by my having supervised the research of former research students, especially Manu Missaghie, Lacra Pavel and Johannes Sveinsson.

The author would appreciate receiving any corrections, comments, or suggestions for future editions should readers wish to do so. This could be done either by post or e-mail: < fairmanf@post.queensu.ca >.

1

Introduction to State Space

1.1 Introduction

A well known behavioral phenomenon of dynamic systems is the appearance of an output in the absence of an input. This effect is explained once it is recognized that the internal storage of energy in the system at the beginning of the response time will produce an output. This kind of behavior is referred to as the system's zero-input response. Alternatively, the production of an output caused solely by an input when there is no energy storage at the start of the response time is referred to as the zero-state response. These two classes of response are responsible for all possible outputs and in the case of linear systems we can always decompose any output into the sum of an output drawn from each of these classes. In this chapter we will use the example of a second order system together with both the zero-input response and the zero-state response to introduce the reader to the use of the state space in modeling the behavior of linear dynamic systems.

1.2 Review of Second Order Systems

A commonly encountered physical process which we will use in the next two sections to introduce the state modeling of linear dynamic systems is the electric circuit formed by connecting an ideal constant resistor R_e, inductor L_e, and capacitor C_e in series in a closed loop as shown in Figure 1.1

Suppose the switch is closed at $t = t_s < 0$ so that there is a current flow $i(t)$, $t \geq 0$, and a voltage across the capacitor $y(t)$, $t \geq 0$. Then applying Kirchhoff's voltage law yields

$$R_e i(t) + L_e \frac{\mathrm{d}i(t)}{\mathrm{d}t} + y(t) = 0$$

where the current in the circuit depends on the capacitor voltage as

$$i(t) = C_e \frac{\mathrm{d}y(t)}{\mathrm{d}t}$$

Combining these equations gives a second order differential equation in the capacitor voltage, $y(t)$,

$$\frac{\mathrm{d}^2 y(t)}{\mathrm{d}t^2} + a_1 \frac{\mathrm{d}y(t)}{\mathrm{d}t} + a_2 y(t) = 0 \tag{1.1}$$

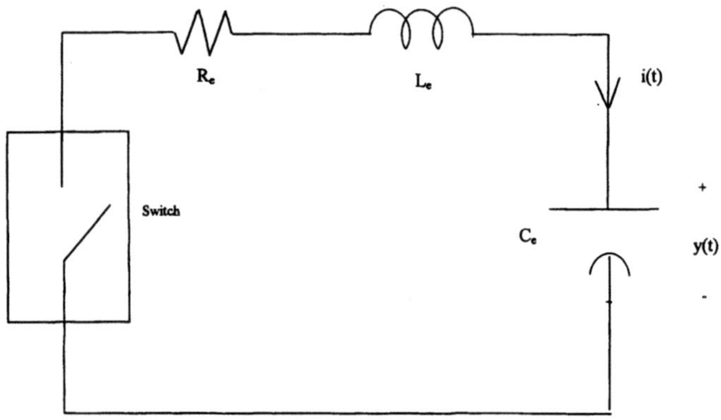

Figure 1.1 Electric circuit with charged capacitor. Switch closed prior to $t = 0$

where

$$a_1 = \frac{R_e}{L_e} \qquad a_2 = \frac{1}{L_e C_e}$$

and we refer to the capacitor voltage as the system's output.

1.2.1 Patterns of behavior

The differential equation (1.1) is said to govern the evolution of the output, $y(t)$, since it acts as a constraint relating $y(t), \frac{dy(t)}{dt}$, and $\frac{d^2y(t)}{dt^2}$ to each other at each instant of time. We will see now that once the initial conditions, i.e., the values of initial output, $y(0)$, and initial derivative of the output, $\dot{y}(0)$, are specified, the differential equation, (1.1), completely determines the output, $y(t)$, for all positive time $t \in (0, \infty)$. We obtain $y(t)$ as follows.

Suppose we have $y(t)$ such that (1.1) is satisfied. Then denoting the derivatives of $y(t)$ as

$$\frac{dy(t)}{dt} = g(t) \qquad \frac{d^2y(t)}{dt^2} = h(t)$$

we see that equation (1.1) becomes

$$h(t) + a_1 g(t) + a_2 y(t) = 0 \tag{1.2}$$

Now the only way this equation can hold for all $t > 0$ is for $h(t)$ and $g(t)$ to be scalar multiples of $y(t)$ where

$$g(t) = sy(t) \qquad h(t) = s^2 y(t) \tag{1.3}$$

Otherwise equation (1.2) can only be satisfied at specific instants of time. Therefore with this assumption assumption (1.2) becomes

$$p(s)y(t) = 0 \tag{1.4}$$

where $p(s)$ is the second degree polynomial

$$p(s) = s^2 + a_1 s + a_2$$

Finally, equation (1.4) holds for all time, when $y(t)$ is not zero for all time, i.e., the trivial solution, if and only if s is any one of the roots, $\{\lambda_i : i = 1, 2\}$ of $p(s)$,

$$p(s) = (s - \lambda_1)(s - \lambda_2) \qquad \lambda_{1,2} = -\frac{a_1}{2} \pm \sqrt{\left(\frac{a_1}{2}\right)^2 - a_2} \tag{1.5}$$

Returning to the requirement that $y(t)$ and its derivatives must be constant scalar multiples of each other, equation (1.3), the function that has this property is the exponential function. This important function is denoted as e^{st} and has series expansion

$$e^{st} = \sum_{i=0}^{\infty} \frac{(st)^i}{i!} \tag{1.6}$$

where $i!$, (factorial i), is the product

$$\begin{aligned} i! &= (i)(i-1)(i-2)\cdots(1) && i > 0 \\ &= 1 && i = 0 \end{aligned}$$

Notice that a bit of algebra shows us that the derivative of e^{st}, equation (1.6), has the desired property of being an eigenfunction for differentiation,

$$\frac{\mathrm{d}e^{st}}{\mathrm{d}t} = se^{st}$$

Now we see from the foregoing that e^{st} satisfies equation (1.1) when $s = \lambda_1$ or λ_2. Therefore any linear combination of $e^{\lambda_1 t}$ and $e^{\lambda_2 t}$ satisfies equation (1.1) so that the output $y(t)$ is given in general as

$$y(t) = k_1 e^{\lambda_1 t} + k_2 e^{\lambda_2 t} \qquad \lambda_1 \neq \lambda_2 \tag{1.7}$$

where the k_is are constant scalars chosen so that $y(t)$ satisfies the initial conditions. We can be do this by solving the equations which result from setting the given values for the initial conditions, $y(0)$ and $\dot{y}(0)$, equal to their values determined from equation (1.7), i.e., by solving

$$\begin{bmatrix} y(0) \\ \dot{y}(0) \end{bmatrix} = \begin{bmatrix} 1 & 1 \\ \lambda_1 & \lambda_2 \end{bmatrix} \begin{bmatrix} k_1 \\ k_2 \end{bmatrix} \tag{1.8}$$

for k_1, k_2. Notice that we can do this only if $\lambda_1 \neq \lambda_2$. In order to proceed when $\lambda_1 = \lambda_2$ we replace equation (1.7) with

$$y(t) = (k_3 t + k_4) e^{\lambda_1 t} \qquad \lambda_1 = \lambda_2 \tag{1.9}$$

and determine the k_is from the appropriate equations to ensure that the initial conditions are satisfied.

Returning to the behavior of the physical process that is under analysis, notice that since R_e, L_e, and C_e are real, the $a_i s$ are real. As a consequence the roots λ_i of $p(s)$, equation (1.5), are both real or are both complex. Moreover when these roots are complex they are conjugates of each other, i.e., $\lambda_1 = \lambda_2^*$.

More generally, if all the coefficients of a polynomial of any degree are real, each complex root must be matched by another root which is its complex conjugate. This property is important in the context of the behavior of linear physical processes since the parameters of these processes, e.g., mass, heat conductivity, electric capacitance, are always real so that the coefficients of $p(s)$ are always real.

Now a plot of the output, $y(t)$, versus time, t, reveals that there are two basic patterns for the behavior of the output depending on whether the $\lambda_i s$ are real or are complex conjugate pairs.

If the $\lambda_i s$ are real, we see from equation (1.8) that the $k_i s$ are also real and the output $y(t) : t \in (0, \infty)$ is given as equation (1.7) or (1.9). In this case we see that the output voltage $y(t)$ exhibits at most one maximum and decays *without oscillation* to the time axis as t tends to infinity. Notice from equation (1.5) that the $\lambda_i s$ are real provided the parameters R_e, L_e, C_e have values such that $(\frac{a_1}{2})^2 \geq a_2$.

Alternatively, if the $\lambda_i s$ are complex, i.e., if $(\frac{a_1}{2})^2 < a_2$, then we see from (1.5) that $\lambda_1 = \lambda_2^*$ and from (1.8) that $k_1 = k_2^*$. Thus $k_1 e^{\lambda_1 t}$ and $k_2 e^{\lambda_2 t}$ are complex conjugates of each other and their sum which gives $y(t)$, equation (1.7), is real. Incorporating these conjugate relations for the $\lambda_i s$ and the $k_i s$ in equation (1.7) allows us to write the output as a *damped oscillation*

$$y(t) = 2|k_1| e^{Re[\lambda_1]t} \cos(\mathrm{Im}[\lambda_1]\mathrm{t} + \theta) \tag{1.10}$$

where

$$k_1 = \mathrm{Re}[k_1] + j\mathrm{Im}[k_1]$$

$$\theta = \tan^{-1}\left(\frac{\mathrm{Im}[k_1]}{\mathrm{Re}[k_1]}\right)$$

Thus we see from (1.10) that the output voltage across the capacitor, $y(t)$, swings back and forth from its initial value to ever smaller values of alternating polarity. This behavior is analogous to the behavior of the position of a free swinging pendulum. The capacitor voltage (pendulum position) eventually goes to zero because of the loss of heat energy from the system resulting from the presence of R_e (friction). In this analogy, voltage and current in the electric circuit are analogous to position and velocity respectively in the mechanical process. The inductance L_e is analogous to mass since the inductance resists changes in the current through itself whereas the inertial effect of mass causes the mass to resist change in its velocity.

In addition, notice from equation (1.10) that the frequency of the oscillation, $\mathrm{Im}[\lambda_1]$, as well as the time constant associated with the decay in the amplitude of the oscillation, $(\mathrm{Re}[\lambda_1])^{-1}$, are each independent of the initial conditions and depend on the system parameters, R_e, L_e, C_e only, i.e., on a_1, a_2 only.

The previous discussion leads to the following characterization of the zero-input response of dynamic processes whose behavior can be modeled by second order differential equations with constant coefficients.

(i) The zero-input response, $y(t) : t > 0$, depends on the set of signals $\{e^{\lambda_i t} : i = 1, 2\}$ referred to as modes of the system where the constants λ_i, (system eigenvalues), are roots of the polynomial $p(s)$, (characteristic polynomial).

(ii) The steady state zero-input response is zero, i.e., $\lim_{t\to\infty} y(t) = 0$, for any initial conditions if and only if all the $\lambda_i s$ are negative or have negative real part, i.e., $\text{Re}[\lambda_i] < 0, \quad i = 1, 2$. In this situation we say that the system is stable.

(iii) We have $\text{Re}[\lambda_i] < 0, \quad i = 1, 2$, if and only if $a_1 > 0$ and $a_2 > 0$. More generally, the condition $a_i > 0, \quad i = 1, 2, \cdots n$ for systems whose behavior is governed by differential equations in the order of $n > 2$, is necessary but not sufficient for the system to be stable, i.e., is necessary but not sufficient for all $\lambda_i s$ to have negative real part

1.2.2 The phase plane

We have just seen that, when there is no input, a second order system having specified $a_i s$ has output, $y(t)$, which is specified completely by the initial conditions, $y(0)$ and $\dot{y}(0)$. This important observation suggests that the same information concerning the behavior of the system is contained in either (a) a plot of $y(t)$ versus t or (b) a plot of $y(t)$ versus $\dot{y}(t)$.

Thus if we make a plot of $y(t)$ versus $\dot{y}(t)$, the point representing $y(t), \dot{y}(t)$ in the $y(t)$ versus $\dot{y}(t)$ plane traces out a curve or trajectory with increasing time.

The two dimensional space in which this trajectory exists is referred to as the state space and the two-element vector consisting of $y(t)$ and $\dot{y}(t)$ is referred to as the state, denoted as $x(t)$ where

$$x(t) = \begin{bmatrix} y(t) \\ \dot{y}(t) \end{bmatrix}$$

This approach to visualizing the behavior of a dynamic process was used by mathematicians at the end of the last century to investigate the solutions for second order nonlinear differential equations, i.e., equations of the form (1.1) but with the $a_i s$ functions of $y(t)$ and/or $\dot{y}(t)$. The term phase plane plot was used to refer to the state trajectory in this case. Since, in general, the dimension of the state space equals the order of the differential equation which governs the output behavior of the process, the state space cannot be displayed for systems of order greater than two. Even so, the mathematical idea of the state space has become of great practical and theoretical importance in the field of control engineering.

Referring to the previous section, we see that the state trajectory for a dynamic process whose behavior can be modeled by a second order differential equation with constant coefficients, can exhibit any one of the following four fundamental shapes.

(i) If the $\lambda_i s$ are complex and $\text{Re}[\lambda_i] < 0$ the system is stable and the state trajectory spirals inwards towards the origin.

(ii) If the $\lambda_i s$ are complex and $\text{Re}[\lambda_i] > 0$ the system is unstable and the state trajectory spirals outwards away from the origin.

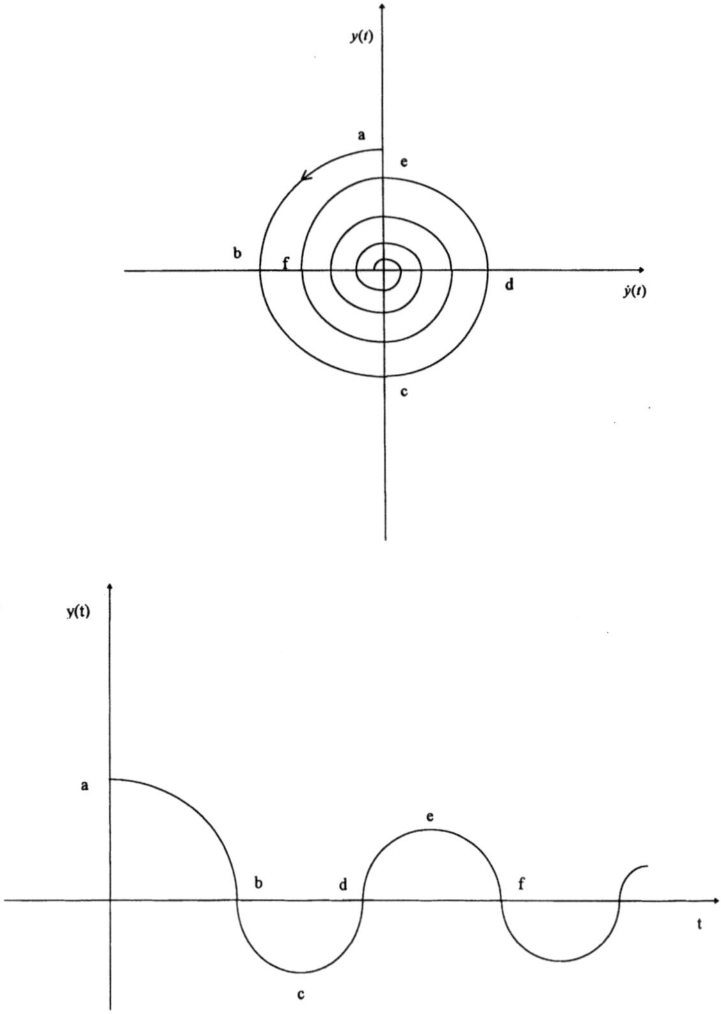

Figure 1.2 Plot of $y(t)$ vs. t and $y(t)$ vs. $\dot{y}(t)$ when λ_is complex

(iii) If the $\lambda_i s$ are real and both $\lambda_i s$ are negative the system is stable and the state trajectory moves towards the origin in an arc.

(iv) If the $\lambda_i s$ are real and one or both are positive the system is unstable and the state trajectory moves away from the origin in an arc.

Notice that state trajectories (ii) and (iv) do not occur in the present example of an electric circuit. This results from the fact that the parameters R_e, L_e, C_e are positive. Thus the coefficients, $a_i : i = 1, 2$ of the characteristic polynomial, equation (1.4) are positive so that the λ_is are negative or have negative real parts. This implies that we are dealing with a stable dynamic process, i.e., state trajectories tend to the origin for all initial states.

So far in this chapter we have used an electric circuit as an example of a system. We used the character of the behavior of this system in response to initial conditions to introduce the concept of the state of a system. In the next section this concept is made more specific by introducing the mathematical characterization of a system referred to as a state model.

1.3 Introduction to State Space Modeling

We saw in the previous section that once a second order system is specified, i.e., once the a_is are given numerical values, the zero-input response is determined completely from the system's initial conditions, $y(0), \dot{y}(0)$. In addition, we noted that the second derivative of the output is determined at each instant from $y(t)$ and $\dot{y}(t)$ through the constraint (1.1). These facts suggest that it should be possible to obtain the zero-input response by solving two first order differential equations involving two signals, $x_1(t), x_2(t)$, which are related uniquely to $y(t), \dot{y}(t)$. One straightforward way of doing this is to identify $\dot{y}(t)$ with $x_1(t)$and $y(t)$ with $x_2(t)$, i.e.,

$$\dot{y}(t) = x_1(t) \tag{1.11}$$

$$y(t) = x_2(t) \tag{1.12}$$

An immediate consequence of this identification is that at every instant the derivative of $x_2(t)$ equals $x_1(t)$

$$\dot{x}_2(t) = x_1(t) \tag{1.13}$$

Moreover, rewriting the second order differential equation, (1.1), as

$$\frac{\mathrm{d}}{\mathrm{d}t}(\dot{y}(t)) = -a_1\dot{y}(t) - a_2 y(t)$$

and using equations (1.11–1.13) gives us the differential equation for $x_1(t)$ as

$$\dot{x}_1(t) = -a_1 x_1(t) - a_2 x_2(t) \tag{1.14}$$

Thus we see from equation (1.13) and (1.14) that the derivative of each of the x_is is a (linear) function of the x_is. This fact is expressed in matrix notation as

$$\dot{x}(t) = Ax(t) \tag{1.15}$$

where

$$A = \begin{bmatrix} -a_1 & -a_2 \\ 1 & 0 \end{bmatrix} \qquad x(t) = \begin{bmatrix} x_1(t) \\ x_2(t) \end{bmatrix}$$

with the vector $x(t)$ being referred to as the state, and the square matrix A being referred to as the system matrix. In addition we see from equation (1.12) that

$$y(t) = Cx(t) \tag{1.16}$$

where

$$C = [0 \quad 1]$$

with C being a row vector referred to as the output matrix.

In summary the second order differential equation (1.1) is equivalent to the vector differential equation (1.15) and the output equation (1.16). These equations, (1.15, 1.16) constitute a state model for the second order system in the absence of an input. Alternatively, the state model can be represented by a block diagram involving the interconnection of blocks which operate as summers, integrators, and scalar multipliers on the components of the state. The Laplace operator $1/s$ is used to indicate integration.

More generally, we can use the foregoing procedure to obtain a state model for the zero-input response of higher order dynamic processes as follows.

Suppose the zero-input response of an nth order process is governed by

$$y^{(n)}(t) + a_1 y^{(n-1)}(t) + a_2 y^{(n-2)}(t) \cdots + a_n y(t) = 0 \tag{1.17}$$

where

$$y^{(i)}(t) = \frac{\mathrm{d}^i y(t)}{\mathrm{d}t^i}$$

Then we proceed as in the second order case to identify components of the state with derivatives of the output as

$$\begin{aligned} x_1(t) &= y^{(n-1)}(t) \\ x_2(t) &= y^{(n-2)}(t) \\ &\vdots \\ x_n(t) &= y(t) \end{aligned} \tag{1.18}$$

Thus using (1.17, 1.18) we obtain a vector differential equation (1.15) having a system

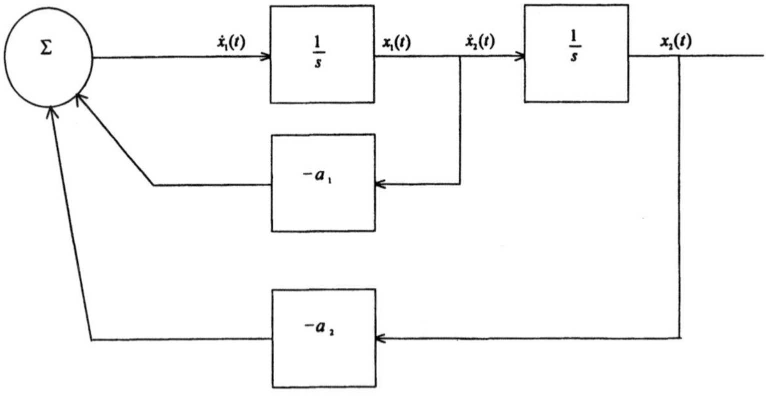

Figure 1.3 Block diagram representation of the state model

matrix A given as

$$A = \begin{bmatrix} -a_1 & -a_2 & -a_3 & \cdots & -a_n \\ 1 & 0 & 0 & \cdots & 0 \\ 0 & 1 & 0 & \cdots & 0 \\ \vdots & \vdots & \vdots & \vdots & \vdots \\ 0 & 0 & 0 & 1 & 0 \end{bmatrix} \tag{1.19}$$

and output equation, (1.16), having an output matrix C given as

$$C = [0 \quad \cdots \quad 0 \quad 1]$$

The pattern of zeros and ones exhibited in A, (1.19), is of particular importance here. Notice that the coefficients of the characteristic polynomial

$$p(s) = s^n + a_1 s^{n-1} + a_2 s^{n-2} + \cdots + a_{n-1}s + a_n$$

appear as the negative of the entries in the first row of A. Matrices exhibiting this pattern are referred to as companion matrices. We will see shortly that given A in any form, the characteristic polynomial is related to A as the matrix determinant

$$p(s) = \det[sI - A]$$

This fact is readily seen to be true in the special case when A is in companion form.

1.4 Solving the State Differential Equation

Recall that the solution to a scalar differential equation, e.g., (1.1), involves the scalar exponential function, $e^{\lambda t}$. In this section we will show that the solution to the state differential equation, (1.15), involves a square matrix, e^{At}, which is referred to as the matrix exponential.

1.4.1 The matrix exponential

Suppose we are given the initial state $x(0)$ and the system matrix A, either constant or time varying. Then we obtain a solution to the state differential equation, (1.15), by finding $\phi(t)$, the square matrix of scalar functions of time, such that

$$x(t) = \phi(t)x(0) \tag{1.20}$$

where $\phi(t)$ is referred to as the transition matrix.

Since the state at each instant of time must satisfy the state differential equation, (1.15), the transition matrix is a matrix function of the system matrix A. In this book A is constant. In this case the dependency of $\phi(t)$ on A is captured by the notation

$$\phi(t) = e^{At} \tag{1.21}$$

where the square matrix e^{At} is referred to as the "matrix exponential of At" since it can be expressed as an infinite series reminiscent of the infinite series for the exponential of a scalar, (1.6), i.e.,

$$e^{At} = I + At + \frac{A^2t^2}{2!} + \frac{A^3t^3}{3!} + \cdots = \sum_{i=0}^{\infty} \frac{A^i t^i}{i!} \tag{1.22}$$

In order to show that the transition matrix given by (1.22) solves the state differential equation, (1.15), we differentiate the foregoing series expansion for the matrix exponential of At to obtain

$$\begin{aligned} \frac{\mathrm{d}e^{At}}{\mathrm{d}t} &= A + \frac{2A^2t}{2!} + \frac{3A^3t^2}{3!} Ae^{At} + \frac{4A^4t^3}{4!} + \cdots \\ &= Ae^{At} \end{aligned}$$

Then using this relation to differentiate the assumed solution

$$x(t) = e^{At}x(0) \tag{1.23}$$

yields

$$\dot{x}(t) = Ae^{At}x(0) = Ax(t)$$

and we see that (1.23) solves the state differential equation, (1.15).

1.4.2 Calculating the matrix exponential

There are many ways of determining e^{At} given A. Some of these approaches are suitable for hand calculation and others are intended for use with a digital computer. An approach of the first kind results from using Laplace transforms to solve the state differential equation. We develop this approach as follows.

We begin by taking the Laplace transform of (1.15) to obtain

$$sX(s) - x(0) = AX(s) \tag{1.24}$$

where A is 2×2 we have

$$X(s) = \begin{bmatrix} X_1(s) \\ X_2(s) \end{bmatrix} \qquad X_i(s) = \int_0^{\infty} x_i(t)e^{-st}\mathrm{d}t$$

Then rewriting (1.24) as

$$(sI - A)X(s) = x(0) \tag{1.25}$$

we see that provided s is such that $(sI - A)$ is invertible, we can solve (1.25) for $X(s)$ as

$$X(s) = (sI - A)^{-1}x(0) \tag{1.26}$$

Now $(sI - A)^{-1}$ can be expressed using Crammer's rule as

$$(sI - A)^{-1} = \frac{\mathrm{adj}[sI - A]}{\det[sI - A]}$$

where when A is an $n \times n$ matrix, the adjugate matrix, $\mathrm{adj}[sI - A]$, is an $n \times n$ matrix of polynomials of degree less than n and $\det[sI - A]$ is a polynomial of degree n.

Finally, taking the inverse Laplace transform of (1.26) yields

$$x(t) = \mathcal{L}^{-1}\left[(sI - A)^{-1}\right]x(0)$$

and we see, by comparing this equation with (1.23), that

$$e^{At} = \mathcal{L}^{-1}\left[(sI - A)^{-1}\right]$$

Now in the case where A is the 2×2 matrix given by (1.15), we have

$$(sI - A)^{-1} = \frac{\mathrm{adj}[sI - A]}{\det[sI - A]} = \begin{bmatrix} s + a_1 & a_2 \\ -1 & s \end{bmatrix}^{-1} \tag{1.27}$$

where

$$\det[sI - A] = s^2 + a_1 s + a_2 = (s - \lambda_1)(s - \lambda_2)$$

$$\mathrm{adj}[sI - A] = \begin{bmatrix} s & -a_2 \\ 1 & s + a_1 \end{bmatrix}$$

Notice from the previous section that $\det[sI - A] = p(s)$, (1.4), is the characteristic polynomial. In general any n by n system matrix A has a characteristic polynomial with roots $\{\lambda_i : i = 1, 2 \cdots n\}$ which are referred to as the eigenvalues of A. The eigenvalues of the system matrix A play an important role in determining a system's behavior.

Returning to the problem of determining the transition matrix for A, (1.15), we apply partial fraction expansion to the expression for $(sI - A)^{-1}$, (1.27), assuming $\det[sI - A]$ has distinct roots, i.e., $\lambda_1 \neq \lambda_2$, to obtain

$$\begin{bmatrix} s + a_1 & a_2 \\ -1 & s \end{bmatrix}^{-1} = \frac{K_1}{s - \lambda_1} + \frac{K_2}{s - \lambda_2} \tag{1.28}$$

where

$$K_1 = \lim_{s \to \lambda_1} \left[(s - \lambda_1) \frac{\mathrm{adj}[sI - A]}{\det[sI - A]} \right] = \frac{\begin{bmatrix} \lambda_1 & -\lambda_1\lambda_2 \\ 1 & -\lambda_2 \end{bmatrix}}{\lambda_1 - \lambda_2}$$

$$K_2 = \lim_{s \to \lambda_2} \left[(s - \lambda_2) \frac{\mathrm{adj}[sI - A]}{\det[sI - A]} \right] = \frac{\begin{bmatrix} \lambda_2 & -\lambda_1\lambda_2 \\ 1 & -\lambda_1 \end{bmatrix}}{\lambda_2 - \lambda_1}$$

Finally, taking the inverse Laplace transform of (1.28), we obtain the transition matrix as

$$e^{At} = (\lambda_1 - \lambda_2)^{-1} \begin{bmatrix} \lambda_1 e^{\lambda_1 t} - \lambda_2 e^{\lambda_2} & -\lambda_1\lambda_2(e^{\lambda_1 t} - e^{\lambda_2 t}) \\ e^{\lambda_1 t} - e^{\lambda_2 t} & -\lambda_2 e^{\lambda_1 t} + \lambda_1 e^{\lambda_2 t} \end{bmatrix} \tag{1.29}$$

We will show in the next section that there are other ways of modeling a dynamic process in the state space. This non-uniqueness in the state model representation of a given dynamic process results from being able to choose the coordinates for expressing the state space. In the next section we will use this fact to simplify the determination of e^{At} by working in co-ordinates where the state model has a diagonal A matrix.

1.4.3 Proper and strictly proper rational functions

Before continuing to the next section, notice that when A is an $n \times n$ matrix, adj$[sI - A]$ is an $n \times n$ matrix of polynomials having degree no larger than $n - 1$. Thus, since the characteristic polynomial for A, det$[sI - A]$, is of degree n, we see from (1.27) that $(sI - A)^{-1}$ is an $n \times n$ matrix of strictly proper rational functions.

In general a rational function

$$r(s) = \frac{n(s)}{d(s)}$$

is said to be;

(i) strictly proper when the degree of its numerator polynomial is less than the degree of its denominator polynomial, i.e.,

$$\deg[n(s)] < \deg[d(s)]$$

(ii) proper when the degree of its numerator polynomial equals the degree of its denominator polynomial, i.e.,

$$\deg[n(s)] = \deg[d(s)]$$

In subsequent chapters we will see that this characterization of rational functions plays an important role in control theory.

1.5 Coordinate Transformation

In Section 1.3 we saw that the zero-input response for a system could be obtained by solving a state vector differential equation where the components of the state were identified with the output and its derivatives. In this section we examine the effect of changing this identification.

1.5.1 Effect on the state model

Referring to the second order process used in the previous section, let $\bar{x}(t)$ denote the state obtained by setting

$$\begin{bmatrix} \dot{y}(t) \\ y(t) \end{bmatrix} = V \begin{bmatrix} \bar{x}_1(t) \\ \bar{x}_2(t) \end{bmatrix} \tag{1.30}$$

where V is any invertible (nonsingular) 2×2 matrix of constants. In the previous section V was the identity matrix.

Now we see from (1.11, 1.12, 1.30) that the state $x(t)$ used in the previous section is related in a one-to-one fashion to the state $\bar{x}(t)$ as

$$x(t) = V\bar{x}(t) \tag{1.31}$$

where we say that $x(t)$ is the state in the old or original coordinates and $\bar{x}(t)$ is the state in the new or transformed coordinates. Then the state model parameters in the old coordinates, (A, C), are transformed by a change of coordinates to $(\bar{A}, \bar{C})$ in the new coordinates as

$$(A, C) \overset{V}{\mapsto} (\bar{A}, \bar{C}) \tag{1.32}$$

where

$$\bar{A} = V^{-1}AV$$
$$\bar{C} = CV$$

We can develop this relation as follows.

First using (1.31) in (1.15) we obtain

$$V\dot{\bar{x}} = AV\bar{x}(t)$$

which, since V is invertible, can be multiplied throughout by V^{-1} to give

$$\dot{\bar{x}}(t) = \bar{A}\bar{x}(t)$$

where

$$\bar{A} = V^{-1}AV$$

Again, using (1.31) in (1.16) we obtain

$$y(t) = \bar{C}\bar{x}(t)$$

where

$$\bar{C} = CV$$

Notice that the transition matrix, $e^{\bar{A}t}$, which applies in the new coordinates is related to the transition matrix, e^{At}, in the original coordinates as

$$
\begin{aligned}
e^{\bar{A}t} &= \sum_{i=0}^{\infty} \frac{(V^{-1}AV)^i t^i}{i!} = V^{-1}\left(\sum_{i=0}^{\infty} \frac{A^i t^i}{i!}\right)V \\
&= V^{-1}e^{At}V
\end{aligned} \tag{1.33}
$$

1.5.2 Determination of e^{At}

The flexibility provided by being able to choose the coordinates for the state model representation of a dynamic process is often of considerable use in the analysis and design of control systems. We can demonstrate this fact by using a change of coordinates to calculate the transition matrix.

Suppose we are given a two dimensional system matrix A having a characteristic polynomial, $\det[sI - A]$, with distinct roots (eigenvalues), i.e., $\lambda_1 \neq \lambda_2$. Then we can always find a coordinate transformation matrix V so that the system matrix $\bar{A}$ in the new coordinates is diagonal and

$$\dot{\bar{x}}(t) = \bar{A}\bar{x}(t) \tag{1.34}$$

where

$$\bar{A} = V^{-1}AV = \begin{bmatrix} \lambda_1 & 0 \\ 0 & \lambda_2 \end{bmatrix}$$

with entries along the diagonal of $\bar{A}$ being the eigenvalues of A.

Now when the system matrix is diagonal, the corresponding transition matrix is also diagonal. We can see this by noting that the state differential equation in these coordinates, (1.34), consists of two scalar first order differential equations which are uncoupled from each other

$$
\begin{aligned}
\dot{\bar{x}}_1(t) &= \lambda_1 \bar{x}_1(t) \\
\dot{\bar{x}}_2(t) &= \lambda_2 \bar{x}_2(t)
\end{aligned}
$$

so that their solution can be immediately written as

$$
\begin{aligned}
\bar{x}_1(t) &= e^{\lambda_1 t}\bar{x}_1(0) \\
\bar{x}_2(t) &= e^{\lambda_2 t}\bar{x}_2(0)
\end{aligned} \tag{1.35}
$$

which in matrix form is

$$\begin{bmatrix} \bar{x}_1(t) \\ \bar{x}_2(t) \end{bmatrix} = \begin{bmatrix} e^{\lambda_1 t} & 0 \\ 0 & e^{\lambda_2 t} \end{bmatrix} \begin{bmatrix} \bar{x}_1(0) \\ \bar{x}_2(0) \end{bmatrix} \tag{1.36}$$

Thus we see that the transition matrix is indeed diagonal

$$e^{\bar{A}t} = \begin{bmatrix} e^{\lambda_1 t} & 0 \\ 0 & e^{\lambda_2 t} \end{bmatrix}$$

Having determined the transition matrix for $\bar{A}$, we can use (1.33) to determine the transition matrix for A as

$$e^{At} = Ve^{\bar{A}t}V^{-1} \tag{1.37}$$

with V being the coordinate transformation matrix which makes $\bar{A}$ diagonal.

Now we will see in the next section that, in general, the coordinate transformation matrix V needed to make $\bar{A}$ diagonal depends on the eigenvectors of A. However in the special case when A is a 2×2 companion matrix, (1.15), with $\lambda_1 \neq \lambda_2$, the required coordinate transformation matrix is simply related to the eigenvalues of A as

$$V = \begin{bmatrix} \lambda_1 & \lambda_2 \\ 1 & 1 \end{bmatrix} \tag{1.38}$$

We can see that this coordinate transformation gives rise to a diagonal system matrix by using

$$V^{-1} = (\lambda_1 - \lambda_2)^{-1} \begin{bmatrix} 1 & -\lambda_2 \\ -1 & \lambda_1 \end{bmatrix} \tag{1.39}$$

to obtain

$$\bar{A} = V^{-1}AV = \frac{1}{\lambda_1 - \lambda_2} \begin{bmatrix} -a_1\lambda_1 - \lambda_1\lambda_2 - a_2 & -a_1\lambda_2 - \lambda_2^2 - a_2 \\ a_1\lambda_1 + \lambda_2^2 + a_2 & a_1\lambda_2 + \lambda_1\lambda_2 + a_2 \end{bmatrix}$$

Then since $s^2 + a_1 s + a_2 = (s - \lambda_1)(s - \lambda_2)$ we have $a_1 = -(\lambda_1 + \lambda_2)$ and $a_2 = \lambda_1\lambda_2$. Therefore the foregoing expression for $\bar{A}$ reduces to

$$\bar{A} = \begin{bmatrix} \lambda_1 & 0 \\ 0 & \lambda_2 \end{bmatrix} \tag{1.40}$$

Finally, the expression obtained for e^{At} using V, (1.38), in (1.37) equals (1.29) which was obtained at the end of Section 1.4 through the use of Laplace transforms.

The foregoing approach to the determination of the transition matrix requires the determination of a coordinate transformation matrix V which diagonalizes the system matrix A. We will see in the next section that the columns of the coordinate transformation matrix required to do this are right-eigenvectors for A.

1.6 Diagonalizing Coordinate Transformation

As mentioned previously, the roots of the characteristic polynomial for a square matrix A are called the eigenvalues of A. In this section we will see that corresponding to each of A's eigenvalues there is at least one right and one left-eigenvector. Moreover we will see that when the eigenvalues of A are distinct, the coordinate transformation V required to make $\bar{A}$ diagonal has columns equal to the right-eigenvectors of A. In addition we will see that V^{-1} has rows which are the transpose of the left-eigenvectors of A.

1.6.1 Right-Eigenvectors

Consider the special case when A is a two-by-two matrix in companion form, (1.15), having unequal eigenvalues $\{\lambda_i : 1, 2\}$. Then writing the characteristic polynomial as $\lambda_i^2 = -a_1\lambda_i - a_2$ we see that

$$\begin{bmatrix} -a_1 & -a_2 \\ 1 & 0 \end{bmatrix} \begin{bmatrix} \lambda_i \\ 1 \end{bmatrix} = \lambda_i \begin{bmatrix} \lambda_i \\ 1 \end{bmatrix} \qquad i = 1, 2 \tag{1.41}$$

or

$$Av^i = \lambda_i v^i \qquad i = 1, 2 \tag{1.42}$$

where

$$v^i = \begin{bmatrix} \lambda_i \\ 1 \end{bmatrix} \qquad i = 1, 2$$

Notice that (1.42) is a general expression relating the ith eigenvalue, right-eigenvector pair (λ_i, v^i) for the any square matrix A, where v^i is said to be the right-eigenvector corresponding to the eigenvalue λ_i. These pairs play a major role in the state analysis of systems. The particular dependence of v^i on λ_i in the present instance is a result of A being in companion form.

Continuing with the construction of V to make $V^{-1}AV$ diagonal, we combine the equations given by (1.42) for $i = 1$ and $i = 2$ to obtain

$$AV = V\Lambda \tag{1.43}$$

where V, Λ are given as

$$V = \begin{bmatrix} v^1 & v^2 \end{bmatrix} \qquad \Lambda = \begin{bmatrix} \lambda_1 & 0 \\ 0 & \lambda_2 \end{bmatrix}$$

Now when A has distinct eigenvalues, i.e., $\lambda_1 \neq \lambda_2$, V is invertible and we can pre-multiply (1.43) by V^{-1} to obtain

$$V^{-1}AV = \Lambda = \bar{A}$$

Thus we see that V is the coordinate transformation matrix required to make $\bar{A}$ diagonal.

More generally, suppose A is any $n \times n$ matrix, not necessarily in companion form, which has distinct eigenvalues, $\lambda_i \neq \lambda_j : i \neq j$. Now it turns out that this condition of distinct eigenvalues is sufficient for the eigenvectors of A to be independent, i.e., v^i and v^j point in different directions. Therefore V has independent columns and is therefore invertible where

$$V = \begin{bmatrix} v^1 & v^2 & \cdots & v^n \end{bmatrix} \tag{1.44}$$

$$Av^i = \lambda_i v^i \qquad i = 1, 2, \cdots, n \tag{1.45}$$

and $V^{-1}AV$ is diagonal.

In the special case when A is in companion form, (1.19), its eigenvalues, $\{\lambda_i : i = 1, 2, \cdots, n\}$ are related to its eigenvectors, $\{v^i : i = 1, 2, \cdots, n\}$ as

$$v^{iT} = \begin{bmatrix} \lambda_i^{n-1} & \lambda_i^{n-2} & \cdots & \lambda_i & 1 \end{bmatrix}$$

In order to see that this result holds, set the last entry in v^i equal to one. Then taking A in companion form, (1.19), solve the last scalar equation in (1.45) and use the result to solve the second to last scalar equation in (1.45). We continue in this way solving successive scalar equations in (1.45), in reverse order, until we reach the first scalar equation. At this stage we will have all the entries in v^i. These entries satisfy the first scalar equation in (1.45) since λ_i is a root of the characteristic polynomial whose coefficients appear with negative signs along the first row of A.

In general, when A is not in any special form, there is no special relation between the eigenvalues and the corresponding eigenvectors. Thus in order to determine the eigenvalue, right-eigenvector pairs when A is not in any special form we need to determine $(\lambda_i, v^i) : i = 1, 2, \cdots n$ so that the equations

$$Av^i = \lambda_i v^i \qquad i = 1, 2, \cdots n \tag{1.46}$$

are satisfied.

1.6.2 Eigenvalue–Eigenvector problem

The problem of determining (λ_i, v^i) pairs which satisfy (1.46) is referred to as the eigenvalue–eigenvector problem. There are well established methods for solving this problem using a digital computer. In order to gain additional insight into the nature of the eigenvalue–eigenvector problem we consider a theoretical approach to finding eigenvalue–eigenvector pairs to satisfy (1.46).

To begin, suppose we rewrite (1.46) as

$$(\lambda I - A)v = \emptyset \tag{1.47}$$

where $\emptyset$ denotes the null vector, i.e., a vector of zeros. Then in order for the solution v to this equation to be non-null we must choose λ so that the matrix $\lambda I - A$ is singular, i.e., does not have an inverse. Otherwise, if $\lambda I - A$ is invertible we can solve (1.47) as

$$v = (\lambda I - A)^{-1}\emptyset$$

and the only solution is the trivial solution $v = \emptyset$. However when $(\lambda I - A)$ is not invertible, (1.47) can be satisfied by $v \neq \emptyset$.

Now from Crammer's rule for matrix inversion we have

$$(\lambda I - A)^{-1} = \frac{\operatorname{adj}[\lambda I - A]}{\det[\lambda I - A]}$$

Therefore $\lambda I - A$ does not have an inverse when $\det[\lambda I - A] = 0$, i.e., $\lambda = \lambda_i$ an eigenvalue of A.

Next recall that singular matrices have dependent columns. Therefore $\lambda_i I - A$ has dependent columns so that we can find scalars $\{v_k^i : k = 1, 2, \cdots, n\}$ not all zero such that

$$\sum_{k=1}^{n} [(\lambda_i I - A)]_k v_k^i = \emptyset \tag{1.48}$$

where $[(\lambda_i I - A)]_k$: $k = 1, 2 \cdots, n$ denote columns of $\lambda_i I - A$. Notice that (1.48) can be rewritten as (1.47) with $\lambda = \lambda_i$ and $v = v^i$ where

$$v^{iT} = \begin{bmatrix} v_1^i & v_2^i & \cdots & v_n^i \end{bmatrix}$$

Since we can always multiply (1.48) through by a nonzero scalar α, the solution, v^i, to (1.48) or (1.47) is not unique since αv^i is another solution. More generally, we say that the eigenvectors of a given matrix are determined to within a scalar multiple, i.e., the directions of the eigenvectors are determined but their lengths are arbitrary.

Assuming that A has a complete set of (n independent) right-eigenvectors, we can decompose any initial state as

$$x(0) = \sum_{i=1}^{n} \alpha_i v^i = V\alpha \tag{1.49}$$

where

$$\alpha^T = [\alpha_1 \quad \alpha_2 \quad \cdots \quad \alpha_n] \qquad\qquad V = \begin{bmatrix} v^1, v^2, \cdots v^n \end{bmatrix}$$

with the $\alpha_i s$ being found as

$$\alpha = V^{-1} x(0)$$

This decomposition of the state into a linear combination of right-eigenvectors of A plays an important role in the analysis of system behavior. This is illustrated in the next section where we will use the eigenvectors of A to reveal certain fundamental properties of state trajectories.

Unfortunately, when an $n \times n$ matrix A has multiple eigenvalues, i.e., when $\det[\lambda I - A]$ does not have n distinct roots, the number of eigenvectors may or may not be equal to n, i.e., A may or may not have a complete set of eigenvectors. When A does not have a complete set of eigenvectors, other related vectors, referred to as generalized eigenvectors, (Appendix), can be used together with the eigenvectors to provide a decomposition of the state space in a manner similar to (1.49). However in this case it is impossible to find a nonsingular matrix V such that $V^{-1}AV$ is diagonal. Then A is said to be not diagonalizable.

In summary, the condition that A has distinct eigenvalues is sufficient but not necessary for A to be diagonalizable. Most of the time, little additional insight into control theory is gained by discussing the case where A does not have a complete set of eigenvectors. Therefore, we are usually able to avoid this complication without loss of understanding of the control theory.

1.6.3 Left-Eigenvectors

Suppose A is any $n \times n$ matrix having n distinct eigenvalues. Then we have seen that the matrix V having columns which are right-eigenvectors of A is invertible and we can write

$$V^{-1}AV = \Lambda \tag{1.50}$$

where $Av^i = \lambda_i v^i : i = 1, 2, \cdots n$ and

$$V = \begin{bmatrix} v^1 & v^2 & \cdots & v^n \end{bmatrix} \qquad \Lambda = \begin{bmatrix} \lambda_1 & 0 & \cdots & 0 \\ 0 & \lambda_2 & \cdots & 0 \\ \vdots & \vdots & \ddots & \vdots \\ 0 & 0 & \cdots & \lambda_n \end{bmatrix}$$

Now suppose we post-multiply both sides of (1.50) by V^{-1} to obtain

$$V^{-1}A = \Lambda V^{-1} \tag{1.51}$$

Then if we denote V^{-1} in terms of its rows as

$$V^{-1} = \begin{bmatrix} w^{1T} \\ w^{2T} \\ \vdots \\ w^{nT} \end{bmatrix}$$

and carry out the matrix multiplication indicated in (1.51) we obtain

$$\begin{bmatrix} w^{1T}A \\ w^{2T}A \\ \vdots \\ w^{nT}A \end{bmatrix} = \begin{bmatrix} \lambda_1 w^{1T} \\ \lambda_2 w^{2T} \\ \vdots \\ \lambda_n w^{nT} \end{bmatrix}$$

Therefore it follows, by equating corresponding rows on either side of this equation, that

$$w^{iT}A = \lambda_i w^{iT} \qquad i = 1, 2, \cdots, n \tag{1.52}$$

which transposing throughout gives

$$A^T w^i = \lambda_i w^i \qquad i = 1, 2, \cdots, n \tag{1.53}$$

Thus we see from equation (1.53) that the column vector w^i is an eigenvector of A^T with corresponding eigenvalue λ_i. However, since the row vector w^{iT} appears on the left side of A in equation (1.52), w^i is referred to as a left-eigenvector of A to distinguish it from the corresponding right-eigenvector of A, namely, v^i.

Notice that, since $V^{-1}V$ is an identity matrix, the left and right-eigenvectors just defined are related to each other as

$$\left.\begin{aligned} w^{iT}v^j &= 0 \quad i \neq j \\ &= 1 \quad i = j \end{aligned}\right\} \tag{1.54}$$

In addition notice that any nonzero scalar multiple of w_i

$$z^i = \gamma_i w^i$$

satisfies (1.52), i.e.

$$z^{iT} A = \lambda_i z^{iT}$$

Therefore z^i is also a left-eigenvector of A and we see from equation (1.54) that in general the left and right eigenvectors are related as

$$\begin{aligned} z^{iT}v^j &= 0 & i \neq j \\ &= \gamma_i \neq 0 & i = j \end{aligned}$$

This basic characteristic of left and right-eigenvectors, referred to as the orthogonality property, is responsible for a number of fundamental facts relating to the behavior of state models.

1.6.4 Eigenvalue invariance

Before we go to the next section, it is important to note the basic fact that the eigenvalues of A and of $\bar{A}$ are the same whenever $\bar{A}$ is related to A as $\bar{A} = V^{-1}AV$ for any invertible matrix V. We can see this by premultiplying both sides of equation (1.45) by V^{-1} and inserting VV^{-1} between A and v^i, viz.,

$$V^{-1}AVV^{-1}v^i = \lambda_i V^{-1}v^i$$

Then setting $t^i = V^{-1}v^i$ and taking $V^{-1}AV = \bar{A}$ gives

$$\bar{A}t^i = \lambda_i t^i$$

which implies that the eigenvalue, right-eigenvector pairs for $\bar{A}$ are (λ_i, t^i). Thus the eigenvalues of $\bar{A}$ equal the eigenvalues of A independent of the coordinate transformation matrix V.

Alternatively, another way we can see this fact is to carry out the following manipulations

$$\begin{aligned}\det[sI - \bar{A}] &= \det\left[sI - V^{-1}AV\right] = \det\left[V^{-1}(sI - A)V\right] \\ &= \det V^{-1} \det[sI - A] \det V \\ &= \det[sI - A]\end{aligned}$$

Thus A and $\bar{A}$ have the same characteristic polynomial, and since the roots of a polynomial are uniquely dependent on the coefficients of the polynomial, A and $\bar{A}$ have the same eigenvalues.

Finally, since the differential equations modeling the behavior of dynamical processes must have real coefficients, we can always work in coordinates where the state model parameters, (A, C), are real. As a consequence, if (λ_i, v^i) is a complex eigenvalue–eigenvector pair for A, then (λ_i^*, v^{i*}) is also an eigenvalue–eigenvector pair for A.

1.7 State Trajectories Revisited

We saw in Section 1.6.2 that assuming A has a complete set of eigenvectors, any initial state can be written in terms of the eigenvectors of A, (1.49). In this section this fact is used to gain additional insight into the nature of a system's state trajectories and zero-input response.

More specifically, under certain conditions on the matrix pair, (A, C), a system can exhibit a null zero-input response, $y(t) = 0$ for all $t > 0$, for some non-null initial state, $x(0) \neq \emptyset$. When this occurs we say that the state trajectory is orthogonal (perpendicular) to C^T, denoted $C^T \perp x(t)$, since the output depends on the state as $y(t) = Cx(t)$. Two vectors α, β are said to be orthogonal if

$$\alpha^T \beta = 0$$

When the state space is n dimensional, a state trajectory which produces no output lies in an $n - 1$ dimensional subspace of state space which is perpendicular to the vector C^T. For example, if the system is second order, this subspace is a straight line perpendicular to C^T; if the system is third order, this subspace is a plane perpendicular to C^T. Thus, in the second order case, we obtain a null output if we can find an initial state such that it produces a straight line trajectory

$$x(t) = \gamma(t)v$$

satisfying

$$Cv = 0$$

where $\gamma(t)$ is a scalar function of time and v is a constant two-element vector. When $n > 2$, any trajectory orthogonal to C^T can be decomposed into a sum of straight line trajectories all lying in the $n-1$ dimensional subspace orthogonal to C^T. Therefore an understanding of straight line trajectories is essential to an understanding of the property posed by certain systems of having a null zero-input response to certain initial states.

1.7.1 Straight line state trajectories: diagonal A

Suppose $\bar{A}$ is a 2×2, real, diagonal matrix. Then the state trajectory is a straight line whenever the initial state lies only on one of the two coordinate axes. We can see this immediately as follows.

Consider the initial states

$$\bar{x}^1(0) = \begin{bmatrix} \bar{x}_1(0) \\ 0 \end{bmatrix} \qquad \bar{x}^2(0) = \begin{bmatrix} 0 \\ \bar{x}_2(0) \end{bmatrix}$$

where $\bar{x}_1(0)$ and, $\bar{x}_2(0)$ are any real nonzero scalars. Then recalling from Section 1.5.2 that the transition matrix $e^{\bar{A}t}$ is diagonal when $\bar{A}$ is diagonal, we see from (1.36) that the state corresponding to each of these initial states is

$$\begin{aligned} \bar{x}(t) &= \begin{bmatrix} \bar{x}_1(0)e^{\lambda_1 t} \\ 0 \end{bmatrix} && \text{for } \bar{x}(0) = \bar{x}^1(0) \\ \bar{x}(t) &= \begin{bmatrix} 0 \\ \bar{x}_2(0)e^{\lambda_2 t} \end{bmatrix} && \text{for } \bar{x}(0) = \bar{x}^2(0) \end{aligned} \tag{1.55}$$

The foregoing suggests that the trajectory for any initial state in these coordinates

$$\bar{x}(0) = \begin{bmatrix} \bar{x}_1(0) \\ \bar{x}_2(0) \end{bmatrix}$$

can be written as

$$\bar{x}(t) = (\bar{x}_1(0)e^{\lambda_1 t})i^1 + (\bar{x}_2(0)e^{\lambda_2 t})i^2 \tag{1.56}$$

where $i^k : k = 1, 2$ are columns from the 2×2 identity matrix, viz.,

$$I = \begin{bmatrix} i^1 & i^2 \end{bmatrix}$$

More generally, when A is a real, diagonal, $n \times n$ matrix, the state trajectory

$$\bar{x}(t) = \bar{x}_k(0)e^{\lambda_k t}i^k \tag{1.57}$$

results when the initial state

$$\bar{x}^T(0) = [\bar{x}_1(0) \quad \bar{x}_2(0) \quad \cdots \quad \bar{x}_n(0)]$$

has components which satisfy

$$\begin{aligned} \bar{x}_i(0) &= 0 && \text{for } i \neq k \\ &\neq 0 && \text{for } i = k \end{aligned}$$

where i^k is the kth column from the $n \times n$ identity matrix.

The foregoing result implies that the zero-input state response for any initial state can be written as

$$\bar{x}(t) = \sum_{k=1}^{n} (\bar{x}_k(0)e^{\lambda_k t})i^k$$

when

$$\bar{x}(0) = \sum_{k=1}^{n} \bar{x}_k(0)i^k$$

In summary, the state trajectory is the vector sum of state trajectories along coordinate axes where each coordinate axis trajectory depends on one of the system modes in the set of system modes, $\{e^{\lambda_k t} : k = 1, 2 \cdots, n\}$.

Now in order to generalize the foregoing result to the case where A is not diagonal, we suppose, in the next section, that the foregoing diagonal case resulted from a coordinate transformation from the original coordinates in which A is given.

1.7.2 Straight line state trajectories: real eigenvalues

Recalling that V, equation (1.44), is the coordinate transformation needed to diagonalize A and taking $\bar{x}(0)$ as in (1.57) we have

$$x(0) = V\bar{x}(0) = \bar{x}_k(0)v^k \tag{1.58}$$

where $Av^k = \lambda_k v^k$. Then, using the series expansion for the matrix exponential, (1.22), we see that when $x(0)$ is given by equation (1.58) we have

$$\begin{aligned} x(t) = e^{At}x(0) &= \left(I + At + \frac{A^2t^2}{2!} \cdots\right)\bar{x}_k(0)v^k \\ &= \bar{x}_k(0)\left(1 + \lambda_k t + \frac{\lambda_k^2 t^2}{2!} \cdots\right)v^k = (\bar{x}_k(0)e^{\lambda_k t})v^k \end{aligned} \tag{1.59}$$

Now with λ_k real we see from equation (1.59) that the point representing the state moves along the eigenvector, v^k, towards the origin when $\lambda_k < 0$ and away from the origin when $\lambda_k > 0$. The case where λ_k is complex is taken up in the next subsection.

More generally, assuming A has a complete set of eigenvectors, we can write any initial state as

$$x(0) = \sum_{i=1}^{n} \gamma_i v^i \tag{1.60}$$

where $\gamma = V^{-1}x(0)$ and

$$V = \begin{bmatrix} v^1 & v^2 & \cdots & v^n \end{bmatrix} \qquad \gamma^T = \begin{bmatrix} \gamma_1 & \gamma_2 & \cdots & \gamma_n \end{bmatrix}$$

Then the state trajectory can be written as

$$x(t) = \sum_{i=1}^{n} (\gamma_i e^{\lambda_i t}) v^i \tag{1.61}$$

Now recall, from Section 1.2, that a system is stable if

$$\lim_{t\to\infty} x(t) = \emptyset \qquad \text{for all } x(0) \tag{1.62}$$

Therefore assuming A has real eigenvalues, we see from equation (1.59) that the system is stable if and only if

$$\lambda_i < 0 \qquad i = 1, 2, \cdots, n$$

It should be emphasized that a system is stable if and only if equation (1.62) is satisfied. Therefore if we need to restrict the initial condition in order to ensure that the zero-input state trajectory goes to the origin, the system is not stable. For example, referring to the expansion (1.60), any initial state, $x(0)$, which is restricted so that its expansion satisfies

$$\gamma_i = 0 \qquad \text{when } \lambda_i \geq 0$$

has a state trajectory which goes to the origin with time even though A has some non-negative eigenvalues.

1.7.3 Straight line trajectories: complex eigenvalues

Since the differential equations governing the input–output behavior of the physical processes we are interested in controlling have real coefficients, we can always choose to work in coordinates so that A, C are real matrices. Then since the eigenvalues, if complex, occur in conjugate pairs we see that the corresponding eigenvectors are also conjugates of each other, i.e., if (λ, v) is a complex, eigenvalue–eigenvector pair, then (λ^*, v^{*T}) is also an eigenvalue-eigenvector pair of A.

Now if the initial state is any scalar multiple of the real or imaginary part of v, the resulting state trajectory lies in a two dimensional subspace of state space composed from the real and imaginary parts of v. More specifically, suppose

$$x(0) = \frac{\gamma}{2}\left[v + v^{*T}\right] = \gamma \mathrm{Re}[v] \tag{1.63}$$

where γ is any real scalar and

$$v = \mathrm{Re}[v] + j\mathrm{Im}[v]$$

Then from (1.59) we obtain

$$x(t) = \alpha_{re}(t)\mathrm{Re}[v] + \alpha_{im}(t)\mathrm{Im}[v] \tag{1.64}$$

where

$$\alpha_{re}(t) = \gamma e^{\mathrm{Re}[\lambda]t}\cos(\mathrm{Im}[\lambda]t)$$
$$\alpha_{im}(t) = -\gamma e^{\mathrm{Re}[\lambda]t}\sin(\mathrm{Im}[\lambda]t)$$

The foregoing generalizes the observation made at the beginning of the chapter that, for second order systems, spiral shaped trajectories result when the roots of the characteristic equation consist of a complex conjugate pair. In the present case where the system order $n > 2$, the spiral shaped trajectory lies in a two-dimensional subspace of the n-dimensional state space.

Notice that the initial state was chosen to ensure that the state $x(t)$ is real. However if we choose the initial state as

$$x(0) = \alpha_1 v + \alpha_2 v^{*T}$$

with the real scalars α_i satisfying

$$|\alpha_1| \neq |\alpha_2|$$

then $x(0)$ would be complex and the resulting trajectory would lie in a 2-dimensional subspace of a complex state space.

1.7.4 Null output zero-input response

Having discussed straight line state trajectories, we return to the problem stated at the beginning of this section concerning the possibility of having initial states which produce null outputs.

Suppose the output matrix happens to satisfy

$$Cv^k = 0 \tag{1.65}$$

for some eigenvector, v^k, of A. Then it follows from (1.59) that if $x(0) = \gamma_k v^k$ then

$$y(t) = Cx(t) = \gamma_k e^{\lambda_k t} C v^k = 0$$

and we see the output is null for all time when the initial state lies along v^k.

This effect of the existence of non-null initial states which do not affect the output, is related to a property of the state model's A and C matrices which is referred to as the system's observability (Chapter 3). More is said about this matter in Section 3.5.2.

In order to obtain some appreciation of the importance of this effect consider a state model with A having all its eigenvalues, except λ_k, in the open left-half plane. Then this state model is internally unstable since $x(0) = v^k$ produces a trajectory which moves away from the origin. However if C satisfies equation (1.65) this trajectory has no effect on the output and in this case we have

$$\lim_{t\to\infty} y(t) = 0 \qquad \text{for all } x(0)$$

and the system is externally stable.

This demonstrates that while the state model is internally unstable its output behavior is stable. However, since it is practically impossible to exactly model a physical process, the foregoing stability of the output in response to initial states exists on paper only and is referred to by saying that the system is not robustly output stable. For this reason, we say that a system is stable if and only if its A matrix has all its eigenvalues in the open left-half plane.

Before we leave this section, it is instructive to consider conditions on C which guarantee that equation (1.65) is satisfied. Recall, from Section 1.6.3, that right and left-eigenvectors corresponding to different eigenvalues are orthogonal,

$$w^{iT} v^j = 0 \qquad i \neq j$$

Therefore suppose A has a complete set of eigenvectors so that we can expand C in terms of the left-eigenvectors of A,

$$C = \sum_1^n \alpha_i w^{iT} \tag{1.66}$$

Then we see that equation (1.65) is satisfied when C is independent of w^k, i.e., when $\alpha_k = 0$ in equation (1.66). This structural characterization of C will be used in Chapter 3 to develop properties of state models which relate to their observability.

1.8 State Space Models for the Complete Response

So far we have used the state space to model a system's zero-input response. In this section we take up the use of the state space in connection with the modeling of a system's zero-state response.

1.8.1 Second order process revisited

Returning to the electric circuit used in Section 1.2, suppose we connect a voltage source as show in Figure 1.4. Then the differential equation governing the output, $y(t)$ (capacitor voltage), becomes

$$\frac{d^2y(t)}{dt^2} + a_1 \frac{dy(t)}{dt} + a_2 y(t) = b_2 u(t) \tag{1.67}$$

where $u(t)$ is the input (source voltage) and a_1, a_2 are as in (1.1) with $b_2 = a_2$.

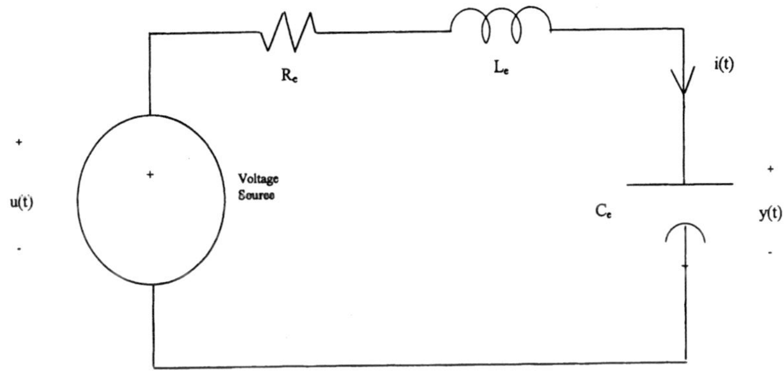

Figure 1.4 Electric circuit with voltage input

Suppose, as in Section 1.3, that we choose the components of the state as

$$\begin{bmatrix} x_1(t) \\ x_2(t) \end{bmatrix} = \begin{bmatrix} \dot{y}(t) \\ y(t) \end{bmatrix} \tag{1.68}$$

so that

$$\dot{x}_2(t) = x_1(t) \tag{1.69}$$

Then we see from (1.67) that

$$\dot{x}_1(t) = \frac{\mathrm{d}^2 y}{\mathrm{d}t^2} = -a_1 x_1(t) - a_2 x_2(t) + b_2 u(t) \tag{1.70}$$

and from (1.68–1.70) that the state differential equation and output equation are given as

$$\dot{x}(t) = Ax(t) + Bu(t) \tag{1.71}$$

$$y(t) = Cx(t) \tag{1.72}$$

where

$$A = \begin{bmatrix} -a_1 & -a_2 \\ 1 & 0 \end{bmatrix} \qquad B = \begin{bmatrix} b_2 \\ 0 \end{bmatrix}$$

$$C = \begin{bmatrix} 0 & 1 \end{bmatrix}$$

with B being referred to as the input matrix. This state model is complete in as much as it can be used to obtain the output caused by any specified combination of initial state, $x(0)$, and input, $u(t)$. We will see that the matrix product of the initial state and transition matrix, which gives the zero-input response, is replaced in the calculation of the zero-state response by an integral, referred to as the convolution integral, of the imput and transition matrix. Before showing this, consider the following modification of the foregoing state model.

Suppose we rewrite (1.67) as

$$\frac{d^2\bar{y}(t)}{dt^2} + a_1 \frac{d\bar{y}(t)}{dt} + a_2\bar{y}(t) = u(t)$$

where

$$\bar{y}(t) = \frac{y(t)}{b_2}$$

and proceed in the manner used to get equation (1.71, 1.72). This gives the alternative state model

$$\begin{aligned} \dot{x}_c(t) &= A_c x_c(t) + B_c u(t) \\ y(t) &= C_c x_c(t) \end{aligned} \tag{1.73}$$

where $A_c = A$ and

$$B_c = \frac{1}{b_2} B = \begin{bmatrix} 1 \\ 0 \end{bmatrix}$$

$$C_c = b_2 C = \begin{bmatrix} 0 & b_2 \end{bmatrix}$$

This state model, equation (1.73), is an example of a controller form state model. Controller form state models are characterized by having B_c as the first column of the identity matrix and A_c in companion form, (1.19). The controller form state model is used in the next chapter to provide insight into the dynamics of closed loop systems employing state feedback.

1.8.2 Some essential features of state models

First notice that when we change coordinates by setting $x(t) = V\bar{x}(t)$ for some constant invertible matrix V, the resulting state model in the new coordinates is given by

$$\dot{\bar{x}}(t) = \bar{A}\bar{x}(t) + \bar{B}u(t)$$

$$y(t) = \bar{C}\bar{x}(t)$$

where the parameters for the original and transformed state models are related as

$$(A, B, C) \overset{V}{\mapsto} (\bar{A}, \bar{B}, \bar{C})$$

where

$$\bar{A} = V^{-1}AV \qquad \bar{B} = V^{-1}B \qquad \bar{C} = CV$$

Second, notice that there are physical processes which transmit the effect of the input directly to the output. In this situation the output equation for the state model has an additional term, $Du(t)$, i.e., the output equation is

$$y(t) = Cx(t) + Du(t)$$

Notice that, unlike the other state model parameters (A, B, C), the D parameter is unaltered by a state coordinate transformation.

Third, some dynamic processes have more than one scalar input and/or more than one scalar output. For instance a dynamic process may have m scalar inputs, $\{u_i(t) : i \in [1, m]\}$ and p scalar outputs, $\{y_i(t) : i \in [1, p]\}$. In this situation the system input, $u(t)$, and system output, $y(t)$, are column vectors of size m and p respectively

$$u^T(t) = \begin{bmatrix} u_1(t) & u_2(t) & \cdots & u_m(t) \end{bmatrix}$$

$$y^T(t) = \begin{bmatrix} y_1(t) & y_2(t) & \cdots & y_p(t) \end{bmatrix}$$

Further, in this case, the state model has an input matrix B with m columns and an output matrix C with p rows. More specifically, the general form for state models used here is given as

$$\left.\begin{aligned} \dot{x}(t) &= Ax(t) + Bu(t) \\ y(t) &= Cx(t) + Du(t) \end{aligned}\right\} \tag{1.74}$$

where $x(t), u(t)$ and $y(t)$ are column vectors of time functions of length n, m and p respectively and the state model parameters, (A, B, C, D), are matrices of constants of size $n \times n$, $n \times m$, $p \times n$, and $p \times m$ respectively.

Finally, systems having $p > 1$ and $m > 1$ are referred to as multiple-input-multiple-output (MIMO) systems and systems having $p = 1$ and $m = 1$ are referred to as single-input-single-output (SISO) systems.

1.8.3 Zero-state response

Recall from Section 1.3 that the zero-input response, $y_{zi}(t)$, depends on the transition matrix and initial state through multiplication as $y_{zi}(t) = Ce^{At}x(0)$. In this section we will show that the zero-state response, $y_{zs}(t)$, depends on the transition matrix and the input through integration as

$$y_{zs}(t) = C\int_0^t e^{A(t-\tau)}Bu(\tau)\mathrm{d}\tau + Du(t)$$

where the integral is known as the convolution integral.

We begin the development of the foregoing relation by assuming that the initial state is null, $x(0) = \emptyset$, and that the state model parameters, (A, B, C, D) are known. Then taking Laplace transforms throughout the state differential and output equations, (1.74), gives

$$sX(s) = AX(s) + BU(s) \tag{1.75}$$

$$Y_{zs}(s) = CX(s) + DU(s) \tag{1.76}$$

Next solving (1.75) for $X(s)$ yields

$$X(s) = (sI - A)^{-1}BU(s)$$

and substituting this expression for $X(s)$ in (1.76) yields

$$Y_{zs}(s) = G(s)U(s) \tag{1.77}$$

where

$$G(s) = C(sI - A)^{-1}B + D \tag{1.78}$$

Now, recalling from Section 1.4.3 that $(sI - A)^{-1}$ is a matrix of strictly proper rational functions, we see that $G_{sp}(s)$ is strictly proper where

$$G_{sp}(s) = C(sI - A)^{-1}B$$

so that

$$\lim_{s\to\infty} G_{sp}(s) = \emptyset$$

Thus $G(s)$, (1.78) is given by

$$G(s) = G_{sp}(s) + D \tag{1.79}$$

and

$$\lim_{s\to\infty} G(s) = D$$

Notice that two state models which are related by a coordinate transformation have the same transfer function. We can see this by calculating the transfer function using the state model in the transformed coordinates, $(\bar{A}, \bar{B}, \bar{C}, D)$

$$\bar{G}(s) = \bar{C}(sI - \bar{A})^{-1}\bar{B} + \bar{D}$$

and substituting $\bar{A} = V^{-1}AV, \bar{B} = V^{-1}B, \bar{C} = CV, \bar{D} = D$ to obtain

$$\begin{aligned}\bar{G}(s) &= CV\left[V^{-1}(sI - A)V\right]^{-1}V^{-1}B + D \\ &= C(sI - A)^{-1}B + D = G(s)\end{aligned}$$

Next recalling the following general property of Laplace transforms

$$Y_{zs}(s) = F_1(s)F_2(s) \Longleftrightarrow y_{zs}(t) = \int_0^\infty f_1(t-\tau)f_2(\tau)\mathrm{d}\tau$$

we see that setting

$$F_1(s) = C(sI - A)^{-1} \qquad F_2(s) = BU(s)$$

so that

$$f_1(t) = Ce^{At} \qquad f_2(t) = Bu(t)$$

gives the inverse transform of (1.77) as

$$y_{zs}(t) = C\int_0^\infty e^{A(t-\tau)}Bu(\tau)\mathrm{d}\tau + Du(t) \tag{1.80}$$

Notice that when D is null and the input is an impulse, $u(t) = \delta(t)$, (1.80) gives

$$y_{zs}(t) = Ce^{At}B \tag{1.81}$$

In addition recalling from section 1.4.2 that

$$\mathcal{L}\left[e^{At}\right] = (sI - A)^{-1}$$

we see that $y_{zs}(t)$, (1.81), has Laplace transform equal to $G_{sp}(s)$ as we expect since the Laplace transform of the impulse response equals the transfer function. In addition we see from (1.81) that the zero-input response equals the impulse response when the initial state is $x(0) = B$.

Now we need to define e^{At} as a null matrix when $t < 0$ in (1.80, 1.81). This is done to match the mathematics to the physical fact that the future input, $u(\tau) : \tau > t$, does not affect the output at time t, i.e., $y_{zs}(t)$ is independent of future values of the input. This property of the transition matrix, i.e., $\phi(t) = \emptyset$ for all $t < 0$, is referred to as the causality constraint and applies when we use the transition matrix in connection with the zero-state response. Thus the causality constraint forces the integrand in (1.80) to be null for $\tau > t$ and enables (1.80) to be rewritten as

$$y_{zs}(t) = C\int_0^t e^{A(t-\tau)}Bu(\tau)\mathrm{d}\tau + Du(t)$$

Notice that when the transition matrix is used in connection with the zero-input response we can interpret $\phi(-t)$ for $t > 0$ as the matrix needed to determine the initial state from the state at time t, i.e., $x(0) = \phi(-t)x(t)$, which implies that $\phi(-t)$ is the inverse of $\phi(t)$.

At this point we can see, by recalling the principle of superposition, that when a system is subjected to both a non-null initial state, $x(0)$, and a non-null input, $u(t)$, we can write the output as

$$y(t) = y_{zi}(t) + y_{zs}(t) \tag{1.82}$$

where

$$y_{zi}(t) = Ce^{At}x(0)$$

$$y_{zs}(t) = C\int_0^t e^{A(t-\tau)}Bu(\tau)d\tau + Du(t)$$

Notice that:

(i) $y_{zi}(t)$, the zero-input response, is caused solely by to $x(0)$
(ii) $y_{zs}(t)$, the zero-state response, is caused solely by to $u(t)$.

In this section we have developed several essential relations involving the state model of a dynamic process. These important relations provide the complete response of a system to any specified input and any specified initial state.

1.9 Diagonal Form State Model

In subsequent chapters we will encounter a number of fundamental properties of state models in connection with their use in the design of feedback control systems. A simple way of beginning to appreciate these properties is to consider state models in coordinates where the system matrix A is diagonal. This type of state model is referred to as a diagonal or normal form state model. We encountered this model earlier in Section 1.5.2 in connection with the determination of the matrix exponential, and in Section 1.7.1 in connection with straight line state trajectories.

1.9.1 Structure

Suppose the state model for a given n^{th} order SISO system has an A matrix which is diagonal. Then the state vector differential equation decomposes into n scalar equations

$$\dot{x}_i(t) = \lambda_i x_i(t) + b_i u(t) \qquad i = 1, 2, \cdots, n \tag{1.83}$$

with the output being a scalar multiple of these components

$$y(t) = \sum_{i=1}^{n} c_i x_i(t) + Du(t) \tag{1.84}$$

We can visualize these equations as a block diagram of the sort introduced in Section 1.3 to model the zero-input response.

Alternatively, when we view this state model in the frequency domain so as to obtain the plant's transfer function, we see that since the state model's system matrix, A, is

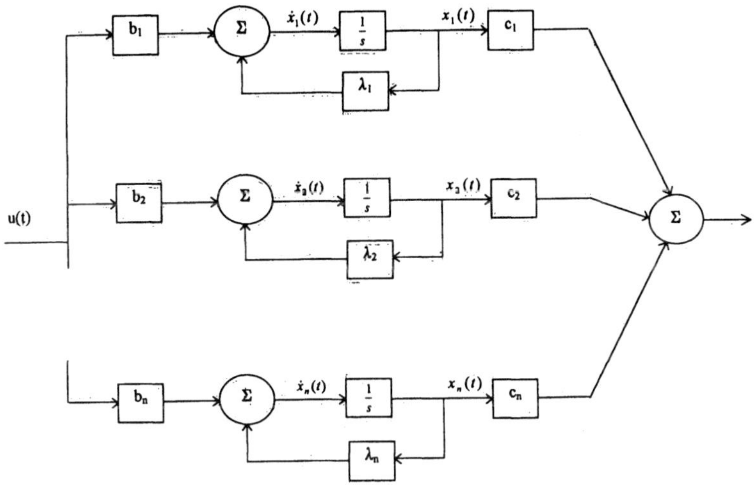

Figure 1.5 Block diagram representation for diagonal form state model

diagonal, we have $(sI - A)^{-1}$ diagonal,

$$(sI - A)^{-1} = \begin{bmatrix} (s-\lambda_1)^{-1} & 0 & \cdots & 0 \\ 0 & (s-\lambda_2)^{-1} & \cdots & 0 \\ \vdots & \vdots & \ddots & \vdots \\ 0 & 0 & \cdots & (s-\lambda_n)^{-1} \end{bmatrix} \tag{1.85}$$

This fact simplifies the dependency of the transfer function on the elements in the B and C matrices since we have

$$\begin{aligned} G(s) &= C(sI - A)^{-1}B + D \\ &= \sum_{i=1}^{n} \frac{c_i b_i}{s - \lambda_i} + D \end{aligned} \tag{1.86}$$

where

$$C = \begin{bmatrix} c_1 & c_2 & \cdots & c_n \end{bmatrix} \qquad B = \begin{bmatrix} b_1 \\ b_2 \\ \vdots \\ b_n \end{bmatrix}$$

1.9.2 Properties

Notice from (1.83) that the evolution of the components of the state, $x_k(t)$, are decoupled from each other, i.e., $x_k(t)$ is independent of $\{x_i(t) : i \in [0, n], i \neq k\}$. This fact plays an important role in the following discussion.

We begin by noting that this decoupling of components of the state from each other implies that any component of the state, say $x_k(t)$, depends on the input through the kth component of the input matrix, b_k, only. Therefore, if one of the components of B is zero, the corresponding component of the state is not affected by the input. Alternatively, notice from (1.86) that when $b_k = 0$, the transfer function, $G(s)$, does not have a pole at λ_k and the zero-state response does not involve the mode $e^{\lambda_k t}$.

These phenomena are related to a property of state models known as controllability. Roughly speaking, a system is said to be controllable if its state can be forced, by the input, to equal any point in state space at some finite time . Therefore if $b_k = 0$, the evolution of the kth component of the state, $x_k(t)$, is independent of the input, $u(t)$, and the state model is not controllable since there is no way of manipulating the input to make $x_k(t_f)$ equal some specified value after the passage of a finite amount of time, t_f. We will see in the next chapter that controllability is a necessary and sufficient condition for being able to arbitrarily assign the closed loop system matrix eigenvalues by using state feedback. Unlike the diagonal form state model, the controllability of a state model in other coordinates, depends on both the A and B matrices.

The foregoing observations were obtained by viewing the system from the input side. We can obtain dual observations by viewing the system from the output side involving the output instead of the input. We develop these observations as follows.

First, since equation (1.83) decouples the evolution the components of the state from each other, we see that the output, $y(t)$, (1.84), depends on the kth component of the state, $x_k(t)$, through c_k only. Therefore, if one of the components of C is zero, the corresponding component of the state does not affect the output. Alternatively, notice from (1.86) that when $c_k = 0$, the transfer function, $G(s)$, does not have a pole at λ_k and the zero-state response does not involve the mode $e^{\lambda_k t}$.

The foregoing phenomena are related to the property of state models known as observability. Recall that we encountered this property in the previous section in connection with the system's zero-input response. In the present instance we see that the observability of a system also relates to the system's zero-state response. Notice that unlike the diagonal form state model, the observability of a state model in other coordinates, depends on both the A and C matrices.

In summary, we see from equation (1.86) that if either b_k or c_k is zero for some integer $k \in [0, n]$, then the system transfer function $G(s)$ does not have a pole at λ_k and the zero-state response does not involve the mode $e^{\lambda_k t}$. Moreover, under these conditions, if we transformed coordinates and recomputed the transfer function from the parameters of the state model in the new coordinates, i.e.,

$$G(s) = \bar{C}(sI - \bar{A})^{-1}\bar{B} + D$$

we would find that the both the numerator polynomial, $\bar{C}\text{adj}(sI - \bar{A})\bar{B}$, and denominator polynomial, $\det(sI - \bar{A})$ (or $\det(sI - A)$ since the characteristic polynomial is invariant to coordinate transformation), would have a zero at λ_k. Since $s - \lambda_k$ would then be a common factor of $\bar{C}\text{adj}(sI - \bar{A})\bar{B}$ and $\det(sI - \bar{A})$, it would cancel from the ratio of these polynomials and be absent from the transfer function. Thus the eigenvalue of A at λ_k is not a pole of $G(s)$ and has no effect on the system's input–output.

Given a transfer function, any state model that is both controllable and observable and has the same input–output behavior as the transfer function, is said to be a minimal realization of the transfer function. Notice that this terminology reflects the fact that lack of either controllability or observability increases the dimension of the state model over the minimal dimension needed to match the zero-state behavior of the state model with that of the transfer function. As we will see, this fact play's an important role in control theory.

1.9.3 Obtaining the diagonal form state model

The relation between the form for the transfer function, equation (1.86) and the diagonal form state model suggests a way of determining a diagonal form state model for a given SISO transfer function, $G(s)$, having simple poles. This is done as follows.

We begin by using one cycle of division when the given transfer function is proper, to separate out the strictly proper part of the transfer function. Thus suppose we are given a transfer function which is proper

$$G_p(s) = \frac{\sum_{i=0}^{n} \beta_{n-i}\, s^i}{s^n + \sum_{i=0}^{n-1} a_{n-i}\, s^i}$$

Then dividing the denominator into the numerator enables $G_p(s)$ to be written as the sum of a strictly proper transfer function plus a constant, i.e.,

$$G_p(s) = G_{sp}(s) \; + \; \beta_0$$

where

$$G_{sp}(s) = \frac{\sum_{i=0}^{n-1} b_{n-i}\, s^i}{s^n + \sum_{i=0}^{n-1} a_{n-i}\, s^i} \qquad\qquad b_i = \beta_i - \beta_0 a_i \quad i = 1, 2, \cdots n$$

Then assuming the strictly proper part, $G_{sp}(s)$, can be expanded in partial fractions as

$$G_{sp}(s) = \sum_{i=1}^{n} \frac{k_i}{s - \lambda_i} \tag{1.87}$$

we can obtain the diagonal form state model by factoring the k_is as

$$k_i = c_i b_i$$

and referring to equation (1.86).

Notice it is only possible to obtain a diagonal form state model if the transfer function can be expanded as equation (1.87), i.e., if the system can be decomposed into first order systems. When $G_{sp}(s)$ has simple poles we can always expand $G_{sp}(s)$, as equation (1.87).

Notice that if $G_{sp}(s)$ has simple poles some of which are complex, the diagonal form state model will be complex. In some situations we may want the benefits of the diagonal form state model, i.e., decoupled state differential equations, but require the state model to be real. In this case a reasonable compromise is to combine complex conjugate pairs of terms in equation (1.87) so that the resulting realization has an A matrix which is block diagonal with each pair of complex conjugate poles corresponding to a 2×2 real block on the diagonal of A. More specifically we expand $G_{sp}(s)$ as

$$G_{sp}(s) = G_{spr}(s) + G_{spc}(s) \tag{1.88}$$

where

$$G_{spr}(s) = \sum_{i=1}^{n_r} \frac{k_i}{s - \lambda_i} \qquad \lambda_i \text{ real}$$

$$G_{spc}(s) = \sum_{i=1}^{n_c} \frac{b_{i1}s + b_{i2}}{s^2 + a_{i1}s + a_{i2}}$$

with the n_c complex pairs of poles of $G_{sp}(s)$ being the poles of $G_{spc}(s)$, i.e.,

$$s^2 + a_{i1}s + a_{i2} = (s - \lambda_{i+n_r})(s - \lambda^*_{i+n_r})$$

Now we can obtain a state model from (1.88) as

$$A = \begin{bmatrix} A_1 & A_2 \\ A_3 & A_4 \end{bmatrix} \qquad B = \begin{bmatrix} B_1 \\ B_2 \end{bmatrix}$$

$$C = \begin{bmatrix} C_1 & C_2 \end{bmatrix}$$

where (A_1, B_1, C_1) constitute a diagonal form state model for $G_{spr}(s)$, i.e., A_1 is diagonal with diagonal elements equal to the real poles and the elements of B_1 and C_1 satisfy $c_{1i}b_{i1} = k_i$ for $i = 1, 2, \cdots n_r$. Both off diagonal blocks of A, namely A_2 and A_3 are null.

The remaining blocks constitute a state model for $G_{spc}(s)$ as

$$A_4 = \begin{bmatrix} A_{41} & \emptyset & \cdots & \emptyset \\ \emptyset & A_{42} & \cdots & \emptyset \\ \vdots & \vdots & \ddots & \vdots \\ \emptyset & \emptyset & \cdots & A_{4n_c} \end{bmatrix} \qquad B_2 = \begin{bmatrix} B_{21} \\ B_{22} \\ \vdots \\ B_{2n_c} \end{bmatrix}$$

$$C_2 = \begin{bmatrix} C_{21} & C_{22} & \cdots & C_{2n_c} \end{bmatrix}$$

where $\{(A_{4i}, B_{2i}, C_{2i}) : i = 1, 2, \cdots n_c\}$ are 2 dimensional real parameter state models which we obtain from $\{G_{ci} : i = 1, 2, \cdots n_c\}$ with

$$G_{ci} = \frac{b_{i1}s + b_{i2}}{s^2 + a_{i1}s + a_{i2}}$$

One such real parameter state model is given as

$$A_{4i} = \begin{bmatrix} -a_{i1} & -a_{i2} \\ 1 & 0 \end{bmatrix} \qquad B_{2i} = \begin{bmatrix} 1 \\ 0 \end{bmatrix} \tag{1.89}$$

$$C_{2i} = \begin{bmatrix} b_{i1} & b_{i2} \end{bmatrix}$$

Finally notice that we encountered a restricted version of the state model (1.89) in Section 1.8.1. In the next Chapter we will make extensive use of this form which is referred to as a controller form. At that time we will provide a means for its determination from a given transfer function.

1.10 Computer Calculation of the State and Output

A fundamental problem encountered in using digital computers to analyses control problems is created by the fact that the dynamic processes needing to be controlled operate in continuous time whereas digital computers perform operations in discrete time. In this section we indicate how this fundamental mismatch is overcome.

In order to be able to use a digital computer to calculate the system output $y(t)$ from (1.82) we need to employ a piecewise constant approximation of the input $u(t)$ over the time interval of interest, $[0, t_N]$. This is done by partitioning the interval into N equal non-overlaping contiguous subintervals each of time duration T. Then the piecewise constant approximation, $\bar{u}(t)$, of the input is given by

$$\bar{u}(t) = u(t_k) \qquad \text{for } t \in [t_k, t_{k+1}) \tag{1.90}$$

where

$$t_k = kT : \ k = 0, 1, 2, \cdots N.$$

The set $\{u(t_k) : \ k = 0, 1, 2, \cdots N\}$ is referred to as the discrete-time equivalent of the continuous-time input, $u(t)$, and the values, $u(t_k)$, of the discrete-time equivalent input are referred to as the sampled values of the input. The continuous time "staircase like" function $\bar{u}(t)$ generated by this approximation is referred to as the sample and hold equivalent of the input. Figure 1.6 shows a plot of a typical input and its sample and hold equivalent.

The output, $y(t)$, at the sample times, $t_k : \ k = 0, 1, 2, \cdots, N$, is computed by repeating the calculation indicated in equation (1.82) for each subinterval with kT considered as the start time for the computation of $y([k+1]T)$. This is carried out in a cyclical fashion with the state that is computed at the end of a given subinterval being used as the initial state for the computation of the state at the end of the next subinterval.

More specifically, the computations are started by specifying the initial state, $x(0)$, and the input, $u(t)$, over the interval of interest, $[0, t_N]$. The computer program then decides on the sample spacing T. This is done so that the sample and hold version of the input is a "close" approximation to the actual input in the time interval $[0, t_N]$. Then taking the

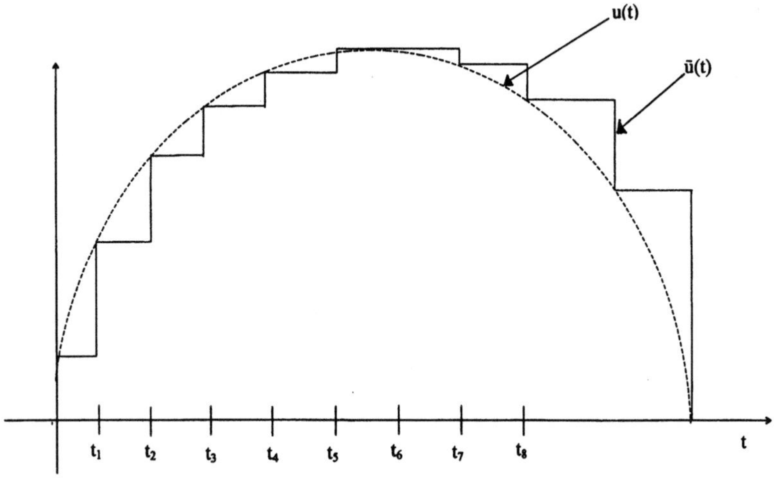

Figure 1.6 Typical input and its piecewise constant approximation

input as $u(0)$ over $[0, T]$ and using (1.82) we compute $y(T)$ from

$$x(T) = e^{AT}x(0) + \int_0^T e^{A(T-\tau)}Bu(0)\mathrm{d}\tau$$

$$y(T) = Cx(T) + Du(T)$$

which can be rewritten as

$$\begin{aligned} x(T) &= Fx(0) + Gu(0) \\ y(T) &= Cx(T) + Du(T) \end{aligned} \tag{1.91}$$

where

$$F = e^{AT}$$

$$G = \int_0^T e^{A(T-\tau)}B\mathrm{d}\tau$$

Then to compute $y(2T)$, we replace $x(0)$ by $x(T)$ and $u(0)$ by $u(T)$ in (1.91). This gives

$$x(2T) = Fx(T) + Gu(T)$$

$$y(2T) = Cx(2T) + Du(2T)$$

This procedure continues with the output at the $(k+1)^{\text{th}}$ sample time being given as

$$x[(k+1)T] = Fx(kT) + Gu(kT)$$

$$y[(k+1)T] = Cx[(k+1)T] + Du[(k+1)T]$$

There are several ways in which the computation of F and G can be carried out. A simple way of doing this is to truncate the series expansion for the matrix exponential, (1.22), with the upper limit, n_0, of the sum being chosen large enough to achieve some specified accuracy. Thus F, G in (1.91) can be expressed as

$$F = \sum_{i=0}^{n_0} \frac{A^i T^i}{i!} + E_F$$

$$G = \left(\int_0^T \sum_{i=0}^{n_0} \frac{A^i \alpha^i}{i!} \mathrm{d}\alpha \right) B + E_G$$

$$= \left(\sum_{i=0}^{n_0} \frac{A^i T^{i+1}}{(i+1)!} \right) B + E_G$$

where the error matrices E_F and E_G are made negligible by choosing n_0 "large enough" so that we can take F and G as

$$F = \sum_{i=0}^{n_0} \frac{A^i T^i}{i!}$$

$$G = \left(\sum_{i=0}^{n_0} \frac{A^i T^{i+1}}{(i+1)!} \right) B$$

Notice that the choice of the number of subintervals needed to cover the main interval, i.e., $[0,t_N]$, also plays a role in the accuracy of the computation, with an increase in accuracy being achieved at the expense of more computation time i.e., the accuracy goes up the smaller T and the larger N.

1.11 Notes and References

There have been a number of excellent textbooks expounding on state space analysis of systems. Some of these books are [5], [4], [23], [6], [9], [15].
Throughout the eighties the book by Kailath [23] was used widely as a text for courses on state space analysis of linear systems. Although Brockett's book [5], is quite old and was written for a more mathematically sophisticated reader, the first half of the book does give the reader a nice exposure to the more general nature of the subject. The treatment of state modeling in the brief book by Blackman [4], shares the view with the present book that understanding basic ideas in the state space can be facilitated by thinking geometrically.

Concerning books that provide insight into matters pertaining to linear algebra, Strange's book [39] gives an excellent introduction. The book by Golub and Van Loan [17] provides a more advanced treatment which is more orientated towards computer computation. In addition the book by Brogan [6], plus the appendix to Kailath's book also provide useful information on linear algebra.

2

State Feedback and Controllability

2.1 Introduction

An important and well known consequence of feedback is behavioral modification. Feedback is present when the input to a system depends on that system's output. When feedback is used to modify a system's behavior the resulting closed loop system is referred to as a feedback control system. For the purposes of discussion we can separate the control system into a plant and a controller. The plant consists of the physical process to be controlled plus any transducers, actuators or sensors, needed to interface the physical process output to the controller and the output of the controller to the input of the physical process. While controllers were implemented in the past by a physical process such as an electric filter, their implementation now is often done by a digital computer. Figure 2.1 shows the general setup for a control system.

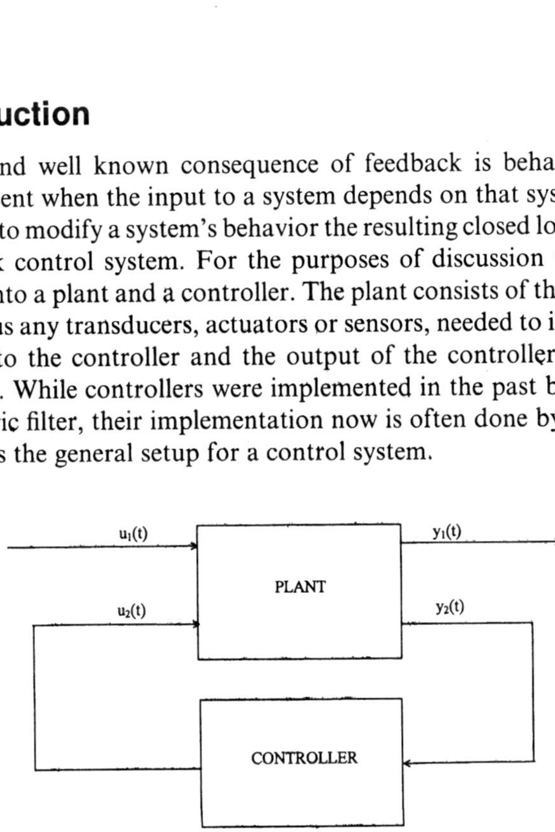

Figure 2.1 Block diagram of the feedback control scheme

$$u_1(t) = \text{external input} \qquad y_1(t) = \text{desired output}$$
$$u_2(t) = \text{controlled input} \qquad y_2(t) = \text{measured output}$$

where $u_1(t), u_2(t), y_1(t), y_2(t)$ have dimensions m_1, m_2, p_1, p_2 respectively.

The general form of the state model for the plant is

$$\left.\begin{aligned}\dot{x}(t) &= Ax(t) + B_1(t)u_1(t) + B_2u_2(t)\\ y_1(t) &= C_1x(t) + D_{11}u_1(t) + D_{12}u_2(t)\\ y_2(t) &= C_2x(t) + D_{21}u_1(t) + D_{22}u_2(t)\end{aligned}\right\} \tag{2.1}$$

This chapter focuses on the effect of using a feedback signal composed from the components of the plant's state as the controlled input,

$$u_2(t) = Kx(t) \tag{2.2}$$

where K has m_2 rows, one row for each scalar signal making up the controlled input. This important kind of feedback is referred to as state feedback and is one of two ingredients needed to complete the design of the controller.

The need for something in addition to state feedback can be appreciated when it is recognized that the plant state is usually unknown. Therefore if we are to use state feedback as a basis for the design of a controller, we must have a means for estimating the plant state. This is done by a state model, referred to as a state estimator, which is designed from knowledge of the plant's state model. The information on the plant state contained in measurements of all or some of the plant's input and output signals is supplied to the state estimator by setting the estimator's input equal to these measurements. The state estimator is designed so that its state mimics the state of the plant. There are several different types of state estimator depending on the assumptions made about the relation between the actual plant signals and their measurements, e.g., exact, corrupted by random measurement noise, or partly unknown. In subsequent chapters we will study both the design of these estimators and the effect on the performance of the feedback control system that is caused by their use in conjunction with state feedback.

In this chapter we assume that the state of the plant's state model is known so that we can use it to generate the state feedback signal given by equation (2.2). Our goal in using state feedback is to produce a closed loop system in which the zero state and zero input behavior is changed in some way which is beneficial to the plant's use. The extent to which the plant's controlled input can affect the plant state, is referred to as a system's controllability. We will see that the controllability of the plant state model determines the extent to which the state model of the closed loop system can differ from the plant's state model when we use state feedback. The important property of state model controllability will be encountered in several different guises throughout this chapter and the rest of the text.

2.2 State Feedback

Applying state feedback, equation (2.2), to the plant, equation (2.1) gives the state model for the feedback system as

$$\left.\begin{aligned}\dot{x}(t) &= (A + B_2K)x(t) + B_1(t)u_1(t)\\ y_1(t) &= (C_1 + D_{12}K)x(t) + D_{11}u_1(t)\\ y_2(t) &= (C_2 + D_{22}K)x(t) + D_{21}u_1(t)\end{aligned}\right\} \tag{2.3}$$

Figure 2.2 gives a block diagram interpretation of state feedback.

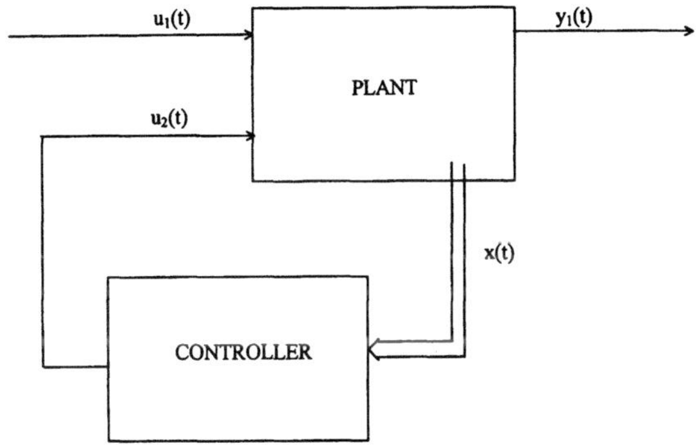

Figure 2.2 Block diagram representation of state feedback.

Notice that the main effect of using state feedback is the transformation of the system matrix from A, when the plant stands alone, to $A + B_2K$, when the plant is embedded in a state feedback loop. Thus we see that the main purpose of state feedback is the assignment of eigenvalues to the closed loop system matrix $A + B_2K$ to achieve a set of specified performance criteria for the behavior of the closed loop system from the external input $u_1(t)$ to the desired output $y_1(t)$. This fact leads immediately to the question: can we choose K so that the eigenvalues of $A + B_2K$ are assigned to any specified set of values? The answer to this question is obviously of great importance to the use of state feedback in the design of feedback control systems. The answer to this question, which is given in the following theorem, involves the use of the left-eigenvectors of A. Since this property applies in general to any state model (A, B, C, D) we will drop the subscript on B, i.e., B_2 will be replaced by B.

Theorem 2.1 Whenever A has a left-eigenvector w^i, i.e., $w^{iT}A = \lambda_i w^{iT}$, such that

$$w^{iT}B = 0$$

the corresponding eigenvalue of A, λ_i, is invariant to state feedback, i.e., $\lambda_i \in \lambda(A + BK)$ for all K.

Proof Suppose the condition in the theorem is satisfied. Then multiplying the closed loop state feedback system matrix, $A + BK$, on the left by w^{iT} gives

$$w^{iT}(A + BK) = w^{iT}A + w^{iT}BK$$

However since $w^{iT}A = \lambda_i w^{iT}$ and $w^{iT}BK = \emptyset$, the previous equation becomes

$$w^{iT}(A + BK) = \lambda_i w^{iT}$$

and we see that λ_i is an eigenvalue of the system matrix for the state model of the feedback control system, since $\lambda_i \in \lambda(A + BK)$ for all K. ■

It turns out that when there are no left-eigenvectors of A which satisfy $w_i^T B = 0$, the eigenvalues of A are all assignable to any specified values by state feedback. This is shown later in Theorem 2.4. Notice that when we have the condition $w_i^T B = 0$ so that the corresponding eigenvalue λ_i cannot be assigned by state feedback, we refer to λ_i as an uncontrollable eigenvalue. Conversely, when we have the condition $w_i^T B \neq 0$ we refer to the corresponding eigenvalue λ_i which can be assigned by state feedback, as a controllable eigenvalue.

Definition 2.1 An eigenvalue λ_i of A is said to be a controllable (uncontrollable) eigenvalue of the pair (A, B) if we can (cannot) find K so that $\lambda_i \notin \lambda[A + BK]$

Uncontrollable (controllable) eigenvalues for a pair (A, B) are invariant to changing coordinates. This can be seen as follows.

Suppose λ_i is an uncontrollable eigenvalue of (A, B). Then we have

$$w^{iT} B = 0 \tag{2.4}$$

$$w^{iT} A = \lambda_i w^{iT} \tag{2.5}$$

Now insert TT^{-1} between w^{iT} and B in equation (2.4) and between w^{iT} and A, in equation (2.5). In addition, postmultiply both sides of equation (2.5) by T. Then we obtain

$$q^{iT} \bar{B} = 0$$

$$q^{iT} \bar{A} = \lambda_i q^{iT}$$

where

$$q^{iT} = w^{iT} T \qquad \bar{B} = T^{-1} B \qquad \bar{A} = T^{-1} A T$$

Thus we see that λ_i remains an uncontrollable eigenvalue of the state model in the transformed coordinates.

Reflection on the foregoing reveals that plant state models which have no eigenvalues of A which are both uncontrollable and unstable have the property that state feedback can be used to make the closed loop system stable. State models having this property are referred to as stabilizable state models. Conversely, a state model is not stabilizable when it has uncontrollable eigenvalues which are unstable. This is an important property since the stability of a control system is almost always a necessary condition for a control system to be of any use.

Definition 2.2 A plant state model is said to be stabilizable if all its uncontrollable eigenvalues are stable.

Recall that the poles of the transfer function for an SISO output feedback control system are constrained to lie on branches of the root locus. In the present case of state feedback, the location of the eigenvalues of the closed loop system matrix $A + BK$ are unconstrained provided, of course, that they are controllable. This additional flexibility of state feedback over output feedback arises from there being one adjustable parameter in K for each component of the state when we use state feedback, whereas there is only one adjustable parameter when we use constant output feedback.

2.3 Eigenvalue Assignment

Recall, Section 1.6.4, that a state model's system matrix has the same eigenvalues independent of the coordinates used to express the state model. In addition, notice that if the state, $x(t)$, in the original coordinates is related to the state, $\bar{x}(t)$, in the new coordinates as

$$x(t) = T\bar{x}(t)$$

then the state feedback matrix is transformed as $\bar{K} = KT$ since

$$u(t) = Kx(t) = KT\bar{x}(t)$$

This fact provides us with the flexibility of being able to carry out the determination of the state feedback matrix in coordinates which are the most suitable for this computation.

We saw in the previous section, that we can choose K to achieve a specified set of closed loop eigenvalues, $\{\mu_i : i = 1, 2, \cdots, n\}$, provided that each uncontrollable eigenvalue of the pair (A, B) equals one of the desired closed loop eigenvalues. Assuming this condition is satisfied, the state feedback matrix, K, which is needed to make the set of closed loop eigenvalues equal the specified set of eigenvalues can be obtained by equating the characteristic polynomial for $A + BK$, denoted $\alpha(s)$, to the characteristic polynomial $\gamma(s)$ having roots equal to the specified set of eigenvalues, $\{\mu_i : \ i = 1, 2, \cdots n\}$, as

$$\alpha(s) = \gamma(s) \tag{2.6}$$

where

$$\alpha(s) = \det[sI - (A + BK)] = s^n + \alpha_1 s^{n-1} + \alpha_2 s^{n-2} + \cdots + \alpha_n$$

$$\gamma(s) = \prod_{i=1}^{n}(s - \mu_i) = s^n + \gamma_1 s^{n-1} + \gamma_2 s^{n-2} + \cdots + \gamma_n$$

Notice that the coefficients of $\alpha(s)$ are functions of the elements of K. Thus we see that equation (2.6) gives rise to n equations in the unknown elements of K, i.e.,

$$\alpha_i = \gamma_i \qquad i = 1, 2, \cdots n$$

In general, these equations in the elements of K are coupled in ways which depend on the A and B matrices. Therefore it becomes difficult to set up a computer program to form and solve these equations for all possible (A, B) pairs. However if the plant's state model is in controller form these equations are uncoupled and therefore easy to solve.

2.3.1 Eigenvalue assignment via the controller form

A state model for an nth order SISO system is said to be in controller form when $A = A_c$ is a companion matrix and $B = B_c$ is the first column of the $n \times n$ identity matrix. More specifically the controller form state model has parameters

$$A_c = \begin{bmatrix} -a_1 & -a_2 & \cdots & -a_{n-1} & -a_n \\ 1 & 0 & \cdots & 0 & 0 \\ 0 & 1 & \cdots & 0 & 0 \\ \cdots & \cdots & \cdots & \cdots & \cdots \\ 0 & 0 & \cdots & 1 & 0 \end{bmatrix} \qquad B_c = \begin{bmatrix} 1 \\ 0 \\ 0 \\ \vdots \\ 0 \end{bmatrix}$$

with D and C being whatever is needed to model the plant's behavior in these coordinates. Use of the controller form state model greatly simplifies the calculation of the state feedback matrix K_c needed to make $A_c + B_cK_c$ have a specified set of eigenvalues. This results from the fact that in these coordinates the state feedback system matrix, $A_c + B_cK_c$, is also in companion form

$$A_c + B_cK_c = \begin{bmatrix} -a_1 + k_{c1} & -a_2 + k_{c2} & -a_3 + k_3 & \cdots & -a_{n-1} + k_{n-1} & -a_n + k_n \\ 1 & 0 & 0 & \cdots & 0 & 0 \\ 0 & 1 & 0 & \cdots & 0 & 0 \\ \cdots & \cdots & \cdots & \cdots & \cdots & \cdots \\ 0 & 0 & 0 & \cdots & 1 & 0 \end{bmatrix}$$

with the $k_{ci}s$ being the elements of the state feedback matrix

$$K_c = \begin{bmatrix} k_{c1} & k_{c2} & k_{c3} & \cdots & k_{cn} \end{bmatrix}$$

Recall from Chapter 1 that since $A_c + B_cK_c$ is a companion matrix, the coefficients of its characteristic polynomial appear, with negative signs, along its first row. Therefore the coefficients of the characteristic polynomial for the closed loop system $\alpha(s)$, equation (2.6), are related to the elements of the state feedback matrix K as

$$\alpha_i = a_i - k_{ci} \qquad i = 1, 2, \cdots n \tag{2.7}$$

Then if we want the closed loop system matrix, $A_c + B_cK_c$, to have eigenvalues at $\{\mu_i : i = 1, 2, \cdots, n\}$, the $\alpha_i s$ are obtained from

$$\prod_{i=1}^{n} (s - \mu_i) = s^n + \alpha_1 s^{n-1} + \alpha_2 s^{n-2} + \cdots + \alpha_n$$

and the elements $\{k_{ci} : i = 1, 2, \cdots, n\}$ of the feedback matrix K are easily determined from equation (2.7) as

$$k_{ci} = a_i - \alpha_i \qquad i = 1, 2, \cdots n \tag{2.8}$$

Before showing how to transform coordinates so that a given state model is transformed to controller form, we consider the problem of determining a controller form state model from a given transfer function.

2.3.2 Realizing the controller form

Having seen that the assignment of eigenvalues is easy to do when the system state model is in controller form, we turn now to the problem of getting a controller form state model for a system specified by a transfer function or differential equation.

Theorem 2.2 Suppose we are given the a_i, b_i parameters in the differential equation or transfer function model of some plant, i.e.,

$$y^{(n)}(t) + \sum_{i=0}^{n-1} a_{n-i}\, y^{(i)}(t) = \sum_{i=0}^{n-1} b_{n-i}\, u^{(i)}(t)$$

or

$$G_{sp} = \frac{\sum_{i=0}^{n-1} b_{n-i}\, s^i}{s^n + \sum_{i=0}^{n-1} a_{n-i}\, s^i} \tag{2.9}$$

where

$$Y(s) = G_{sp}(s)U(s)$$

with $y^{(i)}(0) = u^{(i)}(0) = 0$ for all $i = 0, 1, 2, \cdots, n-1$

Then the controller form state model for this plant has parameters (A_c, B_c, C_c) given by

$$A_c = \begin{bmatrix} -a_1 & -a_2 & -a_3 & \cdots & -a_{n-1} & -a_n \\ 1 & 0 & 0 & \cdots & 0 & 0 \\ 0 & 1 & 0 & \cdots & 0 & 0 \\ \cdots & \cdots & \cdots & \cdots & \cdots & \cdots \\ 0 & 0 & 0 & \cdots & 1 & 0 \end{bmatrix} \qquad B_c = \begin{bmatrix} 1 \\ 0 \\ 0 \\ \vdots \\ 0 \end{bmatrix} \tag{2.10}$$

$$C_c = \begin{bmatrix} b_1 & b_2 & b_3 & \cdots & b_n \end{bmatrix}$$

Proof Factor the transfer function as

$$G_{sp}(s) = N(s)D(s)$$

where

$$N(s) = \sum_{i=0}^{n-1} b_{n-i} s^i \qquad D(s) = \frac{1}{s^n + \sum_{i-0}^{n-1} a_{n-i} s^i}$$

Then we see that

$$Y(s) = N(s)Z(s) \tag{2.11}$$

$$Z(s) = D(s)U(s) \tag{2.12}$$

where $Z(s)$ is the Laplace transform of the signal $z(t)$, i.e.

$$z(t) = \mathcal{L}^{-1}[Z(s)]$$

Since (2.12) gives the zero state response, $z(t)$, of a system having transfer function $D(s)$ and having $u(t)$ as input, we see that (2.12) implies that

$$z^{(n)}(t) + \sum_{i=0}^{n-1} a_{n-i} z^{(i)}(t) = u(t) \tag{2.13}$$

and setting

$$x_i(t) = z^{(n-i)}(t) \qquad\qquad i = 1, 2, \cdots, n \tag{2.14}$$

in (2.13) gives

$$\dot{x}(t) = A_c x(t) + B_c u(t)$$

where the matrices A_c, B_c are as stated in the theorem.

Finally, since the initial conditions are all assumed zero, we see from (2.11) that

$$y(t) = \sum_{i=0}^{n-1} b_{n-i}\, z^{(i)}(t)$$

and using the relation between the components of the state and the derivatives of $z(t)$, (2.14), we see that the foregoing expression for the output can be written as

$$y(t) = C_c x(t)$$

where the matrix C_c is as stated in the theorem. ■

Notice that since $G_{sp}(s)$ is strictly proper, $D_c = 0$. Alternatively when the given transfer function, $G_p(s)$, is proper, we can proceed as in Section 1.9.3 to obtain its controller form (A_c, B_c, C_c, D_c) with A_c, B_c, C_c as in Theorem 2.2 and $D = \beta_0$ where

$$G_p(s) = \frac{\sum_{i=0}^{n} \beta_{n-i}\, s^i}{s^n + \sum_{i=0}^{n-1} a_{n-i}\, s^i} = \frac{\sum_{i=0}^{n-1} b_{n-i}\, s^i}{s^n + \sum_{i=0}^{n-1} a_{n-i}\, s^i} + \beta_0$$

with

$$b_i = \beta_i - \beta_0 a_i \qquad\qquad i = 1, 2, \cdots n$$

In summary, Theorem 2.2 shows that the controller form state model is easily obtained when we are given either the transfer function or the differential equation governing the process behavior.

In some cases we may be given a state model for an SISO plant which is in a form other than the controller form. In this situation it may be possible to simplify the determination of the state feedback matrix K by:

(i) determining the coordinate transformation matrix, T_c, to transform the given plant state model to controller form

(ii) choosing K_c using (2.8) to assign the eigenvalues to the controller form state model

(iii) using the coordinate transformation matrix T_c to obtain the require state feedback matrix as $K = K_c T_c^{-1}$.

This approach, while not of practical significance when applied as just described, does enable the development of a practical computational algorithm for determining K. This algorithm which is referred to as Ackermann's formula will be developed in Section 2.3.4.

At this point we need to develop both a method for determining T_c and conditions on the given state model which enables its transformation to controller form.

2.3.3 Controller form state transformation

Suppose we want to find a coordinate transformation matrix, T_c, which transforms a given state model, (A, B, C), not in controller form to a state model, $(\bar{A}, \bar{B}, \bar{C})$, which is in controller form. Then assuming, for the time being, that such a transformation exists, we can solve this problem by proceeding as follows.

Since, in general A, and $\bar{A} = T^{-1}AT$ have the same characteristic polynomial, we can use the coefficients of the characteristic polynomial for the given A matrix, $\det[sI - A]$, to construct the controller form system matrix $\bar{A} = A_c$ from Theorem 2.2. In addition, since B_c is the first column of the $n \times n$ identity matrix we have $\bar{B} = B_c$ without doing any calculation. Therefore neither A_c nor B_c requires the determination of T_c. However T_c is needed to determine $\bar{C} = CT_c = C_c$. In what follows we use A_c, B_c and their relation to T_c to develop an algorithm for determining T_c, one column at a time.

To begin with, notice that the first column of T_c can be obtained immediately from inspection of the relation $T_c\bar{B} = B$ when T_c is written in terms of its columns,

$$\begin{bmatrix} t^1 & t^2 & \cdots & t^n \end{bmatrix} \begin{bmatrix} 1 \\ 0 \\ \vdots \\ 0 \end{bmatrix} = t^1 = B$$

where t^i is the ith column of T_c.

Having determined the first column of T_c as B, we can see how to determine other columns of T_c by inspection of the relation $T_c\bar{A} = AT_c$ with $\bar{A}$ set equal to A_c and T_c expressed in terms of its columns,

$$\begin{bmatrix} t^1 & t^2 & \cdots & t^n \end{bmatrix} \begin{bmatrix} -a_1 & -a_2 & -a_3 & \cdots & -a_{n-1} & -a_n \\ 1 & 0 & 0 & \cdots & 0 & 0 \\ 0 & 1 & 0 & \cdots & 0 & 0 \\ \cdots & \cdots & \cdots & \cdots & \cdots & \cdots \\ 0 & 0 & 0 & \cdots & 1 & 0 \end{bmatrix} = A\begin{bmatrix} t^1 & t^2 & \cdots & t^n \end{bmatrix}$$

Thus equating like positioned columns on either side of this equation gives

$$\left.\begin{aligned} -a_1 t^1 + t^2 &= At^1 \\ -a_2 t^1 + t^3 &= At^2 \\ &\vdots \\ -a_n t^1 + t^n &= At^{n-1} \end{aligned}\right\} \tag{2.15}$$

$$-a_n t^1 = At^n \tag{2.16}$$

and we see that the remaining columns of T_c can be generated successively from (2.15) by starting with $t^1 = B$ and proceeding as

$$\left.\begin{aligned} t^2 &= a_1 t^1 + At^1 \\ t^3 &= a_2 t^1 + At^2 \\ &\vdots \\ t^n &= a_{n-1} t^1 + At^{n-1} \end{aligned}\right\} \tag{2.17}$$

To recap, we can generate the columns of T_c by

(i) obtaining the coefficients of the characteristic polynomial for the given A
(ii) setting the first column of T_c equal to the given B.
(iii) using (2.17) to determine the remaining columns of T_c in succession.

Notice that so far we have no assurance that the matrix T_c which results from using (2.17) will be invertible. However T_c must be invertible if it is to be a coordinate transformation matrix. We show now that only those state models which have the property of controllability will produce a nonsingular T_c matrix as a result of applying the foregoing algorithm. Before doing this notice that we did not need (2.16) to get (2.17). This extra equation will be used in Section 2.3.5 to derive the Cayley–Hamilton theorem.

2.3.4 Condition for controller form equivalence

In order to determine conditions on the given state model which guarantees that it can be transformed to a controller form, i.e., to insure that the matrix T_c which is generated by (2.17) is invertible, we expand (2.17) by successive substitution to obtain

$$\begin{aligned} t^1 &= B \\ t^2 &= a_1 B + AB \\ t^3 &= a_2 B + a_1 AB + A^2 B \\ &\vdots \\ t^n &= a_{n-1} B + a_{n-2} AB + \cdots + a_1 A^{n-2} B + A^{n-1} B \end{aligned} \tag{2.18}$$

Then writing these equations in matrix form gives

$$T_c = \Omega R \tag{2.19}$$

where

$$\Omega = \begin{bmatrix} B & AB & A^2B & \cdots & A^{n-1}B \end{bmatrix} \qquad R = \begin{bmatrix} 1 & a_1 & a_2 & \cdots & a_{n-1} \\ 0 & 1 & a_1 & \cdots & a_{n-2} \\ 0 & 0 & 1 & \cdots & a_{n-3} \\ \vdots & \vdots & \vdots & \vdots & \vdots \\ 0 & 0 & 0 & \cdots & 1 \end{bmatrix}$$

Now since the product of two square matrices is invertible if and only if each matrix in the product is invertible, T_c is invertible only if both Ω and R are invertible. However since R is an upper-triangular matrix having nonzero elements along its diagonal, R is invertible for any given state model. Therefore T_c is invertible if and only if Ω is invertible.

The matrix Ω, which depends on the interaction between the given state model's A and B matrices, is referred to as the given state model's controllability matrix. For systems having only one input the controllability matrix is square and the state model is said to be controllable if its controllability matrix is invertible. However when the input consists of m scalar inputs, Ω, has nm columns and only n rows. In this case the state model is said to be controllable if its controllability matrix is full rank, i.e., all its rows are independent or equivalently, n of its nm columns are independent.

More generally, the property of controllability of a given state model is preserved under coordinate transformation. We can see this by noting that the controllability matrix $\bar{\Omega}$ of the state model in the transformed coordinates is related to the controllability matrix Ω of the state model in the original coordinates as

$$\begin{aligned} \bar{\Omega} &= \begin{bmatrix} \bar{B} & \bar{A}\bar{B} & \cdots & \bar{A}^{n-1}\bar{B} \end{bmatrix} \\ &= \begin{bmatrix} T^{-1}B & T^{-1}ATT^{-1}B & \cdots & T^{-1}A^{n-1}B \end{bmatrix} \\ &= T^{-1}\Omega \end{aligned} \tag{2.20}$$

Therefore since Ω has fewer rows than columns we have

$$\alpha^T\bar{\Omega} = \beta^T\Omega = \emptyset^T$$

for some $\alpha \neq \emptyset$ only if Ω has dependent rows, since $\beta^T = \alpha^T T^{-1} = \emptyset^T$ only if $\alpha = \emptyset$. This implies that

$$\text{rank}[\bar{\Omega}] = \text{rank}[T^{-1}\Omega] = \text{rank}[\Omega]$$

where the rank of a matrix equals the maximum number of its columns which are independent. However since, for any matrix, the number of independent columns equals the number of independent rows, we see that, in general

$$\text{rank}[\Omega] \leq n$$

since Ω is an $n \times nm$ matrix for any n dimension state model having m scalar inputs. This fact will be used later in the fourth section of this chapter.

When the state model is single input and controllable, $m = 1$ and $\Omega, \bar{\Omega}$ in (2.20) are square and invertible. Therefore we can obtain the coordinate transformation matrix, T, from (2.20) as

$$T = \Omega \bar{\Omega}^{-1} \tag{2.21}$$

Then comparing this expression with (2.19) we see that $\bar{\Omega}^{-1} = \Omega_c^{-1} = R$. Moreover the controllability matrix for a state model in controller form is readily calculated as

$$\Omega_c = \begin{bmatrix} B_c & A_c B_c & \cdots & A_c^{n-1} B_c \end{bmatrix}$$

$$= \begin{bmatrix} 1 & -a_1 & a_1^2 - a_2 & \cdots & * \\ 0 & 1 & -a_1 & \cdots & * \\ 0 & 0 & 1 & \cdots & * \\ \vdots & \vdots & \vdots & & \vdots \\ 0 & 0 & 0 & \cdots & 1 \end{bmatrix} \tag{2.22}$$

which, being upper-triangular with fixed nonzero entries along the diagonal, is invertible. Thus all controller form state models are controllable.

To recap, a given single input state model can be transformed through an appropriate coordinate transformation to a controller form state model if and only if the given system is controllable. Moreover we have just seen that an uncontrollable state model cannot be made controllable by a coordinate transformation and that any controller form state model is controllable. In addition, we will see that there are several other aspects of system controllability which play important roles in control theory. A more basic view of system controllability is developed in the fourth section of this chapter.

2.3.5 Ackermann's formula

As mentioned at the end of Section 2.3.1, we could calculate K to assign the eigenvalues of $A + BK$, for a given single input controllable system by transforming the given system to controller form , finding K_c, and transforming K_c back into the original coordinates through the relation $K = K_c T_c^{-1}$. We will see now that we can use this idea to develop a computational algorithm for obtaining K which is known as Ackermann's formula. In order to do this we will need the following result.

Theorem 2.3 (Cayley–Hamilton) Any square matrix A satisfies its own characteristic equation.

Before proving this theorem, the following explanation may be needed. Recall that the characteristic equation for A,

$$a(s) = \det[sI - A] = s^n + a_1 s^{n-1} + a_2 s^{n-2} + \cdots + a_n = 0 \tag{2.23}$$

is a scalar equation which is satisfied by the eigenvalues of A.

Therefore Theorem 2.3 tells us that if we replace the scalar s in (2.23) by the matrix A, we obtain the result

$$a(A) = A^n + a_1 A^{n-1} + a_2 A^{n-2} + \cdots + a_n I = \emptyset$$

We can readily show the Cayley–Hamilton theorem as follows.

Proof Prcmultiplying (2.18) by A and using (2.16) yiclds

$$(A^n + a_1 A^{n-1} + a_2 A^{n-2} + \cdots a_n I)B = \emptyset \tag{2.24}$$

Now (2.24) holds for any B. Therefore letting $B = I$ in the foregoing equation yields

$$A^n + a_1 A^{n-1} + a_2 A^{n-2} + \cdots a_n I = \emptyset$$

■

Theorem 2.4 (Ackermann' Formula) Given a single input controllable state model (A, B, C, D) and the desired characteristic polynomial, $\alpha(s)$, for the closed loop system

$$\alpha(s) = \det[sI - (A + BK)]$$

then the required feedback matrix K is given by

$$K = -q^T[\alpha(A)]$$

where q^T is the last row of the inverse of the controllability matrix, Ω, for the given pair (A, B).

Proof For convenience, and without loss of generality, suppose the given plant is of dimension $n = 3$. Then the plant state model's system matrix A has characteristic polynomial

$$a(s) = \det[sI - A] = s^3 + a_1 s^2 + a_2 s + a_3$$

Let the desired characteristic polynomial, $\alpha(s)$, for the state feedback system matrix be denoted as

$$\alpha(s) = \det[sI - (A + BK)] = s^3 + \alpha_1 s^2 + \alpha_2 s + \alpha_3$$

Then using the coordinate transformation matrix T_c developed in Section 2.3.3 we transform the given state model for the plant to controller form

$$A_c = \begin{bmatrix} -a_1 & -a_2 & -a_3 \\ 1 & 0 & 0 \\ 0 & 1 & 0 \end{bmatrix} \qquad B_c = \begin{bmatrix} 1 \\ 0 \\ 0 \end{bmatrix}$$

so that the state feedback system matrix in these coordinates is given by

$$A_c + B_c K_c = \begin{bmatrix} -a_1 + k_{c1} & -a_2 + k_{c2} & -a_3 + k_{c3} \\ 1 & 0 & 0 \\ 0 & 1 & 0 \end{bmatrix}$$

where

$$K_c = [k_{c1} \quad k_{c2} \quad k_{c3}]$$

Therefore the desired characteristic polynomial can be written by inspection from the companion form for $A_c + B_c K_c$ as

$$\begin{aligned}\alpha(s) &= s^3 + (a_1 - k_{c1})s^2 + (a_2 - k_{c2})s + (a_3 - k_{c3}) \\ &= a(s) - (k_{c1}s^2 + k_{c2}s + k_{c3}) \end{aligned} \tag{2.25}$$

Next recalling that A and $T^{-1}AT$ have the same characteristic polynomial for any invertible T, we see that $a(s)$ is the characteristic polynomial for both A and A_c. Therefore from the Cayley–Hamilton theorem (Theorem 2.3) we have

$$a(A) = a(A_c) = \emptyset \tag{2.26}$$

Therefore we see from (2.25, 2.26) that

$$\alpha(A_c) = -k_{c1}A_c^2 - k_{c2}A_c - k_{c3}I \tag{2.27}$$

where I is the 3×3 identity matrix

$$I = \begin{bmatrix} i^{1T} \\ i^{2T} \\ i^{3T} \end{bmatrix}$$

However we can readily verify that

$$i^{3T}I = i^{3T} \qquad i^{3T}A_c = i^{2T} \qquad i^{3T}A_c^2 = i^{1T}$$

Therefore we see that pre-multiplying (2.27) by the third row of the identity matrix gives

$$\begin{aligned} i^{3T}\alpha(A_c) &= -k_{c1}i^{1T} - k_{c2}i^{2T} - k_{c3}i^{3T} \\ &= -[k_{c1} \quad k_{c2} \quad k_{c3}] = -K_c \end{aligned} \tag{2.28}$$

Then inserting the identity matrix $T_c^{-1}T_c$ between i^{3T} and $\alpha(A_c)$ on the left side of (2.28) and post-multiplying throughout by T_c^{-1} yields

$$(i^{3T}T_c^{-1})(T_c\alpha(A_c)T_c^{-1}) = -K_cT_c^{-1} \tag{2.29}$$

Moreover using the facts that

$$\begin{aligned} T_c\alpha(A_c)T_c^{-1} &= T_cA_c^3T_c^{-1} + \alpha_1 T_cA_c^2T_c^{-1} + \alpha_2 T_cA_cT_c^{-1} + \alpha_3 I \\ &= A^3 + \alpha_1 A^2 + \alpha_2 A + \alpha_3 I = \alpha(A) \end{aligned} \tag{2.30}$$

and that $K_c T_c^{-1} = K$ we see that (2.29) can be rewritten as

$$i^{3T} T_c^{-1} \alpha(A) = -K \tag{2.31}$$

Finally recalling (2.19) we see that

$$T_c^{-1} = R^{-1}\Omega^{-1} = \begin{bmatrix} 1 & * & * \\ 0 & 1 & * \\ 0 & 0 & 1 \end{bmatrix} \Omega^{-1} \tag{2.32}$$

where

$$\Omega = \begin{bmatrix} B & AB & A^2B \end{bmatrix} \qquad R = \begin{bmatrix} 1 & a_1 & a_2 \\ 0 & 1 & a_1 \\ 0 & 0 & 1 \end{bmatrix}$$

and $*$ are elements of R^{-1} which are not needed. Thus using (2.32) in (2.31) gives

$$K = -(i^{3T}\Omega^{-1})[\alpha(A)]$$

and the theorem is proved. ■

Notice that since only the last row of Ω^{-1} i.e., q^T, is required when we use this result to compute the feedback matrix, K, we need only compute q which satisfies

$$i^n = \Omega^T q$$

with i^n being the last column of the $n \times n$ identity matrix. This avoids having to do the more intensive computation of Ω^{-1}.

2.4 Controllability

So far we have encountered the effects of system controllability twice, once in connection with the condition for eigenvalue assignment by state feedback, (Theorem 2.1), and again in connection with the existence of a coordinate transformation matrix which transforms a given state model to controller form, (Section 2.3.4). In this section we encounter system controllability in the context of the basic problem of the input's ability to manipulate the state of a given state model. The following definition of a state model's controllability is made with this problem in mind.

Definition 2.3 A state model (A, B, C, D), or pair (A, B), is said to be controllable if for every possible initial state, $x(0)$, we can find at least one input $u(t)$, $t \in [0, t_f]$ and some finite final time $t_f < \infty$ so that $x(t_f) = \emptyset$, i.e., so that the state is annihilated by the input in a finite time.

It is important to notice in the foregoing definition, that t_f is required to be finite. This is done to prevent all stable systems from being considered to be controllable. More specifically, since stable systems have the property that

$$\lim_{t\to\infty} x(t) = \emptyset \qquad \text{for any } x(0) \text{ when } u(t) \text{ is null}$$

all stable systems would be controllable if the final time, t_f, in Definition 2.3 were allowed to be infinite.

In order to develop the implications of the foregoing definition we need to recall, from the first chapter, that the state $x(t_f)$ which results from having an initial state, $x(0)$, and an input, $u(t)$, is given by

$$x(t_f) = e^{At_f}x(0) + \int_0^{t_f} e^{A(t_f-\tau)}Bu(\tau)\mathrm{d}\tau \tag{2.33}$$

Then if the state is annihilated at time t_f, i.e., if $x(t_f) = \emptyset$, we see from (2.33) that the input, $u(t)$, must be chosen so that

$$e^{At_f}x(0) = -\int_0^{t_f} e^{A(t_f-\tau)}Bu(\tau)\mathrm{d}\tau \tag{2.34}$$

However from the series expansion for e^{At}, (Section 1.4.1) we see that

$$\left(e^{At_f}\right)^{-1} = e^{-At_f} \qquad \text{and} \qquad e^{A(t_f-\tau)} = e^{At_f}e^{-A\tau}$$

so that we can simplify (2.34) as

$$x(0) = -\int_0^{t_f} e^{-A\tau}Bu(\tau)\mathrm{d}\tau \tag{2.35}$$

This equation is the basic constraint which must be satisfied by any input which drives the system state from $x(0)$ to the origin in state space in finite time, $t_f < \infty$. Notice that the following three questions are immediately evident:

1. For each initial state, $x(0)$, is there at least one input, $u(t)$, which satisfies (2.35)?
2. If it is not possible to satisfy (2.35) by some input for each initial state, how should the initial state be restricted to enable (2.35) to be satisfied by some input $u(t)$?
3. If for a specific initial state, $x(0)$, it is possible to satisfy (2.35), what is the specification of the input $u(t)$ which does so?

Notice that when the answer to the first question is in the affirmative the system is controllable in the sense of Definition 2.3.

2.4.1 Controllable subspace

We will show now that a criterion for a given state model or pair (A, B) to be controllable in the sense of Definition 2.3 is that the controllability matrix be full rank, i.e., $\text{rank}[\Omega] = n$, when the state model is n dimensional where

$$\Omega = \begin{bmatrix} B & AB & \cdots & A^{n-1}B \end{bmatrix}$$

Notice that when rank$[\Omega] = n$, Ω has n independent columns so that we can always find a constant vector, γ, for any initial state, $x(0)$, such that

$$x(0) = \Omega\gamma \tag{2.36}$$

However, if rank$[\Omega] < n$, then Ω has fewer than n independent columns, and it is not possible to find γ to satisfy equation (2.36) for some $x(0)$. However, those $x(0)$ for which we can find γ to satisfy equation (2.36) are said to lie in the range or image of Ω, denoted as

$$x(0) \in \text{range}[\Omega]$$

We we will show, later in this section, that those initial states, $x(0)$, satisfying $x(0) \in \text{range}[\Omega]$, are those initial states for which we can find $u(t)$ to satisfy equation (2.35). This fact answers question 2 and leads to the following definition.

Definition 2.4 Initial states, $x(0)$, for which we can (cannot) find $u(t)$ to satisfy equation (2.35) are said to lie in the controllable subspace $\mathcal{S}_c$ (uncontrollable subspace $\mathcal{S}_{\bar{c}}$) of the state space.

Thus we see from the answer just given to question 2 that

$$\mathcal{S}_c = \text{range}[\Omega]$$

An important property of the controllable subspace is given in the following theorem.

Theorem 2.5 If $x(0) \in\text{range}[\Omega]$ then $Ax(0) \in\text{range}[\Omega]$.

Proof Suppose $x(0) \in \text{range}[\Omega]$. Then we can find a constant vector, γ, to satisfy

$$x(0) = \Omega\gamma = B\gamma_1 + AB\gamma_2 + \cdots A^{n-1}B\gamma_n \tag{2.37}$$

where the γ_is are constant vectors having the same length as $u(t)$. However, we know from the Cayley–Hamilton theorem, (Theorem 2.3), that

$$A^n = -\sum_{i=1}^{n} a_i A^{n-i}$$

where the $a_i s$ are coefficients of the characteristic polynomial

$$\det[sI - A] = s^n + a_1 s^{n-1} + a_2 s^{n-2} \cdots + a_n$$

Therefore multiplying (2.37) by A and using the Cayley–Hamilton theorem gives

$$\begin{aligned} Ax(0) &= AB\gamma_1 + A^2B\gamma_2 + \cdots A^{n-1}B\gamma_{n-1} + \left(\sum_{i=1}^{n} -a_i A^{n-i} B\right)\gamma_n \\ &= \Omega\bar{\gamma} \end{aligned} \tag{2.38}$$

where

$$\bar{\gamma}^T = [\,-\gamma_n a_n \quad \gamma_1 - \gamma_n a_{n-1} \quad \cdots \quad \gamma_{n-2} - \gamma_n a_2 \quad \gamma_{n-1} - \gamma_n a_1\,]$$

and we see from (2.38) that

$$Ax(0) \in \text{range}[\Omega]$$

■

Notice that this theorem implies that if $x(0) \in \text{range}[\Omega]$, then $A^k x(0) \in \text{range}[\Omega]$ for all integers k. This fact enables us to show that $\text{rank}[\Omega] = n$ is a necessary and sufficient condition for $\mathcal{S}_c$ to be the entire state space so that there is an input satisfying (2.35) for all initial states. Thus $\text{rank}[\Omega] = n$ is a necessary and sufficient condition for a state model to be controllable in the sense of Definition 2.3. This fact is shown now in the following theorem.

Theorem 2.6 If a state model (A, B, C, D) or pair (A, B) is controllable in the sense of Definition 2.3 then

$$\text{rank}[\Omega] = n$$

where

$$\Omega = \begin{bmatrix} B & AB & \cdots & A^{n-1}B \end{bmatrix}$$

Proof Suppose (A, B) is controllable in the sense of Definition 2.3. Then substituting the series expansion for $e^{A\tau}$ in (2.35) yields

$$\begin{aligned} x(0) &= -\int_0^{t_f} \left[I - A\tau + \frac{A^2\tau^2}{2} - \frac{A^3\tau^3}{3!} \cdots \right] Bu(\tau)\mathrm{d}\tau \\ &= B\gamma_0 + AB\gamma_1 + A^2B\gamma_2 + \cdots \\ &= \Omega\gamma^0 + A^n\Omega\gamma^1 + A^{2n}\Omega\gamma^2 + \cdots \end{aligned} \tag{2.39}$$

where

$$\gamma^{kT} = \begin{bmatrix} \gamma_{kn}^T & \gamma_{kn+1}^T & \cdots & \gamma_{(k+1)n-1}^T \end{bmatrix} \qquad k = 0, 1, 2 \cdots$$

with

$$\gamma_i = -(-1)^i \int_0^{t_f} \frac{\tau^i}{i!} u(\tau)\mathrm{d}\tau \qquad i = 0, 1, 2 \cdots$$

However from Theorem 2.5 we have

$$A^{kn}\Omega\gamma^k \in \text{range}[\Omega] \qquad k = 0, 1, \cdots \tag{2.40}$$

Therefore the columns on the right side of (2.39) span the entire state space if and only if $\text{rank}[\Omega] = n$. Otherwise when $\text{rank}[\Omega] < n$ there are initial states $x(0) \notin \text{range}[\Omega]$ so that (2.39) is impossible to satisfy by any $u(t)$ and the state model is not controllable in the sense of Definition 2.3. This implies that $\text{rank}[\Omega] = n$ is necessary for the state model to be controllable in the sense of Definition 2.3. ■

2.4.2 Input synthesis for state annihilation

The following theorem shows that rank$[\Omega] = n$ is sufficient for a state model (A, B, C, D) or pair (A, B) to be controllable in the sense of Definition 2.3.

Theorem 2.7 If rank$[\Omega] = n$ then we can satisfy (2.35) for any $x(0)$ by choosing $u(t)$ as

$$u(t) = -B^T e^{-A^T t} W^{-1} x(0)$$

where

$$W = \int_0^{t_f} e^{-A\tau} BB^T e^{-A^T\tau} \mathrm{d}\tau$$

Proof Suppose W is invertible. Then substituting the expression for $u(t)$ given in the theorem into the right side of (2.35) gives

$$\begin{aligned} -\int_0^{t_f} e^{-A\tau} Bu(\tau)\mathrm{d}\tau &= -\int_0^{t_f} e^{-A\tau} B\left[-B^T e^{-A^T\tau}\right]\mathrm{d}\tau W^{-1}x(0) \\ &= WW^{-1}x(0) \\ &= x(0) \end{aligned}$$

which is the left side of (2.35). Thus we have shown that the state model is controllable in the sense of Definition 2.3 when W is invertible.

In order to show that W is invertible, suppose rank$[\Omega] = n$ and W is not invertible. Then we can find $\alpha \neq \emptyset$ such that

$$\alpha^T W = \emptyset^T \tag{2.41}$$

Now since the integrand in the definition of W is quadratic, i.e.,

$$W = \int_0^{t_f} M^T(\tau)M(\tau)\mathrm{d}\tau$$

where

$$M^T(\tau) = e^{-A\tau}B$$

we can only satisfy (2.41) if

$$\alpha^T e^{-A\tau} B = \emptyset^T \qquad \text{for all } \tau \in [0, t_f] \tag{2.42}$$

However using the power series expansion of the matrix exponential we see that

$$e^{-A\tau}B = \Omega\beta^0 + A^n\Omega\beta^1 + A^{2n}\Omega\beta^2 + \cdots$$

where

$$\beta^{kT} = \left[\frac{(-\tau)^{kn}}{(kn)!} \quad \frac{(-\tau)^{kn+1}}{(kn+1)!} \quad \cdots \quad \frac{(-\tau)^{(k+1)n-1}}{([k+1]n-1)!} \right] \qquad k = 0, 1, 2 \cdots$$

Thus (2.42) is satisfied only if

$$\alpha^T \Omega = \varnothing^T \tag{2.43}$$

However since $\text{rank}[\Omega] = n$ and Ω is $n \times nm$, we see that Ω has independent rows and only $\alpha = \varnothing$ satisfies (2.43) and (2.41). This proves that W is invertible when $\text{rank}[\Omega] = n$. ■

In the next section we will see that when $\text{rank}[\Omega] < n$ we can still use the foregoing approach to provide a method for determining an input which annihilates an initial state $x(0)$ in a finite time provided $x(0) \in \text{range}[\Omega]$. This is achieved by using a controllable decomposed form for the system's state model so that the controllable part of the system is immediately evident.

2.5 Controllable Decomposed Form

So far we have considered two special types (canonical forms) of state model, the controller form and the normal or diagonal form. In this section we introduce another form for the state model of an n dimensional system. The purpose of this form of state model is to simplify the specification of the controllable and uncontrollable subspaces, denoted as $\mathcal{S}_c$ and $\mathcal{S}_{\bar{c}}$ respectively, by aligning them with subsets of the coordinate axis. More specifically, when a state model is in controllable decomposed form the state is given by

$$x(t) = \begin{bmatrix} x^1(t) \\ x^2(t) \end{bmatrix}$$

where $x^i(t)$ is of length n_i and

$$x(t) \in \mathcal{S}_c \qquad \text{if } x^2(t) = \varnothing$$

$$x(t) \in \mathcal{S}_{\bar{c}} \qquad \text{if } x^1(t) = \varnothing$$

Then any state, $x(t)$ can be decomposed as

$$x(t) = \alpha_c x^c(t) + \alpha_{\bar{c}} x^{\bar{c}}(t) \tag{2.44}$$

where $x^c(t) \in \mathcal{S}_c$ and $x^{\bar{c}}(t) \in \mathcal{S}_{\bar{c}}$ with $\alpha_c, \alpha_{\bar{c}}$ being scalars. Notice that $\mathcal{S}_c \perp \mathcal{S}_{\bar{c}}$

A state model whose state space is orthogonally decomposed in the foregoing manner is referred to as being in controllable decomposed form. The state model for a system to be in this form is defined as follows.

Definition 2.5 A state model having m scalar inputs is said to be in controllable decomposed form when its A and B matrices have the following block matrix forms

$$A = \begin{bmatrix} A_1 & A_2 \\ \emptyset & A_4 \end{bmatrix} \qquad B = \begin{bmatrix} B_1 \\ \emptyset \end{bmatrix} \tag{2.45}$$

where (A_1, B_1) is a controllable pair; A_1, A_2, A_4 and B_1 have dimensions: $n_1 \times n_1, n_1 \times n_2$, $n_2 \times n_2, n_1 \times m$ respectively with $n = n_1 + n_2$ being the dimension of the state space.

2.5.1 Input control of the controllable subspace

We begin the discussion of the effect of an input on the state when the state model is in controllable decomposed form by noticing that the product of upper-triangular matrices is upper-triangular, and that the coefficients of the series expansion of e^{At} involve powers of A. Therefore the transition matrix, e^{At}, for a state model in controllable decomposed form, (2.45), is upper-triangular,

$$e^{At} = \sum_{i=0}^{\infty} \begin{bmatrix} A_1^i & * \\ \emptyset & A_4^i \end{bmatrix} \frac{t^i}{i!}$$

where the block marked $*$ is not of importance here.

Next recall, from the previous section, that if $u(t)$ drives the state from $x(0)$ to the origin in finite time, t_f, then (2.35) is satisfied. Therefore since the e^{At} is upper-triangular and B has its last n_2 rows null when the state model is in controllable decomposed form, we see that (2.35) can be written as

$$\begin{bmatrix} x^1(0) \\ x^2(0) \end{bmatrix} = -\int_0^{t_f} \left(\sum_{i=0}^{\infty} (-1)^i \begin{bmatrix} A_1^i B_1 \\ \emptyset \end{bmatrix} \frac{\tau^i}{i!} \right) u(\tau) \mathrm{d}\tau \tag{2.46}$$

or

$$x^1(0) = -\int_0^{t_f} e^{-A_1\tau} B_1 u(\tau) \mathrm{d}\tau$$

$$x^2(0) = \emptyset$$

Thus we see that (2.46) can not be satisfied by any input $u(t)$ if $x^2(0) \neq \emptyset$. Alternatively, since (A_1, B_1) is a controllable pair, we can always find $u(t)$ to force $x^1(t_f) = \emptyset$ for any $x^1(0)$ and $x^2(0)$.

More specifically, when the state model is in controllable decomposed form, (2.45), we see that

$$x^2(t) = e^{A_4 t} x^2(0)$$

so that

$$x^1(t_f) = e^{A_1 t_f} x^1(0) + \int_0^{t_f} e^{A_1(t_f - \tau)} \left[A_2 e^{A_4 \tau} x^2(0) + B_1 u(\tau) \right] \mathrm{d}\tau$$

Therefore, proceeding in the same manner as was done to obtain (2.35) we see that if $x^1(t_f)$ is null then we must choose $u(t)$ to satisfy the following equation

$$q = -\int_0^{t_f} e^{-A_1\tau} B_1 u(\tau)\mathrm{d}\tau \tag{2.47}$$

where

$$q = x^1(0) + \int_0^{t_f} e^{-A_1\tau} A_2 e^{A_4\tau} x^2(0)\mathrm{d}\tau$$

Then using the result given in Theorem 2.7 we see that one input which satisfies (2.47) is

$$u(t) = -B_1^T e^{-A_1^T t} W^{-1} q$$

where

$$W = \int_0^{t_f} e^{-A_1\tau} B_1 B_1^T e^{-A_1^T\tau}\mathrm{d}\tau$$

To recap, we have just shown that we can always find an input to annihilate the projection of the initial state onto the controllable subspace. More specifically since any initial state can be written as (2.44) with $t = 0$, we can find $u(t)$ so that $x^c(t_f) = \emptyset$ for any $t_f < \infty$.

2.5.2 Relation to the transfer function

Concerning the input–output behavior, suppose the state model has A,B as specified by (2.45) and C partitioned as

$$C = [\,C_1 \quad C_2\,]$$

with C_i being $p \times n_i, i = 1, 2$ where p is the number of scalar outputs. Then from the block upper-triangular form for A, (2.45), we see that

$$(sI - A)^{-1} = \begin{bmatrix} sI - A_1 & -A_2 \\ \emptyset & sI - A_4 \end{bmatrix}^{-1} = \begin{bmatrix} (sI - A_1)^{-1} & (sI - A_1)^{-1}A_2(sI - A_4)^{-1} \\ \emptyset & (sI - A_4)^{-1} \end{bmatrix}$$

Therefore the system transfer function is given as

$$\begin{aligned} G(s) &= C(sI - A)^{-1}B \\ &= C_1(sI - A_1)^{-1}B_1 = \frac{C_1 \mathrm{adj}[sI - A_1]B_1}{\det[sI - A_1]} \end{aligned} \tag{2.48}$$

Notice that the order of the transfer function is no more than n_1. This reduction in the order of the transfer function results from the eigenvalues of A_4 being uncontrollable. We will see in the next chapter that a further reduction in the number of eigenvalues of A that go over as poles of the corresponding transfer function may occur as a result of the interaction of the state model's (A, C) pair. In order to better appreciate the fact that the eigenvalues of A_4 are uncontrollable, consider the following.

2.5.3 Eigenvalues and eigenvectors of A

Suppose we attempt to determine the left-eigenvectors of A when the state model is given in controllable decomposed form. Then we need to solve

$$w^T A = \lambda w^T$$

for the pair $\{\lambda, w\}$. However from the block structure of A, (2.45), we see that the foregoing equation expands to two equations,

$$w^{1T} A_1 = \lambda w^{1T} \tag{2.49}$$

$$w^{1T} A_2 + w^{2T} A_4 = \lambda w^{2T} \tag{2.50}$$

where

$$w^T = \begin{bmatrix} w^{1T} & w^{2T} \end{bmatrix}$$

with w^i being of length n_i. Then (2.49) is satisfied when we let $w^1 = \emptyset$ and (2.50) becomes

$$w^{2T} A_4 = \lambda w^{2T} \tag{2.51}$$

Therefore for this choice of w^1, we see that λ is an eigenvalue of A_4, w^2 is the corresponding left-eigenvector of A_4, and $w^T = \begin{bmatrix} \emptyset & w_2^T \end{bmatrix}$ is the corresponding left-eigenvector of A. Thus we see that eigenvalues of A_4 are eigenvalues of A. Notice that in this case we have

$$w^T B = \begin{bmatrix} \emptyset & w_2^T \end{bmatrix} \begin{bmatrix} B_1 \\ \emptyset \end{bmatrix} = 0$$

so that eigenvalues of A which are also eigenvalues of A_4 are uncontrollable.

Now the remaining eigenvalues of A are eigenvalues of A_1. We can see this by noting from (2.49) that if $w^1 \neq \emptyset$ then λ is an eigenvalue of A_1 and w^1 is the corresponding left-eigenvector of A_1. Since in controllable decomposed form the pair (A_1, B_1) is controllable, eigenvalues of A which are also eigenvalues of A_1 are controllable. Moreover since the controllable and uncontrollable eigenvalues of any pair (A, B) are disjoint we have that if $\lambda \in \lambda[A_1]$ then $\lambda \notin \lambda[A_4]$ and $\lambda I - A_4$ is invertible. Therefore when w^1 is a left-eigenvector of A_1, we can determine the required w^2 to make $w^T = \begin{bmatrix} w^{1T} & w^{2T} \end{bmatrix}$ a left-eigenvector of A from (2.50) as

$$w^{2T} = w^{1T} A_2 (\lambda I - A_4)^{-1}$$

The foregoing property of the eigenvalues of A is immediately evident when one attempts to use state feedback. Thus if we use state feedback when the state model is in uncontrollable decomposed form, the closed loop system matrix is

$$A + BK = \begin{bmatrix} A_1 + B_1K_1 & A_2 + B_1K_2 \\ 0 & A_4 \end{bmatrix}$$

where the state feedback matrix is partitioned as

$$K = [\,K_1 \quad K_2\,] \tag{2.52}$$

with K_1, K_2 being $m \times n_1$ and $m \times n_2$. Now we just saw that block upper-triangular matrices have the property that their eigenvalues equal the union of the eigenvalues of their diagonal blocks. Thus the eigenvalues of the state feedback system matrix satisfies

$$\lambda[A + BK] = \lambda[A_1 + B_1K_1] \cup \lambda[A_4]$$

and $A + BK$ has a subset of eigenvalues, $\lambda[A_4]$, which is clearly unaffected by state feedback. However since (A_1B_1) is a controllable pair, all the eigenvalues of $A_1 + B_1K_1$ can be assigned by K_1. Notice also, from the definition of stabilizability, Definition 2.2, that a system is stabilizable if, when its state model in controllable decomposed form, the A_4 partition of A is stable.

Finally notice that Ackermann's formula, which we developed to assign the eigenvalues of single input controllable state models, can also be used to assign the controllable eigenvalues of uncontrollable systems. This is done by transforming coordinates to put the given state model in controllable decomposed form. Ackermann's formula is then applied to the controllable pair (A_1, B_1) to determine the K_1 partition of K, (2.52), with K_2 being chosen arbitrarily. Finally, we can obtain the feedback matrix K_o in the original coordinates from $K_o = KT^{-1}$ where T is the coordinate transformation matrix needed to transform the given state model to controllable decomposed form. In the next section we consider how this coordinate transformation matrix can be determined.

2.6 Transformation to Controllable Decomposed Form

In this section we will indicate the role played by the controllability matrix in constructing a coordinate transformation matrix to put any state model in controllable decomposed form.

We begin by noting that when a state model is in controllable decomposed form with the (A, B) pair being given by (2.45), the last n_2 rows of the controllability matrix consist of nothing but zeros, i.e.,

$$\begin{aligned} \Omega &= \begin{bmatrix} B_1 & A_1B_1 & \cdots & A_1^{n_1-1}B_1 & A_1^{n_1}B_1 & \cdots & A_1^{n-1}B_1 \\ \emptyset & \emptyset & \cdots & \emptyset & \emptyset & \cdots & \emptyset \end{bmatrix} \\ &= \begin{bmatrix} \Omega_1 & \Omega_2 \\ \emptyset & \emptyset \end{bmatrix} \end{aligned} \tag{2.53}$$

where

$$\Omega_1 = \begin{bmatrix} B_1 & A_1B_1 & \cdots & A_1^{n_1-1}B_1 \end{bmatrix} \qquad \text{rank}[\Omega_1] = n_1$$

Notice that the columns of Ω_2 depend on the columns of Ω_1, i.e., we can always find a constant matrix Θ to satisfy

$$\Omega_2 = \Omega_1 \Theta$$

Therefore recalling that the subspace spanned by the independent columns of Ω is the controllable subspace, $\mathcal{S}_c$, we see that

$$\text{range}[\Omega] = \text{range}\left(\begin{bmatrix} \Omega_1 \\ \varnothing \end{bmatrix}\right) = \mathcal{S}_c$$

This fact together with computationally robust methods for finding a basis for the range of a matrix, e.g., singular value decomposition, (Chapter 7), provides a means for constructing a coordinate transformation matrix T which takes any state model to controllable decomposed form. The following theorem provides some insight into why a coordinate transformation based on the controllability subspace is able to achieve this result.

Theorem 2.8 A coordinate transformation matrix T transforms a given state model (A, B, C, D) to controllable decomposed form, (Definition 2.5) if

$$\text{range}[T_1] = \text{range}[\Omega] = \mathcal{S}_c$$

where Ω is the controllability matrix for the given state model and

$$T = [\, T_1 \quad T_2 \,]$$

with T_1, T_2 being $n \times n_1$, $n \times n_2$ and with T invertible where $\text{rank}(\Omega) = n_1$.

Proof Let Q^T denote the inverse of T, i.e.,

$$Q^T T = I$$

$$\begin{bmatrix} Q_1^T \\ Q_2^T \end{bmatrix} [\, T_1 \quad T_2 \,] = \begin{bmatrix} I & \varnothing \\ \varnothing & I \end{bmatrix} \tag{2.54}$$

Then the transformed A, B matrices are given by

$$\bar{A} = T^{-1} A T = \begin{bmatrix} Q_1^T A T_1 & Q_1^T A T_2 \\ Q_2^T A T_1 & Q_2^T A T_2 \end{bmatrix} = \begin{bmatrix} \bar{A}_1 & \bar{A}_2 \\ \bar{A}_3 & \bar{A}_4 \end{bmatrix}$$

$$\bar{B} = T^{-1} B = \begin{bmatrix} Q_1^T B \\ Q_2^T B \end{bmatrix} = \begin{bmatrix} \bar{B}_1 \\ \bar{B}_2 \end{bmatrix}$$

Recall that for the transformed state model to be in controllable decomposed form, (Definition 2.5), we need to show that $\bar{A}_3$ and $\bar{B}_2$ are each null.

To show that $\bar{B}_2 = \varnothing$, notice from (2.54) that $Q_2^T T_1 = \varnothing$ so that the columns of T_1 are orthogonal to the columns of Q_2. However $\text{range}(T_1) = \mathcal{S}_c$ so that we have

$$Q_2^T x = \varnothing \qquad \text{when } x \in \mathcal{S}_c \tag{2.55}$$

Now since B appears as the first m columns of Ω,

$$\text{range}[B] \subset \text{range}[\Omega] = \mathcal{S}_c \tag{2.56}$$

and we see from (2.55) that the columns of B are orthogonal to the columns of Q_2 so that

$$\bar{B}_2 = Q_2^T B = \emptyset$$

To show that $\bar{A}_3 = \emptyset$, recall from Theorem 2.5 that if $x \in \mathcal{S}_c$ than $Ax \in \mathcal{S}_c$. Therefore since

$$\text{range}[T_1] = \mathcal{S}_c$$

we have

$$\text{range}[AT_1] \subset \mathcal{S}_c \tag{2.57}$$

Therefore it follows from (2.57, 2.55) that the columns of AT_1 are orthogonal to the columns of Q_2 so that

$$\bar{A}_3 = Q_2^T A T_1 = \emptyset$$

■

Notice, in the foregoing proof, that (2.55–2.57) imply that we can find constant matrices Θ_1, Θ_2 such that

$$B = T_1 \Theta_1$$
$$AT_1 = T_1 \Theta_2$$

and since $Q_2^T T_1 = \emptyset$ we achieve $Q_2 B = \emptyset$ and $Q_2 A T_1 = \emptyset$ as required for T to transform the given system to controllable decomposed form.

2.7 Notes and References

The Ackermann formula was one of the first attempts to develop an algorithm for assigning the eigenvalues to a single input state model by state feedback. More recent work on this problem has concentrated on developing algorithms which are least sensitive to errors caused by the need to round off numbers during the execution of the algorithm using finite precision arithmetic on a digital computer. Further information on this problem can be obtained by consulting [43].

The Cayley–Hamilton theorem plays an important role in calculating matrix functions and can be used to provide an efficient method for calculating e^{At} once the eigenvalues of A are computed. Modern computer oriented algorithms rely on methods which truncate the infinite series expansion for e^{At} after having put A in a certain computationally beneficial form known as real Schur form, [17].

The approach to the problem of transforming a state model to controllable decomposed form which was discussed in Section 2.6 forms the basis for the command CTRBF in MATLAB.

3

State Estimation and Observability

3.1 Introduction

In the previous chapter we saw that state feedback could be used to modify basic aspects of a plant's behavior. However, since the plant state is usually not available, we need a means for obtaining an ongoing estimate of the present value of the state of the plant's state model. In this chapter we will see that this can be done by using measurements of the plant's input and output.

The problem of determining the state of a state model for a plant from knowledge of the plant's input and output is referred to in general as the state estimation problem. There are two classical types of state estimation: deterministic state estimation and stochastic state estimation. Deterministic state estimation, which is the subject of this chapter, assumes that the system input and output are known or measured exactly. The goal of deterministic state estimation is the determination of an estimate of the state having error which tends to decrease with time following the initiation of the estimation procedure. However, in stochastic state estimation the input and output signals are not assumed to be known exactly because of the presence of additive stochastic measurement noise having known statistics. In this situation the state estimation error is always present. The goal of stochastic state estimation is the determination of an estimate of the state so that, in the steady state, the average or expected value of the state estimation error is null while the expected value of the squared error is as small as possible. Stochastic state estimation is discussed in Chapter 6.

Both deterministic and stochastic state estimation are further subdivided according to when the input and output signals are measured relative to when the state estimate is needed. If the plant's input and output are measured over the time interval $[0, T]$ and if we need the state estimate at time t_e we have

1. a prediction problem if $t_e > T$
2. a filtering problem if $t_e = T$
3. a smoothing problem if $t_e < T$

The state estimation problem which is of concern in connection with the implementation of state feedback is a filtering problem since we need an ongoing estimate of the plant's state at the present time. Prediction and smoothing problems, which are not discussed in this chapter, arise, for example, in hitting a moving target with a projectile by aiming at the target's predicted position and in estimating the true value of data obtained from experiments done in the past when the measured data are corrupted by noise.

The computer implementation of a solution to the filtering problem, either deterministic or stochastic, takes the form of a state model which is referred to in general as a filter. The filter's input consists of both the plant's input and output and the filter's output is the state of the filter, i.e., the filter is designed so that its state is an estimate of the plant model's state. In the deterministic case the filter is referred to as an observer. In the stochastic case the filter is referred to as a Kalman filter. The Kalman filter is taken up in Chapter 6.

3.2 Filtering for Stable Systems

Suppose we know the parameters (A, B, C, D) of a plant state model as well as the plant's input and output, $\{u(t), y(t) : t \in [0, t_e]\}$. Then, from Chapter 1, we can express the plant state at any time t_e in terms of the initial state and input as

$$x(t_e) = e^{At_e}x(0) + \int_0^{t_e} e^{A(t_e-\tau)}Bu(\tau)d\tau \tag{3.1}$$

However since the initial plant state is usually unknown we are only able to calculate the zero state response. We use this fact to form an estimate of the plant state at time t_e as

$$\hat{x}(t_e) = \int_0^{t_e} e^{A(t_e-\tau)}Bu(\tau)d\tau \tag{3.2}$$

where from (3.1) we see that the plant state, $x(t_e)$, and plant state estimate, $\hat{x}(t_e)$, differ by the plant state estimation error, $\tilde{x}(t_e)$,

$$\tilde{x}(t_e) = x(t_e) - \hat{x}(t_e) = e^{At_e}x(0) \tag{3.3}$$

Now if the plant's state model is stable, we see from (3.3) that the state estimation error, $\tilde{x}(t_e)$, approaches the null vector with increasing estimation time, t_e,

$$\lim_{t_e \to \infty} \tilde{x}(t_e) = \emptyset \qquad \text{for all } \tilde{x}(0) = x(0)$$

In this situation the state estimate, $\hat{x}(t_e)$, given by (3.2), can be a reasonably good approximation to the plant model's state, $x(t_e)$, provided t_e is large enough to ensure that the effect of the initial state estimation error $\tilde{x}(0)$ is negligible. In this case the estimate is said to be an asymptotic estimate of the state since the state estimate approaches the state being estimated asymptotically with time.

Alternatively, we can view the state estimate obtained in the foregoing approach as the output, $y_e(t)$, from a system called a state estimator having state model (A_e, B_e, C_e). More specifically we have

$$\dot{\hat{x}}(t) = A_e\hat{x}(t) + B_e u(t) \qquad A_e = A \qquad B_e = B \tag{3.4}$$

$$y_e(t) = C_e\hat{x}(t) \qquad C_e = I \tag{3.5}$$

Then since the state of the plant model is governed by

$$\dot{x} = Ax(t) + Bu(t) \tag{3.6}$$

we see by subtracting equation (3.4) from equation (3.6) that the state estimation error is independent of the plant input and is governed by the differential equation

$$\dot{\tilde{x}}(t) = A\tilde{x}(t) \tag{3.7}$$

where

$$\tilde{x}(t) = x(t) - \hat{x}(t)$$

Therefore the state estimation error is independent of the plant input and is given by

$$\tilde{x}(t) = e^{At}\tilde{x}(0) \tag{3.8}$$

Now if the initial state of the plant were known we could take $\hat{x}(0) = x(0)$ as the initial state for the estimator, equation (3.4). Then $\tilde{x}(0) = \emptyset$ and we see from equation (3.8) that we would have the desirable result of exact state estimation, $\hat{x}(t) = x(t)$, for all time. Usually we don't know the initial plant state. In this case we could set $\hat{x}(0) = \emptyset$ in the state estimator, (3.4), so that $\tilde{x}(0) = x(0)$ and provided A is stable, the state estimate would approach the actual plant state asymptotically for any initial plant state.

However, unstable plants are encountered quite frequently, and must be stabilized by feedback, e.g., the feedback control of satellite rocket launchers. In these cases the estimate obtained from the state estimator, (3.4), diverges from the plant state for any $\tilde{x}(0) \neq \emptyset$ since e^{At} becomes unbounded with time and the state estimation error given by equation (3.8) grows indefinitely. Notice that it is impossible to know a physical parameter such as the initial plant state exactly. Therefore in practice it is not possible to set $\hat{x}(t) = x(0)$ so as to obtain $\tilde{x}(t) = \emptyset$ from (3.8). Thus whenever the plant is unstable, we are unable to use (3.4) as an asymptotic state estimator.

Not knowing the initial state and needing to estimate the state to implement stabilizing state feedback for unstable plants, forces us to seek another approach to state estimation.

3.3 Observers

Notice that the foregoing simple approach to the plant state estimation problem ignored the additional information on the plant state which is present in the plant output, i.e.,

$$y(t) = Cx(t) + Du(t) \tag{3.9}$$

In this section we show how to use this information to obtain an asymptotic state estimator for unstable plants.

Suppose, for the moment, that the plant output, $y(t)$, is unknown. Then, assuming we have an estimate of the plant state, $\hat{x}(t)$, we can use it to obtain an estimate of the output, $\hat{y}(t)$, by substituting $\hat{x}(t)$ for the plant state in (3.9), to get

$$\hat{y}(t) = C\hat{x}(t) + Du(t) \tag{3.10}$$

However, we assume in this chapter that the actual plant output, $y(t)$, is known. Therefore we can define a plant output estimation error as

$$\tilde{y}(t) = y(t) - \hat{y}(t) \tag{3.11}$$

Now since we can form $\tilde{y}(t)$ from $u(t), y(t)$ and $\hat{x}(t)$ all of which are known, we can use $\tilde{y}(t)$ to indicate the accuracy of the plant state estimate since $\tilde{y}(t) \neq \emptyset$ is an indication that the state estimate differs from the actual state. More importantly, we can use $\tilde{y}(t)$ to correct future state estimates by letting $\tilde{y}(t)$ affect $\hat{x}(t)$ by subtracting $L\tilde{y}(t)$ to the right side of (3.4) to obtain

$$\dot{\hat{x}} = A\hat{x}(t) + Bu(t) - L\tilde{y}(t) \tag{3.12}$$

which using (3.10, 3.11) can be rewritten as

$$\dot{\hat{x}} = F\hat{x}(t) + G\begin{bmatrix} u(t) \\ y(t) \end{bmatrix} \tag{3.13}$$

where

$$F = A + LC \qquad G = [B + LD \quad -L]$$

When F is stable, the system represented by (3.13) is referred to as an observer. In order to determine the effectiveness of an observer in obtaining an asymptotic estimate of the state, consider the differential equation for the state estimation error, $\tilde{x}(t) = x(t) - \hat{x}(t)$. We obtain this differential equation by subtracting (3.13) from (3.6) to obtain

$$\dot{\tilde{x}}(t) = F\tilde{x}(t) \tag{3.14}$$

where

$$F = A + LC$$

Then the state estimation error is given by

$$\tilde{x}(t) = e^{Ft}\tilde{x}(0)$$

Notice that unlike the simple approach to state estimation presented in the previous section, we now have the possibility of getting an asymptotic estimate of the state, even if the plant is unstable. We do this by choosing L to make F stable. Questions regarding the possibility of doing this are answered by examining certain properties of the right-eigenvectors of A.

Theorem 3.1 Whenever A has a right-eigenvector v^i satisfying

$$Cv^i = 0$$

the corresponding eigenvalue λ_i of A is an eigenvalue of $A + LC$ for all L.

Proof Suppose the condition of the theorem is satisfied. Then multiplying $A + LC$ on the right by v^i gives

$$(A + LC)v^i = Av^i + LCv^i$$

However, since $Cv^i = 0$ and $Av^i = \lambda_i v^i$ we see that the foregoing equation becomes

$$(A + LC)v^i = \lambda_i v^i$$

Thus λ_i is a fixed eigenvalue of the matrix $A + LC$ for all L ■

We will see that if v^i is an eigenvector of A such that $Cv^i \neq 0$ then we can choose L so that the corresponding eigenvalue, λ_i, of A can be assigned to any desired value as an eigenvalue of $A + LC$. This leads to the following definition.

Definition 3.1 An eigenvalue λ_i of A is said to be an observable (unobservable) eigenvalue for the pair (A, C) if we can (cannot) find L such that λ_i is not an eigenvalue of $A + LC$.

Reflection on the foregoing reveals that we can design an observer for a given state model, i.e., we can find L so that $A + LC$ is stable, provided the plant state model has no eigenvalues which are both unstable and unobservable. Plant state models having this property are referred to as being detectable.

Definition 3.2 A plant state model is said to be detectable if all its unobservable eigenvalues are stable.

Thus we can only determine an observer for a given state model if that state model is detectable. This is of obvious importance in the implementation of state feedback using an observer for the stabilization of an unstable plant.

Recall, in the case of state feedback, that the controller form state model facilitates the calculation of K to achieve a specified set of closed loop eigenvalues. In the present situation there is an analogous form for the plant state model which facilitates the calculation L to assign a specified set of eigenvalues to the observer. This form, which we encounter in the next section, is referred to as an observer form.

3.4 Observer Design

Suppose we are given a state model (A, B, C, D) for the plant and we want to determine L so that the observer eigenvalues, $\lambda[A + LC]$, equal a specified set $\{\mu_i : i = 1, 2 \cdots n\}$. Then provided any unobservable eigenvalues of the pair (A, C) are in the specified set, we can determine L so that the observer eigenvalues coincide with the specified set. This could be done by equating coefficients of like powers on either side of the equation

$$\alpha(s) = \gamma(s)$$

where

$$\alpha(s) = \det[sI - (A + LC)] = s^n + \alpha_1 s^{n-1} + \alpha_2 s^{n-2} + \cdots + \alpha_n$$

$$\gamma(s) = \prod_{i=1}^{n}(s - \mu_i) = s^n + \gamma_1 s^{n-1} + \gamma_2 s^{n-2} + \cdots + \gamma_n$$

The resulting equations, $\gamma_i = \alpha_i : i = 1, 2, \cdots n$, in the elements of L are, in general, coupled in a way that depends on the plant's A and C matrices. This makes it difficult to set up a general procedure for determining L from these equations. However, when $y(t)$ is a scalar and the state model for the plant has all its eigenvalues observable, this difficulty can be overcome by using coordinates which put the state model for the plant in observer form.

3.4.1 Observer form

Recall that any state model (A, B, C, D), for a plant is related to the plant transfer function $G(s)$ as

$$G(s) = C(sI - A)^{-1}B + D$$

Then a system having transfer function $G^T(s)$ has state model $(\bar{A}, \bar{B}, \bar{C}, \bar{D})$ where

$$\bar{A} = A^T \qquad \bar{B} = C^T$$

$$\bar{C} = B^T \qquad \bar{D} = D^T$$

State models $(\bar{A}, \bar{B}, \bar{C}, \bar{D})$ and (A, B, C, D) are said to be duals of each other when their parameters are related in this way. When $G(s)$ is symmetric, the dual of any state model for $G(s)$ is also a state model for $G(s)$.

In the SISO case $G(s)$ is a scalar so that $G^T(s) = G(s)$. Therefore, in this case, we can obtain a canonical state model referred to as an observer form, by forming the dual of the controller form state model. Thus if the state model for $G(s)$ is in controller form, (A_c, B_c, C_c, D_c), the dual state model is in observer form, $(A_o = A_c^T,\ B_o = C_c^T,\ C_o = B_c^T,\ D_o = D_c)$, or more specifically

$$A_o = \begin{bmatrix} -a_1 & 1 & 0 & \cdots & 0 & 0 \\ -a_2 & 0 & 1 & \cdots & 0 & 0 \\ \vdots & \vdots & \vdots & \vdots & \vdots & \vdots \\ -a_{n-2} & 0 & \cdots & 0 & 1 & 0 \\ -a_{n-1} & 0 & \cdots & 0 & 0 & 1 \\ -a_n & 0 & \cdots & 0 & 0 & 0 \end{bmatrix} \qquad B_o = \begin{bmatrix} b_1 \\ b_2 \\ \vdots \\ b_{n-2} \\ b_{n-1} \\ b_n \end{bmatrix}$$

$$C_o = \begin{bmatrix} 1 & 0 & \cdots & 0 & 0 & 0 \end{bmatrix} \qquad D_o = D$$

An indication of the importance of this form is evident from the ease with which we can assign observer eigenvalues through L. Thus when the plant is in observer form, $F_o = A_o + L_o C_o$ is in the same form as A_o

$$F_o = \begin{bmatrix} -a_1 + l_{o1} & 1 & 0 & \cdots & 0 & 0 \\ -a_2 + l_{o2} & 0 & 1 & \cdots & 0 & 0 \\ \vdots & \vdots & \vdots & \vdots & \vdots & \vdots \\ -a_{n-2} + l_{on-2} & 0 & \cdots & 0 & 1 & 0 \\ -a_{n-1} + l_{on-1} & 0 & \cdots & 0 & 0 & 1 \\ -a_n + l_{on} & 0 & \cdots & 0 & 0 & 0 \end{bmatrix} \qquad L_o = \begin{bmatrix} l_1 \\ l_2 \\ \vdots \\ l_{n-2} \\ l_{n-1} \\ l_n \end{bmatrix}$$

Therefore if we want the eigenvalues of F to satisfy a specified characteristic polynomial, $\gamma(s)$, we have

$$\det[sI - F] = s^n + \sum_{i=1}^{n} \gamma_i s^{n-i}$$

and the required elements of L_o, $\{l_{oi} : i = 1, 2, \cdots n\}$ are easily obtained from

$$l_{oi} = a_i - \gamma_i \qquad i = 1, 2 \cdots, n$$

In order to preserve the eigenvalues of $A + LC$ under a coordinate transformation we see that if $\bar{L}$ assigns a desired set of eigenvalues to $\bar{A} + \bar{L}\bar{C}$ in the transformed coordinates then we need to replace $\bar{L}$ by $L = T\bar{L}$ to assign the same set of eigenvalues to $A + LC$ in the original coordinates. This is seen by noting that

$$\bar{A} + \bar{L}\bar{C} = T^{-1}AT + \bar{L}CT = T^{-1}(A + T\bar{L}C)T$$

3.4.2 Transformation to observer form

In the previous chapter, (Section 2.3.3), we developed an algorithm for generating a coordinate transformation matrix T so that any single input controllable state model is transformed to a controller form state model. This algorithm can be adapted to provide a coordinate transformation matrix so that a single output observable state model is transformed to an observer form state model. This adaptation is made using the duality between the controller and observer forms. Thus by replacing A, B in the algorithm given in Section 2.3.3 by A^T, C^T we obtain a matrix T which transforms the pair A^T, C^T to controller form. Then T^{-T} is the coordinate transformation matrix needed to transform the given state model to observer form. Just as the transformation to controller form is only possible if (A, B) is a controllable pair so here the pair (A^T, C^T) must be controllable or equivalently the pair (A, C) must be observable, i.e.,

$$\text{rank}\left[\mho\right] = n \tag{3.15}$$

where

$$\mho = \begin{bmatrix} C \\ CA \\ \vdots \\ CA^{n-1} \end{bmatrix}$$

The matrix $\mho$ is called the observability matrix. Notice, in general, that $\mho$ is a $pn \times n$ matrix where p is the number of elements in $y(t)$. Also in general we have $\text{rank}[\mho] \leq n$ since the rank of any matrix is never greater than its smallest dimension.

Notice that if a pair (A, C) is observable, i.e., $\text{rank}[\mho] = n$, we can assign all the eigenvalues of $A + LC$ by choosing L. Otherwise if we are not able to assign all the eigenvalues, one or more of the right-eigenvectors of A satisfies $Cv^i = 0$ (Theorem 3.1) and we can show that

$$\mho v^i = \emptyset \tag{3.16}$$

Therefore at least one of the n columns of $\mho$ is dependent so that $\text{rank}[\mho] < n$ and the pair (A, C) not observable.

Alternatively, if all the eigenvalues of $A + LC$ can be assigned, all the right-eigenvectors of A, v^i satisfy $Cv^i \neq \emptyset$ and (3.16) is not satisfied by any right-eigenvector of A. Then assuming A has a complete set of eigenvectors, any vector q having n elements can be written as

$$q = \sum_{i=1}^{n} \alpha_i v^i$$

and

$$\mho q \neq \emptyset$$

implying that the columns of $\mho$ are independent and $\text{rank}[\mho] = n$. Thus all the eigenvalues of $A + LC$ are assignable if and only if (A, C) is an observable pair.

3.4.3 Ackermann's formula

Recall from the previous chapter that we could determine K to assign the eigenvalues of $A + BK$ for any controllable single input state model by using Ackermann's formula, (Section 2.3.5). Ackermann's formula can also be used to determine L to assign a specified set of eigenvalues to $A + LC$ for any observable single output state model. This is made possible by the fact that any square matrix and its transpose have the same eigenvalues. Therefore $A + LC$ and $A^T + C^T L^T$ have the same eigenvalues. Thus given an observable state model for a single output system and a set of desired eigenvalues for $A + LC$, we can determine L by replacing A and B in Ackermann's formula by A^T and C^T, respectively. Ackermann's formula then produces L^T.

3.5 Observability

So far we have encountered two effects of system observability. Both the capability of assigning all the eigenvalues to $A + LC$ and the capability of transforming coordinates so that the state model is put in observer form requires that the given state model be observable, i.e.,

$$\text{rank}[\mathcal{O}] = n$$

where $\mathcal{O}$ is $pn \times n$ observability matrix, (3.15), with $p = 1$.

It is important to notice from the discussion so far, that we can still design an observer to estimate the state of a given state model when the state model is unobservable, i.e., when $\text{rank}[\mathcal{O}] < n$, provided the given state model is detectable, Definition 3.2. Thus observability is sufficient but not necessary for the existence of an observer for a given plant state model.

In order to gain further insight into the meaning of observability, we are going to consider the problem of determining the state at some estimation time from the derivatives of the input and output at the estimation time. Notice that in practice we avoid using signal differentiation since noise acquired in measuring a signal can appear greatly enlarged in the derivatives of the measured signal. Therefore our intent in discussing the problem of determining the state from the derivatives of the input and output is to provide additional insight into the theoretical nature of observability. We will show that we cannot solve this problem unless the state model is observable.

3.5.1 A state determination problem

Suppose we are given the state model, (A, B, C, D), and the derivatives of the input and output at some estimation time, t_e, i.e.,

$$\{y^{(i)}(t_e), u^{(i)}(t_e) : i = 0, 1, 2, \cdots, n - 1\}$$

Then from the output equation

$$y(t_e) = Cx(t_e) + Du(t_e) \tag{3.17}$$

we have p equations in the n unknown components of $x(t_e)$, where p is the number of elements in $y(t)$. Now if $p = n$, C is square and if the rows of C are independent, or equivalently if the elements in $y(t)$ are not related by constants, i.e., are independent, then we can solve (3.17) for the state at time t_e as

$$x(t_e) = C^{-1}[y(t_e) - Du(t_e)]$$

without requiring derivatives of the input and output signals. However usually $p < n$ and we need to generate equations in addition to those obtained directly from the output equation (3.17). We can do this by using the state differential equation

$$\dot{x}(t_e) = Ax(t_e) + Bu(t_e) \tag{3.18}$$

together with the output equation (3.17).

Suppose $p = 1$. Then we have one equation, (3.17), which we can use in the determination of the n components of the state $x(t_e)$. We can generate a second equation for this purpose by taking the derivative of (3.17)

$$\dot{y}(t_e) = C\dot{x}(t_e) + D\dot{u}(t_e) \tag{3.19}$$

and using (3.18) to obtain

$$\dot{y}(t_e) = CAx(t_e) + CBu(t_e) + D\dot{u}(t_e) \tag{3.20}$$

We can generate a third equation for determining $x(t_e)$ by differentiating (3.20) and substituting (3.18) for the derivative of the state. This gives

$$y^{(2)}(t_e) = CA^2x(t_e) + CABu(t_e) + CBu^{(1)}(t_e) + Du^{(2)}(t_e)$$

Continuing in this way we can generate additional equations in the state until we have the required n equations. These equations can be collected and given as

$$z = \mho x(t_e) \tag{3.21}$$

where

$$z = Y - \Gamma U \qquad Y = \begin{bmatrix} y(t_e) \\ y^{(1)}(t_e) \\ \vdots \\ y^{(n-1)}(t_e) \end{bmatrix} \qquad \mho = \begin{bmatrix} C \\ CA \\ \vdots \\ CA^{n-1} \end{bmatrix}$$

$$U = \begin{bmatrix} u(t_e) \\ u^{(1)}(t_e) \\ \vdots \\ u^{(n-1)}(t_e) \end{bmatrix} \qquad \Gamma = \begin{bmatrix} D & 0 & 0 & \cdots & 0 \\ CB & D & 0 & \cdots & 0 \\ CAB & CB & D & \cdots & 0 \\ \vdots & \vdots & \vdots & \vdots & \vdots \\ CA^{n-2}B & CA^{n-3}B & CA^{n-4}B & \cdots & D \end{bmatrix}$$

Now since we developed (3.21) from a consideration of the effect of the state and input on the output, we have $z \in \text{range}[\mho]$ so that we can always find $x(t_e)$ to satisfy (3.21). When (A, C) is an observable pair, all n columns of $\mho$ are independent, rank $[\mho] = n$, and $\mho^T\mho$ is invertible so that when $p > 1$ we have

$$x(t_e) = \left(\mho^T\mho\right)^{-1}\mho^T z$$

and when $p = 1$ we have

$$x(t_e) = \mho^{-1}z$$

However when rank$[\mho] < n, \mho$ has dependent columns and we have

$$\emptyset = \mho x(t_e) \tag{3.22}$$

for some non-null $x(t_e)$ which we denote by $x^{\bar{o}}(t_e)$. Solutions to (3.22) when rank$[\mho] < n$ are said to lie in the null space of $\mho$ denoted

$$x^{\bar{o}}(t_e) \in \text{null}[\mho] \qquad \text{when} \qquad \mho x^{\bar{o}}(t_e) = \emptyset$$

Hence if $x(t_e)$ is any solution to (3.21) then we have $x(t_e) + x^{\bar{o}}(t_e)$ as another solution to (3.21). Therefore we see that if (A, C) is not observable it is impossible to determine the true state, $x(t_e)$, from the derivatives of the system's input and output.

In summary we have shown that the observability of the state model is a necessary and sufficient condition for being able to determine the actual state of the plant state model at any time from derivatives of the input and output at that time.

3.5.2 Effect of observability on the output

Continuing the discussion begun in Section 1.7.4 we see that when (A, C) is an unobservable pair with right-eigenvector satisfying $Cv^j = \emptyset$ then when $x(0) = v^i$ we obtain the zero-input response as

$$\begin{aligned} y(t) &= Ce^{At}x(0) = C[I + At + \frac{1}{2!}A^2t^2 + \frac{1}{3!}A^3t^3 + \cdots]x(0) \\ &= C[I + \lambda_i t + \frac{1}{2!}\lambda_i^2 t^2 + \frac{1}{3!}\lambda_i^3 t^3 + \cdots]v^i = e^{\lambda_i t}Cv^i = \emptyset \end{aligned} \tag{3.23}$$

We can illustrate this important effect by considering the unobservable state model

$$A = \begin{bmatrix} -3 & -2 \\ 1 & 0 \end{bmatrix} \qquad B = \begin{bmatrix} 1 \\ 0 \end{bmatrix}$$
$$C = \begin{bmatrix} +1 & +1 \end{bmatrix}$$

Then since A is in companion form, (Section 1.6.1), we can express its right-eigenvectors as

$$\left\{ v^i = \begin{bmatrix} \lambda_i \\ 1 \end{bmatrix} : \qquad \lambda_1 = -1, \quad \lambda_2 = -2 \right\}$$

and any initial state given by

$$x(0) = \alpha v^1 = \begin{bmatrix} +\alpha \\ -\alpha \end{bmatrix}$$

gives rise to a trajectory which lies in the null space of C for all time so that $y(t) = 0$ for all t.

Notice also that the transfer function corresponding to the foregoing state model is given by

$$G(s) = \frac{s+1}{s^2+3s+2} = \frac{1}{s+2}$$

and the unobservable eigenvalue of A at -1 is not a pole of the transfer function. This effect was encountered in Section 1.9.2 in connection with diagonal or normal form state models.

We will see, in Chapter 4, that this property of having certain non-null initial states which produce no output when a state model is unobsevable leads to a characterization of observability in terms of the rank of a matrix called the observability Gramian. The observability Gramian is involved in determining the energy in the output caused by a given initial state.

3.6 Observable Decomposed Form

Recall, from the previous section, that we are unable to determine $x^{\bar{o}}(t) \in \text{null}[\mathcal{U}]$ from knowledge of a state model's input and output derivatives at time t. We refer to null$[\mathcal{U}]$ as the state model's unobservable subspace, denoted $\mathcal{S}_{\bar{o}}$, i.e., $\mathcal{S}_{\bar{o}} = \text{null}\ [\mathcal{U}]$. Moreover we refer to range$[\mathcal{U}^T]$ as the state model's observable subspace, $\mathcal{S}_o$. Now it turns out that any solution to (3.21) can be written as

$$x(t) = x^o(t) + x^{\bar{o}}(t)$$

where

(i) $x^{\bar{o}}(t) \in \text{null}[\mathcal{U}] = \mathcal{S}_{\bar{o}}$ and is arbitrary otherwise;
(ii) $x^o(t) \in \text{range}[\mathcal{U}^T] = \mathcal{S}_o$ and depends uniquely on z.

Moreover the observable and unobservable subspaces are orthogonal, $\mathcal{S}_{\bar{o}} \perp \mathcal{S}_o$, i.e.,

$$x^{oT}(t)x^{\bar{o}}(t) = 0 \tag{3.24}$$

for any $x^o(t) \in \text{range}\left[\mathcal{U}^T\right]$ and any $x^{\bar{o}}(t) \in \text{null}[\mathcal{U}]$. This can be seen by using the fact that

$$x^o(t) \in \text{range}\left[\mathcal{U}^T\right]$$

if and only if we can find w to satisfy

$$x^o(t) = \mathcal{U}^T w$$

so that we have

$$x^{oT}(t)x^{\bar{o}}(t) = \left(\mathcal{U}^T w\right)^T x^{\bar{o}}(t) = w^T \mathcal{U} x^{\bar{o}}(t) = 0$$

In this section we show that any unobservable state model can be transformed, using a change of coordinates, so that in the transformed coordinates the observable and unobservable subspaces align with subsets of the coordinate axes. More specifically, when a state model is in observable decomposed form the state is given by

$$x(t) = \begin{bmatrix} x^1(t) \\ x^2(t) \end{bmatrix} \tag{3.25}$$

where $x^i(t)$ is of length n_i and

$$x(t) \in \mathcal{S}_o \qquad \text{if } x^2(t) = \emptyset$$

$$x(t) \in \mathcal{S}_{\bar{o}} \qquad \text{if } x^1(t) = \emptyset$$

Notice that (3.24) is satisfied when the state is given by (3.25). Now the structure required for the state model to be in observable decomposed form is given in the following definition. Notice that this structure is dual to the structure of the controllable decomposed form, Definition 2.5, in the sense described in Section 3.4.1.

Definition 3.3 A state model having p scalar outputs is said to be in observable decomposed form when its A and C matrices are in the following block forms

$$A = \begin{bmatrix} A_1 & \emptyset \\ A_3 & A_4 \end{bmatrix}$$

$$C = \begin{bmatrix} C_1 & \emptyset \end{bmatrix}$$

where (A_1, C_1) is an observable pair and $A_1, A_3, A_4,\ C_1$ have dimensions $n_1 \times n_1, n_2 \times n_1, n_2 \times n_2$ and $p \times n_1$ respectively with $n = n_1 + n_2$ being the dimension of the state space.

3.6.1 Output dependency on observable subspace

One way of seeing that a state model in observable decomposed form decomposes the state space in the manner specified in equation (3.25) is to note that, in these coordinates, the transition matrix is lower triangular,

$$e^{At} = \sum_{i=0}^{\infty} \begin{bmatrix} A_1^i & \emptyset \\ * & A_4^i \end{bmatrix} \frac{t^i}{i!}$$

so that the zero input response is given by

$$y(t) = Ce^{At}x(0) = C_1 \sum_{i-0}^{\infty} A_1^i \frac{t^i}{i!} x^1(0)$$

$$= C_1 e^{A_1 t} x^1(0) \tag{3.26}$$

Thus the zero input response is independent of $x^2(0)$ and $\mathcal{S}_{\bar{o}}$ is as specified in equation (3.25). In addition since (A_1, C_1) is observable, we could determine $x^1(t)$ from derivatives of the output. Thus $\mathcal{S}_o$ is also as specified in equation (3.25).

3.6.2 Observability matrix

Alternatively, we can also see that the decomposition of the state space given by equation (3.25) is achieved when the state model is in observable decomposed form by inspection of the observability matrix.

$$\mho = \begin{bmatrix} C_1 & \emptyset \\ C_1 A & \emptyset \\ \vdots & \vdots \\ C_1 A_1^{n_1 - 1} & \emptyset \\ C_1 A_1^{n_1} & \emptyset \\ \vdots & \vdots \\ C_1 A_1^{n-1} & \emptyset \end{bmatrix} = \begin{bmatrix} \mho_1 & \emptyset \\ \mho_2 & \emptyset \end{bmatrix}$$

where

$$\mho_1 = \begin{bmatrix} C_1 \\ C_1 A_1 \\ \vdots \\ C_1 A_1^{n_1 - 1} \end{bmatrix} \qquad \text{rank}[\mho] = \text{rank}[\mho_1] = n_1$$

and we have

$$x \in \text{null}[\mho] \quad \text{if and only if} \quad x = \begin{bmatrix} \emptyset \\ x^2 \end{bmatrix} \tag{3.27}$$

for any x^2 of length $n - n_1$.

3.6.3 Transfer function

Again, notice that the observable decomposed form gives the transfer function as

$$\begin{aligned} G(s) &= [\,C_1 \quad \emptyset\,] \begin{bmatrix} (sI - A_1)^{-1} & \emptyset \\ * & (sI - A_4)^{-1} \end{bmatrix} \begin{bmatrix} B_1 \\ B_2 \end{bmatrix} \\ &= C_1 (sI - A_1)^{-1} B_1 \end{aligned}$$

with order equal to or less than n_1 depending on the controllability of the pair (A_1, B_1). Notice also that the eigenvalues of A_4 are not poles of the transfer function since they are unobservable.

This latter observation is readily verified from the structure of $(A + LC)$ in these coordinates since

$$A + LC = \begin{bmatrix} A_1 + L_1C_1 & \emptyset \\ L_2C_1 & A_4 \end{bmatrix} \qquad L = \begin{bmatrix} L_1 \\ L_2 \end{bmatrix}$$

and since $A + LC$ is block lower triangular we have

$$\lambda[A + LC] = \lambda[A_1 + L_1C_1] \cup \lambda[A_4]$$

with $\lambda[A_4]$ being eigenvalues of $A + LC$ for all L, i.e., being unobservable eigenvalues of (A, C).

3.6.4 Transformation to observable decomposed form

Now we saw in Chapter 2 that the controllable subspace is A-invariant, i.e., if $x \in$range$[\Omega]$ then $Ax \in$range$[\Omega]$, (Theorem 2.5). The analogous result here is that the unobservable subspace is A-invariant, i.e., if $x \in$null$[\mho]$ then $Ax \in$null$[\mho]$. When the state model is in observable decomposed form we can see this fact directly from (3.27) and the block structure of A. Alternatively, we can see that this result holds more generally in any coordinates by noting that we can always express the last block row of $\mho A$ using the Cayley–Hamilton theorem, Theorem 2.3, as

$$CA^n = -a_1CA^{n-1} - a_2CA^{n-2} \cdots a_nC$$

Therefore the last block row of $\mho A$ depends only on block rows of $\mho$ and thus if $\mho x = \emptyset$ then $\mho Ax = \emptyset$.

Now the A-invariance of the unobservable subspace provides a means for determining the coordinate transformation matrix needed to put a given state model in observable decomposed form.

Theorem 3.2 A coordinate transformation matrix T transforms a given state model (A, B, C, D) having observability matrix $\mho$ with rank$[\mho] = n_1$ to observable decomposed form if

$$\text{range}[T_2] = \text{null}[\mho]$$

where

$$T = [\,T_1 \quad T_2\,]$$

with T_1, T_2 being $n \times n_1, n \times n - n_1$ such that T^{-1} exists.

Proof Let Q^T denote the inverse of T, i.e.,

$$\begin{bmatrix} Q_1^T \\ Q_2^T \end{bmatrix} [\,T_1 \quad T_2\,] = \begin{bmatrix} I & \emptyset \\ \emptyset & I \end{bmatrix} \tag{3.28}$$

Then the transformed A and C matrices are given by

$$\bar{A} = T^{-1}AT = \begin{bmatrix} Q_1^T AT_1 & Q_1^T AT_2 \\ Q_2^T AT_1 & Q_2^T AT_2 \end{bmatrix} = \begin{bmatrix} \bar{A}_1 & \bar{A}_2 \\ \bar{A}_3 & \bar{A}_4 \end{bmatrix}$$
$$\bar{C} = CT = [CT_1 \quad CT_2] = [\bar{C}_1 \quad \bar{C}_2]$$

Now since C appears as the first p rows of $\mathcal{U}$ and T_2 is chosen so that $\mathcal{U}T_2 = \emptyset$, we have $\bar{C}_2 = CT_2 = \emptyset$.

Finally since range$[T_2] = \mathcal{S}_{\bar{o}}$ and Q^T is the inverse of T we see from (3.28) that

$$Q_1^T x = \emptyset \qquad \text{when } x \in \mathcal{S}_{\bar{o}}$$

Then since $Ax \in \mathcal{S}_{\bar{o}}$ if $x \in \mathcal{S}_{\bar{o}}$, we have range$[AT_2] = \mathcal{S}_{\bar{o}}$ and thus $Q_1^T AT_2 = \emptyset$. ■

3.7 Minimal Order Observer

Recall from Section 3.3 that use of the plant output enables the construction of an asymptotic state estimator for unstable plants. This estimator, referred to as an observer, has dimensions equal to the dimension of the plant state model. In this section we further recruit the plant output into the task of estimating the plant state by using it to reduce the dimension of the observer. More specifically, if the plant output $y(t)$ consists of p independent scalars, $\{y_i(t) : i = 1, 2, \cdots p\}$, then $y(t)$ contains information about the state in the form of a projection of the state onto a p dimensional subspace of state space. This fact allows us to concentrate on the design of an estimator for the remaining part of the state consisting of its projection on the remaining $n - p$ dimensional subspace of the state space. In this way we will see that we can obtain a plant state estimator having dimension $n - p$ which is referred to as a minimal or reduced order observer. We can develop this reduced order observer as follows.

3.7.1 The approach

Suppose the model for the plant whose state is to be estimated is given as

$$\dot{x}(t) = Ax(t) + Bu(t) \tag{3.29}$$

$$y(t) = Cx(t) \tag{3.30}$$

where

$$y^T(t) = [y_1(t) \quad y_2(t) \quad \cdots \quad y_p(t)]$$

Notice that there is no loss of generality in assuming that the state model has $D = \emptyset$, since when $D \neq \emptyset$ we can replace the left side of (3.30) by $q(t) = y(t) - Du(t)$ and use $q(t)$ in place of $y(t)$ everywhere in the following.

Now we assume throughout that the components $\{y_i(t) : i = 1, 2, \cdots, p\}$ of the plant output $y(t)$ are independent, i.e., we assume that

$$\sum_{i=1}^{p} \alpha_i y_i(t) = 0 \quad \text{for all } t \in [0, \infty) \qquad \text{only if } \alpha_i = 0 \quad \text{for all } i \in [1, p]$$

This implies that C has independent rows. As mentioned earlier in this chapter, if $p = n$ then C is invertible and we can solve (3.30) for $x(t)$ as

$$x(t) = C^{-1}y(t)$$

Since we usually have $p < n$, C is not invertible. However, even in this case, we can still use $y(t)$ and (3.30) to obtain information on the state. More specifically, we can determine $x^R(t)$,the projection of the state on range$[C^T]$, from

$$x^R(t) = C^{\#}y(t) \tag{3.31}$$

where $C^{\#} = C^T(CC^T)^{-1}$ is referred to as a right inverse of C since $CC^{\#} = I$. Then we can write $x(t)$ as

$$x(t) = x^R(t) + x^N(t) \tag{3.32}$$

where

$$x^R(t) \in \text{range}[C^T] \qquad\qquad x^N(t) \in \text{null}[C]$$

Now since we can be obtain $x^R(t)$ from (3.31), we see from (3.32) that we can determine the complete plant state provided we can develop a method for estimating $x^N(t)$. Before doing this we show that $x^R(t)$ can be determined from (3.31).

3.7.2 Determination of $x^R(t)$

Notice that the independence of the rows of C ensures that CC^T is invertible. We can see this as follows. Suppose CC^T is not invertible. Then we can find $y \neq \emptyset$ such that

$$y^T CC^T y = w^T w = 0$$

where $w = C^T y$. However this is possible only if $w = \emptyset$ and therefore C^T has dependent columns. Since this contradicts the assumption that C has independent rows we must have CC^T is invertible.

Now the expression for $C^{\#}$, (3.31), can be derived as follows. Since $x^R(t) \in \text{range}[C^T]$ it follows that

$$x^R(t) = C^T w(t) \tag{3.33}$$

for an appropriate $w(t)$. Then multiplying this equation through by C and using the invertability of CC^T enables us to determine $w(t)$ as

$$w(t) = (CC^T)^{-1}Cx^R(t) \tag{3.34}$$

However from (3.32) we see that

$$y(t) = Cx(t) = Cx^R(t)$$

and we can rewrite (3.34) as

$$w(t) = (CC^T)^{-1}y(t) \tag{3.35}$$

Finally, premultiplying this equation by C^T and using (3.33) yields (3.31).

3.7.3 A fictitious output

Having determined the projection, $x^R(t)$, of $x(t)$ on range$[C^T]$ along null$[C]$, we need to develop a method for estimating $x^N(t)$. This is done by introducing a fictitious output, $y^F(t)$, and using it with $y(t)$ to form a composite output, $y^C(t)$ as

$$y^C(t) = \begin{bmatrix} y(t) \\ y^F(t) \end{bmatrix} = \begin{bmatrix} C \\ C_F \end{bmatrix} x(t) = C_C x(t) \tag{3.36}$$

where C_F is chosen so that C_C is invertible, i.e., C_F has $n - p$ independent rows each independent of the rows of C. Then we can solve (3.36) for $x(t)$ as

$$x(t) = C_C^{-1} y^C(t) \tag{3.37}$$

Before showing how to generate $y_F(t)$, it should be pointed out that we can ensure the invertability of C_C by choosing C_F so that

$$\text{range}[C_F^T] = \text{null}[C] \tag{3.38}$$

We can see this by using the general result for matrices which is developed at the beginning of Section 3.6 in terms of the matrix $\mho$, i.e.,

$$\text{range}\left[\mho^T\right] \perp \text{null}[\mho]$$

Therefore we have

$$\text{range}\left[C_F^T\right] \perp \text{null}[C_F] \tag{3.39}$$

and it follows from (3.38, 3.39) that

$$\text{null}[C] \cap \text{null}[C_F] = \{\emptyset\}$$

so that

$$\begin{bmatrix} C \\ C_F \end{bmatrix} x = \emptyset \qquad \text{only if} \qquad x = \emptyset$$

which implies that C_C is invertible.

Now assuming we chose C_F to satisfy (3.38) we can readily show that

$$C_C^{-1} = \begin{bmatrix} C^{\#} & C_F^{\#} \end{bmatrix} \tag{3.40}$$

where

$$C^{\#} = C^T(CC^T)^{-1} \qquad C_F^{\#} = C_F^T(C_F C_F^T)^{-1}$$

This enables (3.37) to be rewritten as

$$x(t) = C^{\#} y(t) + C_F^{\#} y^F(t) \tag{3.41}$$

3.7.4 Determination of the fictitious output

Having obtained the state of the plant model in terms of the known plant output and the fictitious plant output, (3.41), we need a means for generating the fictitious output. This is done by using the plant state differential equation to set up a state model which has the plant's input and output, $(y(t), u(t))$, as input and has state which can be used as an estimate of the fictitious output. The derivation of this state model is given as follows.

We begin by multiplying the plant state differential equation (3.29) by C_F to obtain

$$C_F \dot{x}(t) = C_F A x(t) + C_F B u(t) \tag{3.42}$$

Then differentiating $x(t)$, (3.41), and substituting the result in (3.42) gives

$$C_F C^{\#} \dot{y}(t) + C_F C_F^{\#} \dot{y}^F(t) = C_F A C^{\#} y(t) + C_F A C_F^{\#} y^F(t) + C_F B u(t) \tag{3.43}$$

However from (3.38, 3.40) we have $C_F C^{\#} = \emptyset$ and $C_F C_F^{\#} = I$. Therefore (3.43) can be written as

$$\dot{y}^F(t) = C_F A C_F^{\#} y^F(t) + C_F A C^{\#} y(t) + C_F B u(t) \tag{3.44}$$

Now equation (3.44) suggests that an estimator for $y^F(t)$ has differential equation

$$\dot{\hat{y}}^F(t) = C_F A C_F^{\#} \hat{y}^F(t) + C_F A C^{\#} y(t) + C_F B u(t) \tag{3.45}$$

where $\hat{y}^F(t)$ will be an asymptotic estimate of $y^F(t)$ provided $C_F A C_F^{\#}$ is stable. This follows from the differential equation for the estimation error, $\tilde{y}^F(t)$, which we can determine by subtracting equation (3.44) from equation (3.45) to obtain

$$\dot{\tilde{y}}^F(t) = F_R\tilde{y}^F(t)$$

where

$$F_R = C_F A C_F^{\#} \qquad \tilde{y}^F(t) = \hat{y}^F(t) - y^F(t)$$

Thus we have

$$\tilde{y}^F(t) = (e^{F_R t})\tilde{y}^F(0)$$

so that

$$\lim_{t\to\infty} \tilde{y}^F(t) = \emptyset \qquad \text{for all } \tilde{y}^F(0) \tag{3.46}$$

if and only if F_R is stable.

3.7.5 Assignment of observer eigenvalues

Now the eigenvalues of $F_R = C_F A C_F^{\#}$ are fixed by the plant state model. Therefore when some of these eigenvalues are unstable we cannot satisfy (3.46). Recall we encountered the same sort of difficulty in Section 3.2 in connection with trying to estimate the state of a plant state model without using the plant output. We can overcome the present problem when the state model, whose state is to be estimated, is detectable by replacing C_F everywhere in the foregoing development by

$$T = C_F + L_R C \tag{3.47}$$

where L_R is used in the same way that L is used in Section 3.3 to assign the eigenvalues of an observer's system matrix, $A + LC$. This replacement is effected by replacing $y^F(t)$ by $z(t) = Tx(t)$ in (3.36) to obtain

$$\begin{bmatrix} y(t) \\ z(t) \end{bmatrix} = C_T x(t) \tag{3.48}$$

where

$$C_T = \begin{bmatrix} C \\ T \end{bmatrix} = QC_C \qquad Q = \begin{bmatrix} I & \emptyset \\ L_R & I \end{bmatrix}$$

Notice that since C_C is invertible and Q is invertible, independent of L_R, C_T is invertible for any L_R. Therefore we can determine $x(t)$ from (3.48) for any L_R as

$$x(t) = My(t) + Nz(t) \tag{3.49}$$

where

$$C_T^{-1} = [M \quad N] \tag{3.50}$$

Now by forming $C_T C_T^{-1} = I$ we can show that M and N must satisfy

$$\begin{aligned} CM &= I & \qquad CN &= \emptyset \\ TM &= \emptyset & \qquad TN &= I \end{aligned} \tag{3.51}$$

Therefore recalling from (3.38, 3.40) that

$$\begin{bmatrix} C \\ C_F \end{bmatrix} [C^{\#} \quad C_F^{\#}] = \begin{bmatrix} I & \emptyset \\ \emptyset & I \end{bmatrix} \tag{3.52}$$

we see that the constraints (3.51) on M and N are satisfied when

$$M = C^{\#} - C_F^{\#} L_R \qquad\qquad N = C_F^{\#}$$

In order to obtain a differential equation for $z(t)$, we follow the approach used to get a differential equation for $y_F(t)$, (3.44). Therefore multiplying equation (3.29) by T and substituting for $x(t)$ from (3.49) yields

$$\dot{z}(t) = TANz(t) + TAMy(t) + TBu(t) \tag{3.53}$$

where $z(t) = Tx(t)$ and

$$TAN = C_F A C_F^{\#} + L_R CAC_F^{\#}$$

$$TAM = C_F AC^{\#} + L_R CAC^{\#} - C_F AC_F^{\#} L_R - L_R CAC_F^{\#} L_R$$

This differential equation suggests that an estimator, having the same form as an observer, (3.13), can be formulated for $z(t)$ as

$$\dot{\hat{z}}(t) = F_T \hat{z}(t) + G_T \begin{bmatrix} u(t) \\ y(t) \end{bmatrix} \tag{3.54}$$

where

$$F_T = \text{TAN} \qquad\qquad G_T = [TB \quad TAM]$$

Then the estimation error for $z(t)$ is given by

$$\tilde{z}(t) = (e^{F_T t})\tilde{z}(0)$$

where $\tilde{z}(t) = z(t) - \hat{z}(t)$ and we have an asymptotic estimate of $z(t)$ if and only if F_T is stable.

Thus provided F_T is stable, we can use $\hat{z}(t)$ generated by equation (3.54) in place of $z(t)$

in (3.49) to obtain an asymptotic estimate of $x(t)$ as

$$\hat{x}(t) = My(t) + N\hat{z}(t) \tag{3.55}$$

Notice that the dependency of $F_T = TAN$ on L_R, (3.53), implies that we are only able to choose L_R so that TAN is stable if the pair $(C_F AC_F^{\#},\ CAC_F^{\#})$ is detectable. We show now that this detectability condition is satisfied when the plant state model or pair (A, C) is detectable.

Theorem 3.3 The pair $(C_F AC_F^{\#}, CAC_F^{\#})$ is detectable if and only if the pair (A, C) is detectable.

Proof Suppose (λ, v) is an eigenvalue, right-eigenvector pair for A. Let $\mathbf{\Gamma}$ be a coordinate transformation matrix given as

$$\mathbf{\Gamma} = \begin{bmatrix} C^{\#} & C_F^{\#} \end{bmatrix} \qquad \mathbf{\Gamma}^{-1} = \begin{bmatrix} C \\ C_F \end{bmatrix} \tag{3.56}$$

Then premultiplying the eigenvalue–eigenvector equation $Av = \lambda v$ by $\mathbf{\Gamma}^{-1}$ and inserting $\mathbf{\Gamma}\mathbf{\Gamma}^{-1}$ between A and v gives

$$\bar{A}q = \lambda q \tag{3.57}$$

where $q = \mathbf{\Gamma}^{-1}v$ and

$$\bar{A} = \mathbf{\Gamma}^{-1}A\mathbf{\Gamma} = \begin{bmatrix} \bar{A}_1 & \bar{A}_2 \\ \bar{A}_3 & \bar{A}_4 \end{bmatrix} = \begin{bmatrix} CAC^{\#} & CAC_F^{\#} \\ C_F AC^{\#} & C_F AC_F^{\#} \end{bmatrix}$$

Then from (3.52) we see that C is transformed as

$$\bar{C} = C\mathbf{\Gamma} = C\begin{bmatrix} C^{\#} & C_F^{\#} \end{bmatrix} = \begin{bmatrix} I & \emptyset \end{bmatrix}$$

Now suppose λ is an unobservable eigenvalue of the pair (A, C). Then recalling that unobservable eigenvalues are invariant to coordinate transformation, we see that q, (3.57), satisfies

$$\bar{C}q = \emptyset$$

which from the form for $\bar{C}$ requires q to have its first p elements zero

$$q = \begin{bmatrix} \emptyset \\ q^2 \end{bmatrix}$$

where q^2 has $n - p$ components. Then using the form just obtained for $\bar{A}$, we see that the eigenvalue–eigenvector equation, (3.57) becomes

$$\begin{bmatrix} \bar{A}_2 q^2 \\ \bar{A}_4 q^2 \end{bmatrix} = \begin{bmatrix} \emptyset \\ \lambda q^2 \end{bmatrix}$$

implying that λ is an unobservable eigenvalue of the pair $(\bar{A}_4, \bar{A}_2)$ which equals the pair $(C_F A C_F^{\#}, CAC_F^{\#})$ specified in the statement of the theorem. This shows that the observability or detectability, (if $\text{Re}[\lambda] \geq 0$), of (A, C) is necessary for the observability or detectability of the pair $(C_F A C_F^{\#}, CAC_F^{\#})$. Sufficiency can be shown by reversing the foregoing argument. ■

Notice that when the plant state model's (A, C) pair is observable the coordinate transformation Γ, (3.56), just used in the proof of Theorem 3.3, can be used to determine L_R, (3.47), to achieve some specified set of eigenvalues for the minimal order observer's system matrix, F_T, (3.54). One way of doing this would be to use Ackermann's formula, introduced in the previous chapter, on the pair $(\bar{A}_4, \bar{A}_2)$, (3.57).

The proof of Theorem 3.3 suggests that if we work in coordinates where the state model's C matrix is

$$\bar{C} = [I \quad \emptyset]$$

then a minimal order observer to estimate the state in the transformed coordinates can be determined by taking T, (3.47), as

$$T = [L_R \quad Q]$$

where Q is any nonsingular matrix of appropriate size. This ensures that C_T, (3.48), is invertible independent of L_R. Notice that Q plays the role of a coordinate transformation matrix for the coordinates of the minimal order observer and therefore Q does not affect the minimal order observer eigenvalues. Therefore referring to (3.54, 3.55) and taking $Q = I$ for simplicity, we obtain a minimal order observer for the estimation of the transformed state as

$$\hat{q}(t) = My(t) + N\hat{z}(t)$$

$$\dot{\hat{z}}(t) = T\bar{A}N\hat{z}(t) + T\bar{A}My(t) + T\bar{B}u(t)$$

where

$$M = \begin{bmatrix} I \\ -L_R \end{bmatrix} \qquad N = \begin{bmatrix} \emptyset \\ I \end{bmatrix}$$

$$T\bar{A}N = \bar{A}_4 + L_R\bar{A}_2$$

$$T\bar{A}M = L_R\bar{A}_1 - \bar{A}_4L_R - L_R\bar{A}_2L_R + \bar{A}_3$$

In summary, we have seen that a minimal order observer for an n-dimensional plant state model is $(n-p)$-dimensional, where p is the number of independent scalar plant outputs available for use in estimating the plant state. As in the case of the full order observer introduced in the first few sections of this chapter, unobservable eigenvalues of the plant state model's pair (A, C) are fixed eigenvalues for the minimal order observer's system matrix F_T, (3.54). Thus as in the case of the full order observer we are unable to design an asymptotic observer when the plant state model is not detectable.

3.8 Notes and References

The observer was originally proposed by David G. Luenberger in his Ph.D. thesis in 1964. For this reason the observer is sometimes referred to as a Luenenberger observer. Previously, in 1960, the use of a state model having the same structure as an observer was proposed by R. E. Kalman for the purposes of least squares signal estimation in the presence of additive noise. We will encounter the Kalman filter in Chapter 6. For further discussion of observers the reader is referred to Chapter 4 of [23]

Finally, the term " output injection" is used by some authors to refer to the system signal manipulation we use to obtain an observer having system matrix $A + LC$. This terminology should be compared with the use of the term "state feedback" in connection with the formation of a system having system matrix $A + BK$.

4

Model Approximation via Balanced Realization

4.1 Introduction

In previous chapters it was noted that whenever a state model (A, B, C) is either not controllable or not observable the transfer function resulting from the calculation $C(sI - A)^{-1}B$ has pole-zero cancellations. Thus system uncontrollability and/or unobservability gives rise to a transfer function with order less than the dimension of the state space model. Therefore systems that are "almost uncontrollable" or "almost unobservable", should be able to be approximated by a transfer function model of order less than the dimension of the state model. Reduced order model approximation can be of importance in the implementation of feedback control systems. For instance, model order reduction may be beneficial when the large dimension of the full state model gives rise to computational problems during the operation of the control system.

This chapter treats the problem of extracting a subsystem of a given full order system which works on that part of the state space which is most involved in the input–output behavior of the original system. This is accomplished by introducing two matrices called the observability Gramian and the controllability Gramian. It is shown that these matrices can be used to quantify the energy transfer from the input to the state and from the state to the output. In this way it is possible to identify which part of the state space is controlled most strongly and which part is most strongly observed. Then in order to construct a reduced order model by extracting a subsystem from the given system, it will be shown that the strongly controlled subspace must coincide with the strongly observed subspace. To achieve this coincidence we change coordinates, if necessary, so that the Gramians become equal and diagonal. The resulting state model is said to be a balanced realization. The reduced order model is then obtained by extracting a subsystem from the balanced realization.

4.2 Controllable-Observable Decomposition

The idea for model order reduction being pursued in this chapter grew out of the basic fact that only the jointly controllable and observable part of a state model is needed to model the input–output behavior.

The controllable and observable part of the state space can be viewed as one of four possible subspaces for the state space of any given state model. More specifically, consider the four dimensional state model in diagonal form which is shown in Figure 4.1.

Inspection of the block diagram reveals that the four dimensional state model has a transfer function which is only first order since only the first component of the state is both affected by the input and affects the output. Thus that part of the state space which is both controllable and observable forms a one dimensional subspace $\mathcal{S}_{co}$ which is specified as

$$x(t) \in \mathcal{S}_{co} \qquad \text{if } x_i(t) = 0 \text{ for } i = 2, 3, 4$$

where

$$x^T(t) = [\,x_1(t) \quad x_2(t) \quad x_3(t) \quad x_4(t)\,]$$

On further inspection of Figure 4.1 we see that there are three other subspaces $\mathcal{S}_{c\bar{o}}$, $\mathcal{S}_{\bar{c}o}$, $\mathcal{S}_{\bar{c}\bar{o}}$ specified as

$$\begin{aligned} x(t) &\in \mathcal{S}_{c\bar{o}} && \text{if } x_i(t) = 0 \text{ for } i = 1, 3, 4 \\ x(t) &\in \mathcal{S}_{\bar{c}o} && \text{if } x_i(t) = 0 \text{ for } i = 1, 2, 4 \\ x(t) &\in \mathcal{S}_{\bar{c}\bar{o}} && \text{if } x_i(t) = 0 \text{ for } i = 1, 2, 3 \end{aligned}$$

where $\mathcal{S}_{co}$, $\mathcal{S}_{c\bar{o}}$, $\mathcal{S}_{\bar{c}o}$, $\mathcal{S}_{\bar{c}\bar{o}}$ are referred to as follows.

$\mathcal{S}_{co}$ is the controllable and observable subspace
$\mathcal{S}_{c\bar{o}}$ is the controllable and not observable subspace
$\mathcal{S}_{\bar{c}o}$ is the not controllable and observable subspace
$\mathcal{S}_{\bar{c}\bar{o}}$ is the not controllable and not observable subspace

More generally when we are given a higher dimension state model, (A, B, C), which is

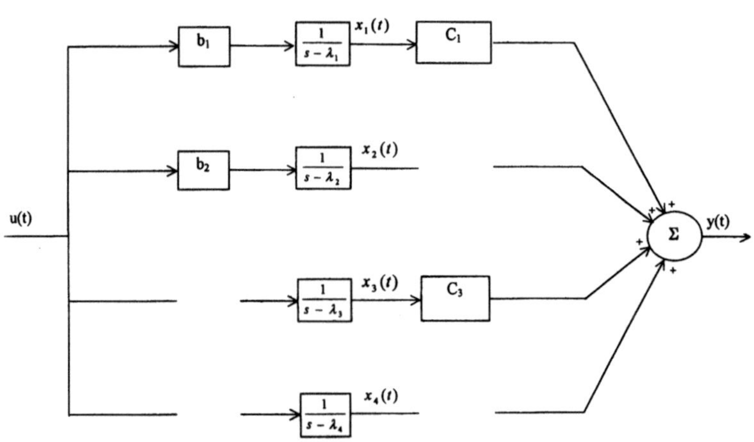

Figure 4.1 Example of controllable-observable decomposition

not in any special form, we can extract a subsystem from the given system which has its state space coincident with the controllable and observable subspace. This can be done by using two successive coordinate transformations as described now.

First we change coordinates so that the state model is in a controllable decomposed form

$$A_c = \begin{bmatrix} A_{c1} & A_{c2} \\ \emptyset & A_{c4} \end{bmatrix} \qquad B_c = \begin{bmatrix} B_{c1} \\ \emptyset \end{bmatrix} \qquad C_c = \begin{bmatrix} C_{c1} & C_{c2} \end{bmatrix} \tag{4.1}$$

with (A_{c1}, B_{c1}) being a controllable pair. The transformation matrix to do this was discussed in Chapter 2.

Notice that, in the new coordinates, we have the full n dimensional state space split into controllable and uncontrollable subspaces, $\mathcal{S}_c, \mathcal{S}_{\bar{c}}$. It remains to split $\mathcal{S}_c$ into observable and unobservable subspaces, i.e., to split $\mathcal{S}_c$ into $\mathcal{S}_{co}$ and $\mathcal{S}_{c\bar{o}}$.

Next we carry out a second coordinate transformation which affects only the controllable part of the controllable decomposed state model, (4.1). This is done using a coordinate transformation matrix, T, which is constructed as

$$T = \begin{bmatrix} T_o & \emptyset \\ \emptyset & I \end{bmatrix}$$

with T_o being chosen so that the controllable state model (A_{c1}, B_{c1}, C_{c1}) is transformed to observable decomposed form, i.e., so that

$$\bar{A}_{c1} = \begin{bmatrix} \bar{A}_{o1} & 0 \\ \bar{A}_{o3} & \bar{A}_{o4} \end{bmatrix} \qquad \bar{B}_{c1} = \begin{bmatrix} \bar{B}_{o1} \\ \bar{B}_{o2} \end{bmatrix} \qquad \bar{C}_{c1} = \begin{bmatrix} \bar{C}_{o1} & \emptyset \end{bmatrix} \tag{4.2}$$

with $(\bar{A}_{o1}, \bar{C}_{o1})$ being an observable pair. The transformation T_o to do this was discussed in Chapter 3. Since coordinate transformation preserves controllability, the state model (4.2) is controllable. Therefore the subsystem of the observable part of the state model (4.2), i.e., $(\bar{A}_{o1}, \bar{B}_{o1}, \bar{C}_{o1})$ is both controllable and observable. Thus the transfer function for the entire system has order equal to the dimension of this subsystem and is given by

$$G(s) = \bar{C}_{o1}(sI - \bar{A}_{o1})^{-1}\bar{B}_{o1}$$

Notice that the foregoing model reduction procedure applies to full order state models which are either not controllable or not observable or both not controllable and not observable. More important, the reduced order model we obtain has exactly the same input–output behavior as the full order model.

In this chapter we are interested in obtaining a reduced order model when the full order state model is completely controllable and observable. In this situation any reduced order model will have input–output behavior which differs from the input–output behavior of the full order state model. Our goal is to develop a method for achieving a reduced order state model having input–output behavior which approximates the input–output behavior of the full order state model. We will see that we can develop an approach for solving this model approximation problem by replacing the concepts of controllability and observability by concepts of energy transfer. We will see that the distribution of energy among components of the state is related to properties of Gramian matrices which replace the controllability and observability matrices in this analysis. Just as coordinate transformations are important by extracting the controllable and observable part of a state model, so too are coordinate transformations important to being able to separate out that part of a controllable and observable state model which is most important in the transfer of energy from the system's input to the system's output. The state model in coordinates which enable the identification of this part of the system is called a balanced realization.

4.3 Introduction to the Observability Gramian

In order to define the observability Gramian we need to consider the related idea of output energy. Recall that if the state model (A, B, C) is known, then the zero input response, $y(t)$, caused by any specified initial state, $x(0)$, is given as

$$y(t) = Ce^{At}x(0) \tag{4.3}$$

The output energy, E_o, for each initial state, $x(0)$, is then defined as the integral of a nonnegative, real, scalar function $v(t)$ of $y(t)$, i.e., $v(t) \geq 0$ for all $t \in [0, \infty)$

$$E_o = \int_0^\infty v(t)\mathrm{d}t$$

Notice that since $v(t)$ is non-negative real for all t, the output energy E_0 for any initial state $x(0)$ must satisfy

$$E_o \geq 0 \quad \text{for all } x(0)$$
$$E_o = 0 \quad \text{only if } v(t) = 0 \text{ for } t \geq 0$$

A simple way to choose $v(t)$ so that E_o has these properties is as follows.

if $y(t)$ is a real scalar	then choose	$v(t) = y^2(t)$
if $y(t)$ is a complex scalar	then choose	$v(t) = \lvert y(t)\rvert^2$
if $y(t)$ is a real vector	then choose	$v(t) = y^T(t)y(t)$
if $y(t)$ is a complex vector	then choose	$v(t) = y^*(t)y(t)$

Notice that when $y(t)$ is a scalar, $y^*(t)$ denotes the complex conjugate of $y(t)$ and $|y(t)|^2 = y^*(t)y(t)$. Alternatively when $y(t)$ is a complex vector, $y^*(t)$ denotes the transpose of $y(t)$ and the complex conjugation of each entry in the resulting row vector. Thus if $y(t)$ is a complex vector of length p then

$$y^*(t)y(t) = \sum_{i=1}^{p} |y_i(t)|^2$$

More important, notice that each of the foregoing choices is a special case of the last choice. Therefore, to avoid special cases, we define the output energy E_o, in general as

$$E_o = \int_0^\infty y^*(t)y(t)\mathrm{d}t \tag{4.4}$$

so that after substituting from (4.3) we obtain

$$E_o = x^*(0)\left(\int_0^\infty e^{A^*t}C^*Ce^{At}dt\right)x(0) \tag{4.5}$$

$$r = x^*(0)W_ox(0) \tag{4.6}$$

where the constant $n \times n$ matrix W_o is called the observability Gramian.

Notice that in order for the foregoing integral to converge, or equivalently for W_o to exist, the observable eigenvalues of (A, C) must be in the open left half plane. This can be seen by referring to section 7 of Chapter 1. Alternatively, if (A, C) is observable W_o exists only if A is stable. Notice also that W_o must be

1. an Hermitian matrix, i.e., $W_o = W_o^*$,
2. a non-negative matrix, denoted $W_o \geq 0$.

Requirement number one results from the integral defining W_o, (4.5). Hermitian matrices have entries in mirror image positions about the diagonal which are conjugates of each other, i.e., $(W_o)_{ij} = (W_o)_{ji}^*$. When W_o is real we require $W_o = W_o^T$ and W_o is referred to as a symmetric matrix . Hermitian (symmetric) matrices have eigenvalues and eigenvectors with some rather striking properties which we will develop in the next section.

Requirement number two results from the output energy being nonnegative for all initial states since a non-negative matrix, W_o, has the property that

$$x^*(0)W_ox(0) \geq 0$$

for all vectors $x(0)$ of appropriate length. Now it turns out that we can always find a matrix M such that any Hermitian, nonnegative matrix W_o can be factored as $W_o = M^*M$. This fact allows us to write

$$\begin{aligned} x(0)W_ox(0) &= x(0)^*M^*Mx(0) \\ &= z^*z = \sum_{i=1}^{n}|z_i|^2 \end{aligned}$$

where $z = Mx(0)$. Since $\sum_{i=1}^n |z_i|^2 \geq 0$ for any z we have $W_o \geq 0$. Moreover since $\sum_{i=1}^n |z_i|^2 = 0$ if and only if $z = \emptyset$, we have $x(0)^* W_ox(0) = 0$ for some $x(0) \neq \emptyset$ if and only if M and hence W_o is singular, i.e., if and only if W_o has one or more zero eigenvalues. More is said about this in the next section.

Finally the observability of a state model is closely tied to the nonsingularity of its observability Gramian. This can be seen by noting from (4.4) that $E_o = 0$ if and only if $y(t) = \emptyset$ for all time. However we saw in Chapter 3 that the zero input response is null, $y(t) = \emptyset$, for some $x(0) \neq \emptyset$ if and only if (A, C) is unobservable. Therefore it follows that

(A, C) is unobservable if and only if W_o is singular. Further insight into this fact is given in subsequent sections where it is shown that W_o can be obtained by solving a linear matrix equation which is referred to as the Lyapunov equation and which involves A and C in a simple fashion.

4.4 Fundamental Properties of W_o

In the previous section we saw that the observability Gramian is a member of the class of matrices which are characterized as being Hermitian and nonnegative. In this section we give some of the essential features of these kinds of matrices.

4.4.1 Hermitian matrices

Recall that a matrix W_o is said to be Hermitian if it equals its conjugate transpose, $W_o = W_o^*$. This implies the ij^{th} and ji^{th} entries are conjugates of each other. The eigenvalues and eigenvectors of Hermitian matrices have some rather striking properties which are given as follows.

Theorem 4.1 If W_o is Hermitian then:

(i) W_o has real eigenvalues
(ii) W_o has right eigenvectors which satisfy $v^{i*}v^j = 0$, when $\sigma_{oi} \neq \sigma_{oj}$

Proof (i) Let (σ_o, v) be any eigenvalue–eigenvector pair for W_o, i.e.,

$$W_o v = \sigma_o v \tag{4.7}$$

Then premultiplying (4.7) by v^* gives

$$v^* W_o v = \sigma_o v^* v \tag{4.8}$$

and taking the conjugate transpose throughout this equation gives

$$v^* W_o^* v = \sigma_o^* v^* v \tag{4.9}$$

Then subtracting (4.9) from (4.8) we obtain

$$v^*(W_o - W_o^*)v = (\sigma_o - \sigma_o^*)v^* v \tag{4.10}$$

Now since W_o is Hermitian, the left side of (4.10) is zero. Therefore either $\sigma_o = \sigma_o^*$ or $v^*v = 0$. However $v^*v = 0$ holds only if $v = \phi$, and since v is an eigenvector of W_o, we have $v^*v \neq 0$. Therefore (4.10) can only be satisfied if $\sigma_o = \sigma_o^*$ which shows that σ_o must be real. ■

Proof(ii) Let $(\sigma_{oi}, v^i : i = 1, 2)$ be two eigenvalue–eigenvector pairs for W_o with $\sigma_{o1} \neq \sigma_{o2}$. Then we have

$$W_o v^1 = \sigma_{o1} v^1 \tag{4.11}$$

$$W_o v^2 = \sigma_{o2} v^2 \quad \text{or} \quad v^{2*} W_o^* = \sigma_{o2}^* v^{2*} \tag{4.12}$$

Next operate on these equations as follows:

(i) premultiply (4.11) by v^{2*}
(ii) postmultiply the second form for (4.12) by v^1
(iii) subtract the result of (ii) from (i)

This yields

$$v^{2*}(W_o - W_o^*)v^1 = (\sigma_{o1} - \sigma_{o2}^*)v^{2*}v^1 \tag{4.13}$$

However, since W_o is Hermitian, the left side of (4.13) is zero. Moreover we are assuming that $\sigma_{o1} \neq \sigma_{o2}^*$. Therefore (4.13) is satisfied only if $v_1^* v_2 = 0$. ■

Notice that when W_o is real, i.e., symmetric, eigenvectors corresponding to different eigenvalues are orthogonal, i.e., $v^{iT}v^j = 0 : i \neq j, \sigma_{oi} \neq \sigma_{oj}$.

Recall that if a matrix has distinct eigenvalues, it has a complete set of eigenvectors and can be diagonalized by a matrix having columns made up from its eigenvectors. Thus when W_o has distinct eigenvalues we can use its eigenvectors $\{v^i : i = 1, 2, \cdots n\}$ to form a matrix V_0 which diagonalizes W_o

$$V_o^{-1} W_o V_o = \Sigma_o \tag{4.14}$$

where

$$\Sigma_o = \begin{bmatrix} \sigma_{o1} & 0 & \cdots & 0 \\ 0 & \sigma_{o2} & \cdots & 0 \\ \vdots & \vdots & \ddots & \vdots \\ 0 & 0 & \cdots & \sigma_{on} \end{bmatrix} = \text{diag}[\sigma_{o1}, \sigma_{o2}, \cdots, \sigma_{on}]$$

$$V_o = [v^1 \quad v^2 \quad \cdots \quad v^n]$$

However if the eigenvectors are normalized,

$$v^{i*}v^i = 1 : i = 1, 2, \cdots, n$$

then we see from Theorem 4.1 that

$$V_o^* V_o = \begin{bmatrix} v^{1*} \\ v^{2*} \\ \vdots \\ v^{n*} \end{bmatrix} [v^1 \quad v^2 \quad \cdots \quad v^n] = I \tag{4.15}$$

Therefore $V_o^{-1} = V_o^*$ so that $V_o^* W_o V = \Sigma_o$. Matrices satisfying the condition $V_o^{-1} = V_o^*$ are referred to as unitary matrices.

Alternatively, when W_o is real then (4.15) becomes

$$V_o^T V_o = I$$

and therefore $V_o^{-1} = V_o^T$ so that $V_o^T W_o V_o = \Sigma_o$. Matrices satisfying the condition

$V_o^{-1} = V_o^T$ are referred to as orthogonal matrices. It can be shown that any Hermitian matrix can be diagonalized by a unitary matrix even if the Hermitian matrix has multiple eigenvalues.

To recap, we have encountered the following matrix types.

Definition 4.1

If M is complex then

$$\begin{array}{lll} M \text{ is Hermitian} & \text{if} & M = M^* \\ M \text{ is unitary} & \text{if} & M^{-1} = M^* \end{array}$$

If M is real then

$$\begin{array}{lll} M \text{ is symmetric} & \text{if} & M = M^T \\ M \text{ is orthogonal} & \text{if} & M^{-1} = M^T \end{array}$$

4.4.2 Positive definite and non-negative matrices

Suppose that we change coordinates, viz., $x(0) = Tz(0)$, with T nonsingular. Then for each initial state $x(0)$ in the old coordinates, there is a unique initial state $z(0)$ in the new coordinates and vice versa. Thus we see from the expression for the output energy, (4.6), that the output energy can be rewritten in terms of the initial state in the new coordinates as

$$E_o = z^*(0)\tilde{W}_o z(0) \tag{4.16}$$

where $\tilde{W}_o = T^* W_o T$.

Now if we choose T as the unitary matrix V_o, (4.14), then

$$\tilde{W}_o = V_o^* W_o V_o = \Sigma_o$$

and (4.16) becomes

$$\mathrm{E}_o = z^*(0)\Sigma_o z(0) = \sum_{i=1}^{n} \sigma_{oi}|z_i(0)|^2 \tag{4.17}$$

However since there is a unique $z(0)$ for each $x(0)$ we have that $E_o \geq 0$ for all $x(0)$ if and only if $E_o \geq 0$ for all $z(0)$. Thus we see from (4.17) that $E_o \geq 0$ for all $x(0)$ if and only if all eigenvalues of W_o are non-negative, i.e.,

$$\sigma_{oi} \geq 0 \qquad i = 1, 2, \cdots n$$

Otherwise, if W_o were to have some negative eigenvalues, we could choose $z(0)$ so that

$$z_i(0) \neq 0 \text{ if } \sigma_{0i} < 0 \quad \text{and} \quad z_i(0) = 0 \text{ if } \sigma_{0i} > 0 \qquad i = 1, 2, \cdots, n$$

which would give $E_o < 0$, for the initial state $x(0) = Tz(0)$ and the requirement for E_o to be non-negative for all initial states would be violated.

Finally, we can show by a similar argument that $E_0 > 0$ for all $x(0) \neq \emptyset$ if and only if all eigenvalues of W_o are positive.

To recap, we have defined the matrix properties of positive definiteness and non-negativeness as follows.

Definition 4.2 W_o said to be:

(i)	positive definite, denoted as $W_o > 0$	when $x^* W_o x > 0$	for all $x \neq \emptyset$
(ii)	non-negative, denoted as $W_o \geq 0$	when $x^* W_o x \geq 0$	for all x

Moreover these properties impose certain restrictions on the eigenvalues of W_o namely

(i)	if $W_o > 0$	then $\sigma_{oi} > 0$	for $\sigma_{oi} \in \lambda[W_o]$
(ii)	if $W_o \geq 0$	then $\sigma_{oi} \geq 0$	for $\sigma_{oi} \in \lambda[W_o]$

4.4.3 Relating E_o to $\lambda[W_o]$

In order to begin to appreciate the role played by the eigenvalues of W_o in achieving a lower order state model approximation we consider the problem of finding the initial state, $x_{\max}(0)$, which produces the most output energy, $E_{0\max}$, when the initial state is constrained so that

$$x^*_{\max}(0)x_{\max}(0) = 1 \tag{4.18}$$

When $x(0)$ is constrained in this way we say $x(0)$ is normalized to one. Notice that this constraint on the initial state removes the possibility of obtaining an unbounded output energy by choosing an initial state with an unbounded component.

We begin the development of a solution to this problem by noticing that if we use a coordinate transformation matrix T which is unitary, the constraint (4.18) is preserved, i.e., if T satisfies $T^*T = I$ then we have

$$x^*(0)x(0) = z^*(0)T^*Tz(0) = z^*(0)z(0) \tag{4.19}$$

Therefore if $x^*(0)x(0) = 1$ so does $z^*(0)z(0)$.

Now we have seen that using the unitary matrix $T = V_o$ as a coordinate transformation matrix makes the observability Gramian diagonal in the new coordinates. Therefore our problem can be solved by first solving the problem in coordinates in which the observability Gramian is diagonal and then transforming the solution, $z_{\max}(0)$ in these coordinates back into the original coordinates.

Suppose we are in coordinates where the observability Gramian is a diagonal matrix, Σ_o. Suppose also that we have reorder the components of the state in these coordinates so that the largest entry on the diagonal of Σ_o is first, i.e., $\sigma_{o1} \geq \sigma_{oi} : \ i = 2, 3, \cdots, n$. Then we see from equation (4.17) that the output energy satisfies

$$E_o = \sum_{i=1}^{n} \sigma_{oi}|z_i(0)|^2 \leq \sigma_{o1} \sum_{i=1}^{n} |\, z_i(0)\, |^2 \tag{4.20}$$

However since we are assuming that $x(0)$ is normalized to one, we see from (4.19) that $z(0)$ is also normalized to one

$$z^*(0)z(0) = \sum_{i=1}^{n} | z_i(0) |^2 = 1 \tag{4.21}$$

Therefore under this constraint we see from (4.20) that the output energy, E_o, is bounded by the observability Gramian's largest eigenvalue, i.e.,

$$E_o \leq \sigma_{o1} \tag{4.22}$$

Notice that if we choose

$$z^T(0) = [1 \quad 0 \quad \cdots \quad 0]$$

then (4.21) is satisfied and the output energy becomes

$$E_o = \sum_{i=0}^{n} \sigma_{oi}|z_i(0)|^2 = \sigma_{o1}$$

and we achieve the upper bound on E_o, (4.22). Thus the normalized initial state in the original coordinates which maximizes the output energy is obtained as

$$\begin{aligned} x_{\max}(0) &= V_o z(0) \\ &= \begin{bmatrix} v^1 & v^2 & \cdots & v^n \end{bmatrix} \begin{bmatrix} 1 \\ 0 \\ \vdots \\ 0 \end{bmatrix} = v^1 \end{aligned} \tag{4.23}$$

Therefore our problem is solved by setting the initial state equal to the eigenvector of W_o corresponding to the largest eigenvalue of W_o when that eigenvector is normalized to one. Moreover, it is not hard to see from the foregoing that, in general, we have

$$E_o = \sigma_{ok} v^{k*} v^k$$

when

$$x(0) = v^k$$

The foregoing analysis indicates that the relative importance of the components of $z(0)$ in producing output energy depends on the size of an eigenvalues of W_o relative other eigenvalues of W_o. We can use this observation to suggest how we are going to be able to obtain a reduced order state model approximation. We do this as follows.

Suppose we choose the coordinates for the state space so that the eigenvalues of W_o are in descending size, i.e., $\sigma_{o1} \geq \sigma_{o2} \geq \cdots \geq \sigma_{on}$ with W_0 diagonal. Suppose we consider $z(0)$

to have all its elements equal, $z_i(0) = z_j(0)$. Then the output energy is given by

$$E_o = z^*(0)\begin{bmatrix} \sigma_{o1} & 0 & \cdots & 0 \\ 0 & \sigma_{o2} & \cdots & 0 \\ \vdots & \vdots & \ddots & \vdots \\ 0 & 0 & \cdots & \sigma_{on} \end{bmatrix} z(0) = \gamma \sum_{i=1}^{n} \sigma_{oi}$$

where

$$\gamma = |z_i(0)|^2$$

Moreover, we can always choose an integer $n_1 \in [0, n]$ large enough so that

$$E_{o1} \gg E_{o2} \tag{4.24}$$

where

$$E_{o1} = \gamma \sum_{i=1}^{n_1} \sigma_{oi} \qquad E_{o2} = \gamma \sum_{i=n_1+1}^{n} \sigma_{oi}$$

Therefore in this situation a good approximation to the output energy is

$$E_o \simeq E_{o1}$$

Since the last $n - n_1$ components of the initial state are less important to the zero input response than the first n_1 components under the foregoing condition of an equal element initial state, a reduced order state model in these coordinates could be formed as $(\bar{A}_1, \bar{B}_1, \bar{C}_1)$ where $\bar{A}_1$ is the $n_1 \times n_1$ partition in the top left corner of $\bar{A}$ and $\bar{B}_1, \bar{C}_1$ are corresponding partitions of $\bar{B}, \bar{C}$.

However, our interest in this chapter is in obtaining a reduced dimension state model whose input–output behavior approximates the input–output behavior of the full dimension model. The foregoing argument for neglecting the last $n - n_1$ components of $z(t)$ focuses only on the zero input response. Therefore what is missing from the previous argument, if we are interested in modeling the input–output behavior, is the effect the input has on the neglected components of $z(t)$. The effect of the input on different components of the state is related to properties of a matrix known as the controllability Gramian.

4.5 Introduction to the Controllability Gramian

The controllability Gramian, which we introduce in this section, is involved in characterizing the energy delivered to the state when the input is a unit impulse and the initial state is null. Since the Laplace transform of the zero state response to this input is the given system's transfer function the distribution of energy among the components of the state in this situation is appropriate to our goal of obtaining a reduced order transfer function approximation. We begin the development of the controllability Gramian as follows.

Suppose we are given a state model, (A, B, C). Then the state in response to a unit impulse at time zero when the initial state is null is given by

$$x(t) = \int_0^t e^{A(t-\tau)} B\delta(\tau)\mathrm{d}\tau = e^{At}B$$

Then the energy, E_c, transferred from this input to the state is given by

$$E_c = \int_0^\infty x^*(t)x(t)\mathrm{d}t = \int_0^\infty B^* e^{A^* t} e^{At} B\mathrm{d}t \tag{4.25}$$

Notice that the integrand obtained here is strikingly different from the one obtained in the integral for the output energy, (4.5). We proceed now to express E_c (4.25) as an integral which appears as the dual of the integral expression of E_o. More specifically, we show now that E_c can be expressed as the trace of the integral obtained earlier for E_o, (4.5) with A replaced by A^* and C replaced by B^*. The trace of a matrix is a scalar which is defined as follows.

Definition 4.3 The trace of any square matrix M of size n, is defined as the sum of its diagonal entries,

$$\text{trace}[M] = \sum_{i=1}^{n} m_{ii}$$

where m_{ij} is the ij^{th} entry in M.

In order to proceed we need the following result involving the trace of the product of two, not necessarily square, matrices

Theorem 4.2 Given two matrices M, N of size $p \times n$, and $n \times p$ then

$$\text{trace}[MN] = \text{trace}[NM]$$

Proof Let $MN = Q$. Then

$$\text{trace}[MN] = \sum_{i=1}^{p} q_{ii}$$

where q_{ii} is the result of multiplying the ith row of M by the ith column of N,

$$q_{ii} = \sum_{j=1}^{n} m_{ij}n_{ji}$$

Therefore we can rewrite the trace of the matrix product as

$$\text{trace}[MN] = \sum_{i=1}^{p}\sum_{j=1}^{n} m_{ij}n_{ji}$$

Next interchanging the order of the summations gives

$$\text{trace}[MN] = \sum_{j=1}^{n}\sum_{i=1}^{p} n_{ji}m_{ij}$$

and we see that

$$\sum_{i=1}^{p} n_{ji}m_{ij} = s_{jj}$$

where s_{jj} is the jth entry on the diagonal of $S = NM$, and the result follows ■

An immediate consequence of this theorem is that the trace of any square matrix is the sum of its eigenvalues. More specifically if $M = V^{-1}\Lambda V$, then

$$\text{trace}[M] = \text{trace}\left[V^{-1}\Lambda V\right] = \text{trace}\left[\Lambda V V^{-1}\right] = \sum_{i=1}^{n} \lambda_i$$

where $\lambda_i \in \lambda[M]$.

Returning to the problem of re-expressing the energy E_c, (4.25), we use Theorem 4.2 with

$$M = B^* e^{A^* t} \qquad\qquad N = e^{At} B$$

to obtain

$$E_c = \text{trace}[W_c] \tag{4.26}$$

where W_c, the controllability Gramian, is given by

$$W_c = \int_0^\infty e^{At} BB^* e^{A^* t} \mathrm{d}t \tag{4.27}$$

Notice that W_c is Hermitian and that W_c is non-negative since the integrand in (4.27) is quadratic. Notice also that if we replace A by A^* and B by C^* in the foregoing definition of the controllability Gramian, W_c, (4.27), we obtain the observability Gramian, W_o, (4.6). Thus all the properties discussed in connection with W_o hold for W_c.

More specifically, since W_c is Hermitian, W_c has real eigenvalues and can be diagonalized by a unitary matrix, V_c, i.e.,

$$V_c^* W_c V_c = \Sigma_c \tag{4.28}$$

where $V_c^{-1} = V_c^*$ and Σ_c is real and diagonal. Notice that V_c has columns which are normalized right eigenvectors of W_c and the diagonal entries in the diagonal matrix Σ_c are the eigenvalues of W_c. In addition recalling the discussion following the proof of Theorem 4.2 that the trace of a square matrix equals the sum of its eigenvalues, we have

$$E_c = \text{trace}[\Sigma_c]$$

Now we will see in the next section that we can always transform coordinates so that the controllability Gramian is diagonal. Assuming we have done this transformation, we permute components of the state, if need be, so that the entries along the diagonal of Σ_c appear in descending size, i.e., $\sigma_{ci} \geq \sigma_{c(i+1)} : i = 1, 2, \cdots n-1$. Then we can always choose n_1 so that

$$E_{c1} \gg E_{c2}$$

where

$$E_c = E_{c1} + E_{c2} \qquad \text{with} \quad E_{c1} = \sum_{i=1}^{n_1} \sigma_{ci} \qquad \text{and with} \quad E_{c2} = \sum_{i=n_1+1}^{n} \sigma_{ci}$$

This inequality implies that the first n_1 components of the state in these coordinates receive most of the energy from the input. If in addition, in these coordinates, the observability Gramian happened to be diagonal with entries on the diagonal being in descending size, the first n_1 components of the state would also be relatively more important than the last $n - n_1$ components in the transfer of energy to the output. We could then obtain an approximation to the full order transfer function for the system as

$$G_{\text{apx}}(s) = C_{b1}(sI - A_{b1})^{-1}B_{b1}$$

where the full state model in these special coordinates has its system parameters, (A_b, B_b, C_b), partitioned to reflect the partitioning of the state , i.e.,

$$A_b = \begin{bmatrix} A_{b1} & A_{b2} \\ A_{b3} & A_{b4} \end{bmatrix} \qquad B_b = \begin{bmatrix} B_{b1} \\ B_{b2} \end{bmatrix} \begin{matrix} \updownarrow n_1 \\ \updownarrow n - n_1 \end{matrix}$$

$$C_b = \begin{bmatrix} C_{b1} & C_{b2} \end{bmatrix}$$

In the next section we will see that it is always possible to transform the coordinates so that any controllable and observable state model has controllability and observability Gramians which are equal and diagonal, i.e., so that in the new coordinates

$$W_c = W_o = \Sigma_b$$

with Σ_b having entries in descending size along its diagonal. A state model having this property is referred to as a balanced realization. A reduced order approximation, $G_{\text{apx}}(s)$, to the full order system is obtained from the balanced realization in the manner just described.

Finally, we are assuming the pair (A, B) is controllable so that every component of the state receives some of the energy supplied by the input. Therefore, from the foregoing, we have all eigenvalues of W_c bigger than zero or $W_c > 0$ when (A, B) controllable. In addition, it turns out that

$$\text{rank}[W_c] = \text{rank}[\Omega]$$

4.6 Balanced Realization

Suppose we are given the controllability and observability Gramians, W_c, W_o, for a controllable and observable state model, $\{A, B, C, D\}$, of the system to be approximated. Then in this section we will show how to construct a coordinate transformation matrix so that the state model in the new coordinates is a balanced realization. Just as in Section 4.2 we saw that a succession of two state coordinate transformations are required to achieve a controllable-observable decomposed form, so too in this section we will see that we need a succession of two transformations to achieve a balanced realization. The first coordinate transformation gives a new state model with controllability Gramian equal to an identity matrix. The second coordinate transformation keeps the controllability Gramian diagonal and diagonalizes the observability Gramian obtained from the first transformation so that the resulting Gramians are equal and diagonal.

To develop these transformations we need the fact that, if the coordinates are changed using coordinate transformation matrix T, then the Gramians $(\tilde{W}_c, \tilde{W}_o)$for the state model in the new coordinates are related to the Gramians (W_c, W_o) for the state model in the original coordinates as

$$\tilde{W}_c = T^{-1} W_c T^{-*} \qquad \tilde{W}_o = T^* W_o T \tag{4.29}$$

where $x(t) = T\tilde{x}(t)$. These relations can be obtained by replacing $(\tilde{A}, \tilde{B}, \tilde{C})$ in the integrals for $\tilde{W}_c$ and $\tilde{W}_o$ by $(T^{-1}AT, T^{-1}B, CT)$. We use (4.29) now to develop a the sequence of three coordinate transformations $T_i : i = 1, 2$ and P so that $T_b = T_1 T_2 P$ transforms the given controllable and observable state model to a balanced realization, where P is the permutation matrix required to put the diagonal elements of the Gramians in decreasing size. We will show now that these coordinate transformation matrices are determined from the Gramians for the given state model. In the next section, we will show that the Gramians for the given state model can be determined by solving two linear matrix equations known as Lyapunov equations.

For the time being suppose we are given the controllability and observability Gramians, W_c, W_o for the state model we wish to approximate. Then we form the first coordinate transformation as

$$T_1 = V_c \Sigma_c^{\frac{1}{2}}$$

where $V_c^* W_c V_c = \Sigma_c$ with $V_c^* = V_c^{-1}$and $\Sigma_c^{\frac{1}{2}} > 0$ is diagonal with $\Sigma_c^{\frac{1}{2}}\Sigma_c^{\frac{1}{2}} = \Sigma_c$. Then using (4.29) we see that the Gramians which result from the first coordinate transformation are given as

$$\tilde{W}_c = T_1^{-1} W_c T_1^{-*} = I \qquad \tilde{W}_0 = T_1^* W_o T_1$$

Next, we take the second coordinate transformation as

$$T_2 = \tilde{V}_0 \tilde{\Sigma}_0^{-\frac{1}{4}}$$

where $\tilde{V}_0^* \tilde{W}_0 \tilde{V}_0 = \tilde{\Sigma}_0$ and $\tilde{\Sigma}_0^{\frac{1}{4}} > 0$ with $(\tilde{\Sigma}_0^{\frac{1}{4}})^4 = \tilde{\Sigma}_0$. Then using (4.29) we see that the Gramians which result from the second transformation are given by

$$\bar{W}_c = T_2^{-1} \tilde{W}_c T_2^{-*} = \tilde{\Sigma}_0^{\frac{1}{2}}$$
$$\bar{W}_o = T_2^* \tilde{W}_o T_2 \quad = \tilde{\Sigma}_0^{\frac{1}{2}}$$

and we have the Gramians equal and diagonal.

It remains to permute the components of the state so that the entries on the diagonal of the Gramians are ordered in decreasing size. This corresponds to a third coordinate transformation, this time using a permutation matrix P as the coordinate transformation matrix so that

$$P\bar{W}_c P = P\bar{W}_o P = \Sigma_b$$

where

$$\Sigma_b = \mathrm{diag}[\sigma_{b1}, \sigma_{b2}, \cdots \sigma_{bn}] \qquad \text{with } \sigma_{b1} \geq \sigma_{b2} \cdots \geq \sigma_{bn}$$

and the σ_{bi}s are referred to as the Hankel singular values for the system. More is said about this in Section 4.9

At this stage we can obtain the balanced realization, (A_b, B_b, C_b), from $(T^{-1}AT, T^{-1}B, CT)$ with $T = T_1T_2P$. Then partitioning the balanced realization as

$$A_b = \begin{bmatrix} A_{b1} & A_{b2} \\ A_{b3} & A_{b4} \end{bmatrix} \quad B_b = \begin{bmatrix} B_{b1} \\ B_{b2} \end{bmatrix} \begin{matrix} \updownarrow n_1 \\ \updownarrow n - n_1 \end{matrix} \tag{4.30}$$

$$C_b = [\,C_{b1} \quad C_{b2}\,]$$

with n_1 large enough so that

$$\sum_{i=1}^{n_1} \sigma_{bi} \gg \sum_{i=n_1+1}^{n} \sigma_{bi}$$

we obtain the reduced order state model as $(A_{b1}, B_{b1}, C_{b1}, D)$ and the reduced order transfer function model as

$$G_{\mathrm{apx}}(s) = C_{b1}(sI - A_{b1})^{-1}B_{b1}$$

Recall that we are assuming that the full order state model is stable, i.e., A_b is stable. Therefore we get a bounded output for every bounded input to the full order, unreduced system. Thus if $G_{\mathrm{apx}}(s)$ is to be a meaningfully reduced order approximation to the full order transfer function, we expect $G_{\mathrm{apx}}(s)$ to be stable. Otherwise the behavior of the reduced order system would be dramatically different from the behavior of the original full order system. The following theorem specifies conditions which guarantee that the reduced order model is stable.

Theorem 4.3 If in (4.30) we have $\sigma_{bn_1} \neq \sigma_{bn_1+1}$ then A_{b1} is stable.

Proof Appendix

In the next section we show that the Gramians for a given state model can be determined by solving linear matrix equations called Lyapunov equations. In this way the calculation of the integrals in the definition of the Gramians is avoided. We will also see that further relations between the Gramians for a state model and the observability, controllability, and stability of that state model are uncovered by considering the Gramians to be solutions to Lyapunov equations.

4.7 The Lyapunov Equation

We begin this section by generalizing the output energy as follows. Recall that in the absence of an input the observability Gramian gives the output energy caused by any initial state as

$$E_o = x^*(0)W_o x(0)$$

Now suppose we replace this constant quadratic function of the initial state by $e_o(t)$, a time varying quadratic function of the state given as

$$e_o(t) = x^*(t)Px(t) \tag{4.31}$$

where P is a positive definite Hermitian matrix and $x(t) = e^{At}x(0)$. Notice that this function is always positive

$$e_o(t) > 0 \qquad \text{for all } t \in (0, \infty) \text{ and all } x(0) \neq \emptyset \tag{4.32}$$

Next assume that $e_o(t)$ has a derivative which is a negative quadratic function of the state

$$\frac{de_o(t)}{dt} = -x^*(t)Qx(t) < 0 \qquad \text{for all } t \in (0, \infty) \text{ and all } x(0) \neq \emptyset \tag{4.33}$$

where Q is a positive definite Hermitian matrix. Then we see from (4.32, 4.33) that $e_o(t)$ is a positive, decreasing function of time which is bounded below by zero. This implies that

$$\lim_{t\to\infty} e_o(t) = 0 \tag{4.34}$$

However since P is invertible we have $e_o(t) = 0$ only if $x(t) = \emptyset$ and (4.34) implies that

$$\lim_{t\to\infty} x(t) = \emptyset \qquad \text{for all } x(0) \tag{4.35}$$

Therefore equations (4.31), and (4.33) imply that the state model is internally stable, i.e., the state model's A matrix has all its eigenvalues in the open left half plane.

Now if we differentiate (4.31) and use $\dot{x}(t) = Ax(t)$ we can express the time derivative of $e_o(t)$, as

$$\dot{e}_o(t) = x^*(t)(A^*P + PA)x(t)$$

These observations lead us to formulate the following theorem concerning the stability of A.

Theorem 4.4 If for any Hermitian $Q > 0$ we obtain $P > 0$ as a solution to the linear matrix equation

$$A^*P + PA = -Q \tag{4.36}$$

then A is stable.

Proof Suppose (λ, v) is any eigenvalue, right-eigenvector pair for A. Then pre and post multiplying (4.36) by v^* and v respectively gives

$$v^*(A^*P + PA)v = -v^*Qv$$

which recalling $v^*A^* = \lambda^*v^*$becomes

$$(\lambda^* + \lambda)v^*Pv = -v^*Qv \tag{4.37}$$

However since $v^*Pv > 0$ and $-v^*Qv < 0$ for all $v \neq \phi$, we see that (4.37) is satisfied only if $(\lambda^* + \lambda) = 2\text{Re}[\lambda] < 0$. This shows that all eigenvalues of A lie in the open left half plane and therefore A is stable. ■

The linear matrix equation (4.36) is called a Lyapunov equation after the Russian mathematician who originated this approach in the study of stability.

4.7.1 Relation to the Gramians

We show next that when we take $Q = C^*C$ the solution P to the Lyapunov equation, (4.36), is the observability Gramian, i.e., W_o satisfies the Lyapunov equation

$$A^*W_o + W_oA = -C^*C \tag{4.38}$$

Analogously, the controllability Gramian, W_c, is the solution the Lyapunov equation

$$AW_c + W_cA^* = -BB^* \tag{4.39}$$

The foregoing relations between the Gramians and the solutions to Lyapunov equations are shown in the following theorem.

Theorem 4.5 If F is a square stable matrix and if

$$\int_0^\infty e^{Ft}Qe^{F^*t}\mathrm{d}t = W \tag{4.40}$$

then

$$FW + WF^* = -Q \tag{4.41}$$

Proof We begin the proof by defining the matrix function

$$Z(t) = e^{Ft}Qe^{F^*t}$$

Then we differentiate $Z(t)$ to obtain

$$\frac{\mathrm{d}Z(t)}{\mathrm{d}t} = Fe^{Ft}Qe^{F^*t} + e^{Ft}QF^*e^{F^*t}$$

Next recalling the series expansion of e^{F^*t}, we see that F^* and e^{F^*t} commute. Therefore the foregoing equation can be written as

$$\frac{dZ(t)}{dt} = FZ(t) + Z(t)F^* \tag{4.42}$$

Finally, since A is being assumed stable, we have $Z(\infty) = \emptyset$. Therefore integrating (4.42) from 0 to ∞ gives

$$\int_0^\infty dZ(t) = \int_0^\infty [FZ(t)\ Z(t)F^*]dt$$

$$Z(\infty) - Z(0) = F\left[\int_0^\infty e^{Ft}Qe^{F^*t}dt\right] + \left[\int_0^\infty e^{Ft}Qe^{F^*t}dt\right]F^*$$

$$-Q = FW + WF^*$$

■

Now there are two important observations we can make when we compare the Lyapunov equation (4.36) with the Lyapunov equations for the Gramians (4.38, 4.39).

First, notice that unlike Q on the right side of (4.36), C^*C and BB^* in (4.38, 4.39) are only non-negative.

Second, (4.38, 4.39) are linear equations in the elements of the matrices W_o, W_c. Therefore we may be able to solve (4.38, 4.39) for matrices W_o and W_c when A is not stable. However since the observability and controllability Gramians (4.5, 4.27) are only defined when A is stable, the matrices W_o, W_c obtained by solving (4.38, 4.39) are system Gramians only if A is stable.

The implications of these observations are taken up next.

4.7.2 Observability, stability, and the observability Gramian

What follows are theorems which describe how the concepts named in the title of this section are interrelated.

Theorem 4.6 If (A, C) is observable and if $P > 0$ satisfies the Lyapunov equation

$$A^*P + PA = -C^*C \tag{4.43}$$

then A is stable.

Proof Suppose, contrary to the theorem, that A is not stable, i.e., $Av = \lambda v$ with $\text{Re}[\lambda] \geq 0$ for some $\lambda \in \lambda[A]$. Then proceeding as in the proof of Theorem 4.4 we obtain

$$2\text{Re}[\lambda](v^*Pv) = -v^*C^*Cv \tag{4.44}$$

Now since (A, C) is observable we have $Cv = z \neq \emptyset$ and $v^*C^*Cv > 0$. However we are assuming that $\text{Re}[\lambda] \geq 0$. Therefore (4.44) is satisfied under these conditions only if we have $v^*Pv < 0$ which contradicts the condition $P > 0$ specified in the Theorem. Therefore our assumption that A is unstable is false. ■

Notice, in connection with the foregoing proof, that when A is unstable and (A, C) is observable we have

$$W_o = \lim_{T\to\infty} \int_0^T e^{A^*t}C^*Ce^{At}dt \tag{4.45}$$

is unbounded and the observability Gramian, W_o, is undefined. However this does not mean that we are unable to find P to satisfy (4.43) in this case.

The foregoing theorem has the following corollary.

Theorem 4.7 If A is stable and P is positive definite and satisfies (4.43) then (A, C) is observable.

Proof Suppose, contrary to the theorem, that (A, C) is not observable. Then we must have a (λ, v) pair, $Av = \lambda v$ and $Cv = \emptyset$. Therefore proceeding as in the proof of Theorem 4.6 we obtain

$$2\text{Re}[\lambda](v^* P v) = -v^* C^* C v = 0 \tag{4.46}$$

Now in order to satisfy (4.46) we must have $\text{Re}[\lambda] = 0$ or $v^* P v = 0$ or both. The condition $\text{Re}[\lambda] = 0$ is impossible since we are assuming that A is stable. The condition $v^* P v = 0$ is also impossible since we are assuming that $P > 0$. Thus (4.46) can not be satisfied and therefore (A, C) must be an observable pair. ■

To recap, the foregoing pair of theorems state that A is stable when (4.43) is satisfied by $P > 0$ if and only if (A, C) is an observable pair. Since the observability Gramian W_o, (4.45) is defined when A is stable and positive definite when (A, C) is an observable pair, if the solution P to (4.43) satisfies $P > 0$ when (A, C) is not an observable pair P is not the observability Gramian. The following theorem characterizes the effect of having a $P > 0$ as a solution to the Lyapunov equation (4.43) when the pair (A, C) is not observable.

Theorem 4.8 If the pair (A, C) is not observable and $P > 0$ satisfies the Lyapunov equation (4.43) then A has imaginary axis eigenvalues.

Proof Since the pair (A, C) is not observable, we can choose a (λ, v) pair so that $Av = \lambda v$ and $Cv = \emptyset$. Then we see from (4.46), that

$$2\text{Re}[\lambda](v^* P v) = 0 \tag{4.47}$$

However since $P > 0$, (4.47) is satisfied only if $\text{Re}[\lambda] = 0$. ■

To recap, we have shown in the foregoing theorems that A is stable when the Lyapunov equation

$$A^* P + PA = -C^* C$$

is satisfied by $P > 0$ if and only if the pair (A, C) is observable. Alternatively, if this Lyapunov equation is satisfied by $P > 0$ when (A, C) is not observable, the unobservable eigenvalues of A lie on the imaginary axis. Notice from the proof of Theorem 4.7 that eigenvalues of A that are not on the imaginary axis are observable and lie in the left half plane.

Analogously, A is stable when the Lyapunov equation

$$AP + PA^* = -BB^*$$

is satisfied by $P > 0$ if and only if the pair (A, B) is controllable. Alternatively, if this Lyapunov equation is satisfied by $P > 0$ when (A, B) is not controllable, the uncontrollable eigenvalues lie on the imaginary axis and the controllable eigenvalues are all in the left half plane.

4.8 Controllability Gramian Revisited

In this section we develop another use for the controllability Gramian which provides us with a different way of viewing the process of model reduction and a new way of thinking about a system's input–output behavior.

4.8.1 *The least energy input problem*

An alternative use for the controllability Gramian, W_c, arises in connection with the solution to the problem of finding the input $u_0(t) : t \in (-\infty, 0]$ which expends the least energy in taking the null state at time $t = -\infty$ to some specified state at time $t = 0$.

In order to solve this problem we need to recall, from Chapter 1, that the basic equation relating the initial state and input to the final state is

$$x(t_f) = e^{At_f}x(-\infty) + \int_{-\infty}^{t_f} e^{A(t_f-\tau)}Bu(\tau)\mathrm{d}\tau \tag{4.48}$$

Then, in the present situation, we have $x(-\infty) = \emptyset$, and $t_f = 0$ so that (4.48) becomes

$$x(0) = \int_{-\infty}^{0} e^{-A\tau}Bu(\tau)\mathrm{d}\tau \tag{4.49}$$

Now we want to choose $u(t)$ to satisfy (4.49) so that the input energy, E_{cu}, given by

$$E_{cu} = \int_{-\infty}^{0} u^*(\tau)u(\tau)\mathrm{d}\tau \tag{4.50}$$

achieves its minimum value, E_{cu_0}, determined from

$$E_{cu_0} = \min_{u(t)} E_{cu}$$

It turns out that one input that solves this problem is given as

$$u_0(t) = B^*e^{-A^*t}W_c^{-1}x(0) \tag{4.51}$$

where W_c is the controllability Gramian, (4.27). Notice that this input satisfies (4.49) and has energy determined as

$$\begin{aligned} E_{cu} &= \int_{-\infty}^{0} u^*(\tau)u(\tau)\mathrm{d}\tau \\ &= x^*(0)W_c^{-1}\left(\int_{-\infty}^{0} e^{-A\tau}BB^*e^{-A^*\tau}\mathrm{d}\tau\right)W_c^{-1}x(0) \\ &= x^*(0)W_c^{-1}x(0) = E_{cu_0} \end{aligned}$$

Notice that $u_0(t)$ exists provided W_c is invertible. This requirement that the (A, B) pair be controllable is a necessary condition for being able to satisfy (4.49) for any $x(0)$, (see Chapter 2 Section 2.4).

Recall that the coordinate transformation matrix, V_c, (4.28), that diagonalizes W_c is unitary, $V_c^{-1} = V_c^*$. Thus we see that this coordinate transformation also diagonalizes

W_c^{-1} since

$$(V_c^* W_c V_c)^{-1} = V_c^* W_c^{-1} V_c = \Sigma_c^{-1}$$

Therefore we can use this fact to rewrite E_{cu_0} as

$$\begin{aligned} E_{cu0} &= x^*(0) V_c V_c^* W_c^{-1} V_c V_c^* x(0) \\ &= z^*(0) \Sigma_c^{-1} z(0) = \sum_{i=1}^{n} \frac{| z_i(0) |^2}{\sigma_{bi}} \end{aligned} \tag{4.52}$$

where

$$z(0) = V_c^* x(0)$$

Then if the components of $z(0)$ are equal and have been permuted so the entries along the diagonal of Σ_c are ordered, $\sigma_{bi} \geq \sigma_{b(i+1)} : i = 1, 2, \cdots n-1$, we see from (4.52) that most of the energy in the input is needed to achieve components in the lower partition of $z(0)$. This implies that the lower partition of $z(t)$ is harder to control than the upper partition.

Now in the following we make use of the foregoing alternative interpretation of the controllability Gramian to provide further insight into the use of a balanced realization to achieve model reduction.

Recall that a balanced realization has both Gramians equal to the same diagonal matrix with the entries encountered in descending the diagonal forming a nonincreasing sequence. Then we saw, in Section 4.4.3, that most of the energy supplied to the output by an initial state with equal component values comes from components in the upper partition of the initial state. However, we have just seen that the components in the upper partition of the state are more easily manipulated by the input than components in the lower partition. Therefore, in balanced coordinates, the subspace of the state space corresponding to the upper partition of the full state is more involved in the input–output behavior than the subspace corresponding to the lower partition of the state. This suggests that we can capture most of the full order system's input–output behavior by using a reduced order state model formed by truncating the balanced realization in the manner given at the end of Section 4.6.

4.8.2 Hankel operator

We can obtain further insight into a system's input–output behavior by using the present interpretation of the controllability Gramian. This is done by noting that:

(i) the controllability Gramian is used to characterize the system's transformation of $u(t) : t \in (-\infty, 0]$ into $x(0)$;
(ii) the observability Gramian is used to characterize the system's transformation of $x(0)$ into $y(t) : t \in [0, \infty)$ when $u(t) = \emptyset : t \in [0, \infty)$.

This suggests that, in addition to the concept of a system implied by its transfer function, $G(s)$, i.e., that the system transforms or maps system inputs, $u(t) : t \in [0, \infty)$, to system outputs, $y(t) : t \in [0, \infty)$, we can view the system as a map from past inputs, $u(t) : t \in (-\infty, 0]$ to future outputs, $y(t) : t \in [0, \infty)$. This map is referred to as the system's Hankel operator and is denoted as Γ_G. When a system is stable its Hankel operator has an operator norm which is referred to as the Hankel norm. Operators and operator norms will be developed in more depth in Chapter 7. For the time being we want to show that the Hankel norm of a system equals the largest entry on the diagonal of the Gramians when the state model for the system is a balanced realization.

Suppose we are given a strictly proper transfer function, $G(s)$, and minimal state model, (A, B, C), for some stable system. Then the Hankel operator, Γ_G, for this system has output $y(t) : t \in [0, \infty)$ produced by input $u(t) : t \in (-\infty, 0]$ when $x(-\infty) = \emptyset$ and $u(t) = \emptyset : t \in [0, \infty)$.

Now we define the system's Hankel norm, denoted $\|G(s)\|_H$ as

$$\|G(s)\|_H = \left(\max_{\substack{u(t)\\ u(t)=\emptyset:\\ t\in[0,\infty)\\ x(-\infty)=\emptyset}} \left[\frac{\int_0^\infty y^*(t)y(t)\mathrm{d}t}{\int_{-\infty}^0 u^*(t)u(t)\mathrm{d}t} \right] \right)^{\frac{1}{2}} \tag{4.53}$$

where $(\cdot)^{\frac{1}{2}}$ indicates the positive square root. There is a close connection between a system's Hankel norm and that system's balanced realization as shown in the following theorem.

Theorem 4.9 The Hankel norm $\|G(s)\|_H$ for a stable system having transfer function $G(s)$ is given as

$$\|G(s)\|_H^2 = \lambda_{\max}[W_c W_o] = \sigma_{b1}^2 \tag{4.54}$$

where σ_{b1} is the largest entry on the diagonal of the Gramians in balanced coordinates.

Proof Recalling (4.29) we see that the eigenvalues of the product of the controllability and observability Gramians are invariant to a coordinate transformation since we have

$$\tilde{W}_c \tilde{W}_o = T^{-1} W_c T^{-*} T^* W_o T = T^{-1} W_c W_o T \tag{4.55}$$

Thus when T is the coordinate transformation which puts the system in balanced form we have

$$\lambda[\tilde{W}_c \tilde{W}_o] = \lambda[\Sigma_b^2] = \{\sigma_{bi}^2 : i = 1, 2, \cdots n\}$$

where $\{\sigma_{bi} : i = 1, 2, \cdots n\}$ are referred to as the Hankel singular values.

Next recall from Section 4.4.3 that when the state model is a balanced realization, the maximum output energy, $E_{o\max}$, produced by an initial state, $z(0)$, constrained to satisfy $z^*(0)z(0) = \alpha^2$, when $u(t) = \emptyset : t \in [0, \infty)$, is given by

$$E_{o\max} = \max_{\substack{z(0)\\ z^*(0)z(0)=\alpha^2}} [E_o] = \left[z^{\max}(0)\right]^* \Sigma_b [z^{\max}(0)] = \alpha^2 \sigma_{b1}$$

where

$$[z^{\max}(0)]^T = [\alpha \quad 0 \quad \cdots \quad 0]$$

Alternatively, recalling the solution to the least energy input problem, we see from (4.52) that using the least energy input $u_0(t) : t \in (-\infty, 0]$ when $z(-\infty) = \emptyset$ to produce $z^{\max}(0)$ gives the least energy as

$$E_{c\min} = \min_{\substack{u(t):t\in(-\infty,0] \\ z(-\infty)=\emptyset \\ z(0)=z^{\max}(0)}} [E_{cu}] = [z^{\max}(0)]^* \Sigma_b^{-1} z^{\max}(0) = \frac{\alpha^2}{\sigma_{b1}}$$

Therefore we see that the maximum value of the ratio of energy out to energy in is given as

$$\frac{E_{o\max}}{E_{c\min}} = \sigma_{b1}^2 = \lambda_{\max}^2 [W_c W_o]$$

which from (4.53, 4.55) yields (4.54). ■

4.9 Notes and References

Hermitian matrices play an important role in many branches of science and engineering, e.g., quantum mechanics. In general a square matrix having repeated eigenvalues may not be able to be diagonalized by a similarity transformation, i.e., we may not be able to find V such that $V^{-1}MV$ is diagonal. However a Hermitian matrix can always be diagonalized even if it has repeated eigenvalues, [39].

The term "Gramian" arose originally in connection with the problem of determining if the vectors in a set of vectors $\{v_i : i = 1, 2, \cdots, n\}$ were linearly independent. The Gramian G was defined as the scalar whose value was determined as $G = \det[M]$, where $(M)_{ij} = v_i{}^* v_j$. The independence of the $v_i s$ required G to be nonzero, [5]. As used in system theory, the term Gramian describes a matrix, not a scalar, which is formed in manner reminiscent of the foregoing matrix G.

Balanced realization was first used in the area of digital filters to combat inaccuracies (roundoff) resulting from computation errors brought about by the requirement that calculations be done using finite precision arithmetic [29]. The use of a balanced realization to obtain a reduced order model approximation and the method given here for calculating a balanced realization was originally presented in [27]. Theorem 4.8 and the proof of Theorem 4.3 (Appendix) were originally given in [34]. The idea of a balanced realization was extended to unstable linear systems, [22] and to a class of nonlinear state models, [38].

Finally, we need to take note of the following important aspect of model reduction. We will see in the remainder of this book that optimal controller design techniques, e.g., quadratic, Gaussian, and H_∞ feedback control theories, make use of an observer as part of the feedback controller. Since the observer order matches the plant order, there can be problems in implementing controllers for use with high order plants. Unfortunately, the stability of the closed loop system may not be preserved when the controller designed using the optimal control theory is replaced by a reduced order model. A good summary of approaches to this problem up until 1989 is given in [2]. A more recent approach to solving this problem, involving H_∞ robust control theory is given in [30] for linear plants and in [32] for a class of nonlinear plants.

5

Quadratic Control

5.1 Introduction

In the preliminary part of this chapter, we introduce the basic approach to feedback control known as observer based feedback control. In this scheme the plant state $x(t)$ in the state feedback control signal $u(t) = Kx(t) + v(t)$ (Chapter 2) is replaced by the estimated plant state $\hat{x}(t)$ obtained from an observer (Chapter 3). As we will see in the remainder of this book, this scheme has enabled the development of several optimization based controller design techniques. In addition to quadratic control treated in this chapter, we will see that observer based feedback control is used in linear quadratic Gaussian, LQG, control and in H_∞ control.

Following this, we take up the use of quadratic control to determine the K and L matrices in an observer based feedback control setup. Quadratic control arises when there is a need to maintain the control system's steady-state output constant in response to a constant control input. In this situation the control scheme is said to operate as a regulator. In practice, impulse like disturbance signals, i.e., signals having large amplitude and short time duration, enter the control loop and cause the output to jiggle about its desired level. A well known example of this is the regulation of the pointing direction of an outdoor dish antenna. Unfortunately the dish acts like a sail and experiences torque disturbances because of wind gusts which tend to change its pointing direction. Therefore we need to choose K and L so as to alter the closed-loop system dynamics, i.e., assign the eigenvalues of the closed-loop system matrix, in a way which reduces the effect on the output of these impulse like disturbance signals. In quadratic control this choice of K and L is done so as to minimize the energy in the output signal caused by the disturbance.

We will see that the determination of the optimal K and L matrices needed to achieve this minimum requires the solution to two nonlinear matrix equations known as algebraic Riccati equations. Special solutions to these equations, referred to as stabilizing solutions are required. We will develop conditions on the plant and optimization criteria which will ensure the existence of such solutions. We will encounter the Riccati equation again in connection with other optimal control schemes developed in this book. More specifically, we will see that the algebraic Riccati equation is involved in the solution of the disturbance attenuation problem when the disturbances are viewed as random signals with known statistics (Chapter 6) and when the disturbance signals are restricted to the so-called L_2 class of signals (Chapters 9 and 10).

5.2 Observer Based Controllers

Although errors in the estimated plant state in an observer based control system produce effects at the control system's output, we will see that the dynamics of the observer do not contribute to the control system's zero-state response. More specifically, we will show that the control system's transfer function equals the transfer function for the closed loop feedback system when exact state feedback is used, i.e., when the estimated state equals the plant state.

Suppose the plant state model parameters (A, B, C, D) are given. Then recalling the development of the observer, (Chapter 3), we see that the plant state, $x(t)$, and the observer state, $\hat{x}(t)$, are governed by the following vector differential equations

$$\dot{x}(t) = Ax(t) + Bu(t) \tag{5.1}$$

$$\dot{\widehat{x}}(t) = (A + LC)\widehat{x}(t) + (B + LD)u(t) - Ly(t) \tag{5.2}$$

Now recall from Chapter 2 that to apply state feedback we set $u(t) = Kx(t) + v(t)$. Lacking exact knowledge of the plant state we replace it by its estimate, $\hat{x}(t)$, to get a control input

$$u(t) = K\widehat{x}(t) + v(t) \tag{5.3}$$

Then substituting (5.3) in (5.1, 5.2) and the plant's output equation, we see that the control system is a $2n$ dimensional system having state model given as

$$\dot{x}_c(t) = A_c x_c(t) + B_c v(t) \tag{5.4}$$

$$y(t) = C_c x_c(t) + D_c v(t) \tag{5.5}$$

where

$$A_c = \begin{bmatrix} A & BK \\ -LC & A + LC + BK \end{bmatrix} \qquad B_c = \begin{bmatrix} B \\ B \end{bmatrix} \qquad x_c(t) = \begin{bmatrix} x(t) \\ \hat{x}(t) \end{bmatrix}$$

$$C_c = \begin{bmatrix} C & DK \end{bmatrix} \qquad D_c = D$$

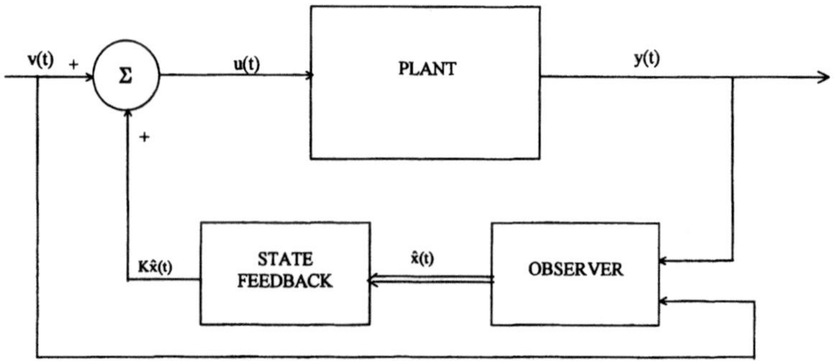

Figure 5.1 Setup for observer based feedback control

Notice that since the plant and observer are each of dimension n, the closed-loop system state model is of dimension $2n$. However even if the plant state model is minimal, i.e., observable and controllable, the closed-loop state model, (5.4, 5.5), is not controllable. We can see this by changing coordinates so that the resulting state model is in controllable decomposed form, (Section 2.5). We do this as follows.

Recall that controllable decomposed form requires the transformed input matrix $\bar{B}_c$ to have a lower partition which is null. Since B_c is partitioned into two equal halves, a coordinate transformation matrix, T, which does this is given by

$$T = \begin{bmatrix} I & \emptyset \\ I & -I \end{bmatrix}$$

Therefore, after using this coordinate transformation matrix the control system state model parameters, (5.4, 5.5), become

$$\bar{A}_c = \begin{bmatrix} \bar{A}_{c1} & \bar{A}_{c2} \\ \emptyset & \bar{A}_{c4} \end{bmatrix} \qquad \bar{B}_c = \begin{bmatrix} B \\ \emptyset \end{bmatrix} \qquad \bar{x}_c = \begin{bmatrix} x(t) \\ \tilde{x}(t) \end{bmatrix} \tag{5.6}$$
$$\bar{C}_c = [\, C + DK \quad -DK \,] \qquad \bar{D}_c = D$$

where

$$\bar{A}_{c1} = A + BK \qquad \bar{A}_{c2} = -BK \qquad \bar{A}_{c4} = A + LC$$
$$\tilde{x}(t) = x(t) - \hat{x}(t)$$

Notice, from the null block structure of $\bar{A}_c$ and $\bar{B}_c$, that the plant state estimation error, $\tilde{x}(t)$, is uncontrollable and that the control system transfer function is given as

$$\begin{aligned} G_c(s) &= \bar{C}_c(sI - \bar{A}_c)^{-1}\bar{B}_c + \bar{D}_c \\ &= [\, C + DK \quad -DK \,]\begin{bmatrix} (sI - \bar{A}_{c1})^{-1} & * \\ \emptyset & (sI - \bar{A}_{c4})^{-1} \end{bmatrix}\begin{bmatrix} B \\ \emptyset \end{bmatrix} + D \\ &= (C + DK)[sI - (A + BK)]^{-1}B + D \end{aligned} \tag{5.7}$$

Therefore provided the plant state model is minimal, i.e., (A, B) controllable and (A, C) observable, we have $(A + BK, B)$ controllable and $(A + BK, C + DK)$ observable so that there are no pole-zero cancellations in the foregoing transfer function and the control system and plant have the same order. Notice this control system transfer function is the same as we obtained in Chapter 2 when we used exact state feedback. Therefore we can conclude that the control system's input-output or zero-state behavior is unaffected by the presence of the observer. However, the control system's zero-input response is affected by the observer. We can show this as follows.

From the control system's state model, (5.6), we see that the differential equations governing the plant state and plant state estimation error are

$$\dot{x}(t) = \bar{A}_{c1}x(t) + \bar{A}_{c2}\tilde{x}(t) + Bv(t) \tag{5.8}$$

$$\dot{\tilde{x}}(t) = \bar{A}_{c4}\tilde{x}(t)$$

Then the state estimation error is independent of the input and depends only on its initial value as

$$\tilde{x}(t) = e^{(A+LC)t}\tilde{x}(0) \tag{5.9}$$

Therefore substituting this expression for $\tilde{x}(t)$ in (5.8) yields

$$\dot{x}(t) = (A + BK)x(t) - BKe^{(A+LC)t}\tilde{x}(0) + Bv(t) \tag{5.10}$$

and we see that the evolution of the plant state depends on the state estimation error.

However, this effect which is caused by the presence of the observer in the feedback loop, disappears with time. More specifically, recall from Chapter 3 that L must always be chosen to make $A + LC$ is stable. Therefore we see from (5.9) that

$$\lim_{t\to\infty} \tilde{x}(t) = \emptyset \qquad \text{for any } \tilde{x}(0)$$

Thus, for some time following the initiation of the feedback control, the plant state, $x(t)$, and plant output, $y(t)$, are both affected by the plant state estimation error, $\tilde{x}(t)$. Eventually the term $\bar{A}_{c2}\tilde{x}(t)$ can be neglected in comparison to other terms on the right side of (5.8) so that for large t the derivative of the plant state can be approximated as

$$\dot{x}(t) = (A + BK)x(t) + Bv(t) \tag{5.11}$$

Notice that (5.11) is the differential equation for the plant state when exact state feedback is used.

Finally, recalling that the eigenvalues of the system matrix A_c are preserved under coordinate transformation, (Section 1.6.4), and that block upper-triangular matrices have eigenvalues which are the union of the eigenvalues of the diagonal blocks, (Appendix), we see from (5.6) that

$$\lambda[A_c] = \lambda[\bar{A}_c] = \lambda[A + BK] \cup \lambda[A + LC]$$

Thus half the closed loop system eigenvalues depend on K and are independent of L, and the other half depend on L and are independent of K. This spliting of the closed loop eigenvalues is responsible, in part, for being able to break the optimal control problems treated in this book into two simpler problems:

1. the state feedback control problem
2. the state estimation problem

This fact, referred to as the separation principle, has been a major theme running through the development of linear control theory since its discovery in the 1960s.

In summary, the use of an observer-estimated plant state in place of the true plant state in a state feedback control configuration leads to a feedback control system having transfer function (steady-state, input-output behavior) which is the same as we would obtain using exact state feedback. However, initially at the start-up of the control system, any departure of the observer-estimated state from the true plant state will affect the control system output. This effect diminishes with time as the observer-estimated state approaches the true plant state.

Recall, from Chapters 2 and 3, that we needed the specification of eigenvalues for $A + BK$ and $A + LC$ in order to determine K for the design of the state feedback control and L for the design of the state estimator, respectively. However, in practice, we are only given some desired goals for the behavior of the control system. Therefore, in this situation, we need methods for turning specified goals for the behavior of the control system into values for the K and L matrices. In the remainder of this chapter, and in subsequent chapters, we will show in each case that by interpreting the specified goal for the control system's behavior as a mathematical optimization problem, we can develop equations for determining the required K and L.

5.3 Quadratic State Feedback Control

There are many possible control system design objectives in addition to the mandatory requirement that a control system be stable. One of these additional requirements arises when there are large-amplitude, short-duration disturbance signals entering the control loop and we want to maintain the output constant. In this situation we need to design the controller both to stabilize the closed-loop system and to diminish, as much as possible, the effect of the disturbance signals on the plant output. The solution to the optimal quadratic control problem attempts to achieve this goal by determining K and L both to stabilize the closed-loop system and to minimize the energy in the departure of the control system output from a desired steady-state value. As mentioned in the previous section, we can use the separation principle to split this problem into two simpler subproblems:

1. the quadratic state feedback control problem and
2. the quadratic state estimation problem

with the solution to 1. determining K, and the solution to 2. determining L.

In this section we develop the design equations for K which solves the quadratic state feedback control problem. When K satisfies these design equations, the sum of the energies in the output and feedback signal caused by the disturbance signal, is minimized. One of these design equations is a nonlinear matrix equation known as the quadratic control algebraic Riccati equation (QCARE). The theoretical background needed to solve this equation is provided in the next section, Section 5.4.

In Section 5.5 we develop the design equations for L to solve the quadratic state estimation problem. Again we will see that this involves the solution of an algebraic Riccati equation. This equation is closely related to the QCARE and is known as the quadratic filtering algebraic Riccati equation (QFARE). The required solution to the QFARE can be obtained using procedures developed in Section 5.4 for solving the QCARE.

5.3.1 Motivating the problem

Let the disturbance input to the plant be denoted as $u_1(t)$; let the controlled input involved in implementing the state feedback be denoted by $u_2(t)$. Then the plant state model is given by the equations

$$\dot{x}(t) = Ax(t) + B_1u_1(t) + B_2u_2(t) \tag{5.12}$$

$$y(t) = Cx(t) + D_1u_1(t) + D_2u_2(t) \tag{5.13}$$

and setting $u_2(t) = Kx(t) + v(t)$, where $v(t)$ is the command input, we obtain a state feedback control system having state model given as

$$\dot{x}(t) = \bar{A}x(t) + B_1u_1(t) + B_2v(t) \tag{5.14}$$

$$y(t) = \bar{C}x(t) + D_1u_1(t) + D_2v(t) \tag{5.15}$$

where

$$\bar{A} = A + B_2K \qquad \bar{C} = C + D_2K$$

$u_1(t)$ = disturbance input $\quad$ $y(t)$ = control system output

$u_2(t)$ = controlled input $\quad$ $v(t)$ = control system command input

However, if

(i) the disturbance input is null, $u_1(t) = \emptyset$
(ii) the control system command input is constant, $v(t) = v_s$,
(iii) the feedback matrix K makes the control system stable so that all eigenvalues of $\bar{A}$ lie in the open left half plane,

then the constant steady-state control system state, x_s, is obtained from the control system state differential equation, (5.14) by setting $\dot{x}(t) = \emptyset$. Then using x_s in (5.15) gives the steady-state control system output $y_s$as

$$y_s = \bar{C}x_s + D_2v_s \qquad x_s = -\bar{A}^{-1}B_2v_s$$

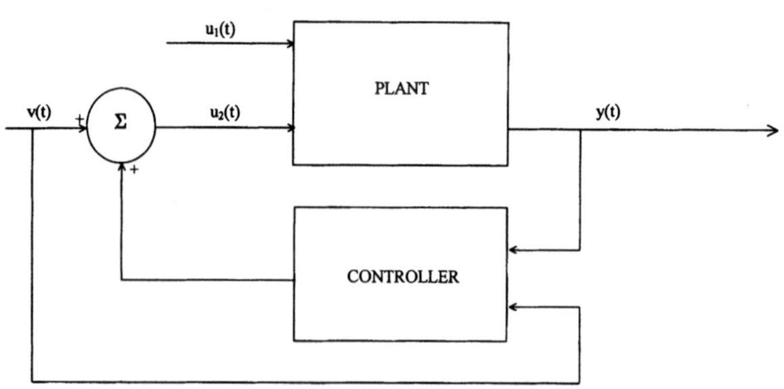

Figure 5.2 Setup for disturbance input attenuation

where

$$x_s = \lim_{t\to\infty} x(t)$$

Now suppose the control system is operating at this steady-state level at time $t = t_1$ when a short-duration disturbance signal, lasting T seconds, arrives at the disturbance input. Then the disturbance input can be written as

$$\begin{aligned} u_1(t) &\neq \emptyset && \text{for } t \in [t_1, t_1 + T] \\ &= \emptyset && \text{otherwise} \end{aligned}$$

Therefore, shifting the time origin to the instant immediately following the end of the disturbance, i.e., to time $t_1 + T$, we are left with the state immediately following the end of the disturbance being given by

$$x(0) = x_s + x_d(0)$$

However since $A + BK$ is assumed stable we have

$$\lim_{t\to\infty} x_d(t) = \emptyset$$

and the state of the plant returns to its steady-state value with time,

$$\lim_{t\to\infty} x(t) = x_s$$

Our goal in choosing K is to decrease the effect of the disturbance input on the control system output during the return of the output to its steady-state value. Since superposition applies, we can assume, without loss of generality, that the control system command input is null, $v(t) = v_s = \emptyset$. Then the desired steady-state output is null and any deviation of the output away from being null is caused by the presence of the disturbance input. Thus we want to choose K so that the action of the feedback signal attenuates these deviations as much as possible.

5.3.2 Formulating the problem

Recall, from the previous chapter, that we used the observability Gramian and its relation to the output energy to obtain a measure of the zero-input response. We use this idea here to reformulate the problem just discussed as the problem of choosing K to minimize the effect of a non-null initial plant state on the sum of the energy in the output and the energy in the feedback signal, i.e.,

$$\min_K J_{QC} = J_{QCo}$$

where

$$\begin{aligned} J_{QC} &= \int_0^\infty (y^*(t)y(t) + \rho u_2^*(t)u_2(t))\mathrm{d}t \\ &= \int_0^\infty [x^*(t)\bar{C}^*\bar{C}x(t) + u_2^*(t)[\rho I]u_2(t)]\mathrm{d}t \end{aligned} \tag{5.16}$$

with ρ being a positive scalar and J_{QC} being referred to as the cost or the performance index for the quadratic control problem.

Notice that the reason for including the feedback signal energy in the performance index, i.e., for including $u_2(t)$ in the integrand, is to limit the feedback signal $u_2(t) = Kx(t)$ to a level that can be physically implemented. A more detailed explanation of this is given as follows.

When we choose K to minimize J_{QC}, we discover that the smaller ρ the larger the elements in K. If we were to let ρ be zero so that there would be no contribution to the cost, J_{QC}, from the feedback control signal, $u_2(t)$, we would find that the K required to minimize J_{QC} would have unbounded entries. Thus in this case the required feedback control signal $u_2(t)$ would be an impulse. This physically unrealizable requirement prohibits the use of $\rho = 0$. However, while increasing the size of ρ has the beneficial effect of reducing the size of the required feedback signal, the effect of the disturbance on the output is increased. Thus a beneficial choice for the size of ρ is a compromise between limiting the size of the control signal, $u_2(t)$ and limiting the effect of the disturbance on the output.

In order to simplify the development to follow, we will assume that there is no direct feedthrough of the feedback controlled input $u_2(t)$ to the plant output, i.e.,

$$\text{we assume } D_2 = 0$$

In addition, since it is always possible to choose coordinates for the state space so that the parameters of the state model are real, we will assume that these parameters are real.

Finally, for generality in what follows, we will replace $\bar{C}^*\bar{C}$ and ρI in (5.16) by Q_c and R_c respectively so that the performance index becomes

$$J_{QC} = \int_0^\infty [x^*(t)Q_c x(t) + u_2^*(t)R_c u_2(t)]\mathrm{d}t \tag{5.17}$$

where Q_c, R_c are real, symmetric matrices with $Q_c \geq 0$ and $R_c > 0$ implying that for all $x(t)$ and $u(t)$ we have

$$x^*(t)Q_c x(t) \geq 0$$

$$u_2^*(t)R_c u_2(t) > 0$$

These restrictions on Q_c and R_c are needed to ensure that the contributions to J_{QC} from both $u_2(t)$ and $x(t)$ are never negative and so that R_c is invertible.

In summary, we have formulated the problem of designing a state feedback controller to combat the effect of unknown short-duration disturbance inputs as an optimization problem. The solution to this optimization problem requires that K be determined so that a quadratic performance index in the disturbed state and the controlled input is minimized.

5.3.3 Developing a solution

From the discussion earlier in this section, we suppose that the command input is null, $v(t) = 0$, and that the time origin is taken at the instant a short-duration disturbance has just ended. This leaves the plant state at some non-null value, $x(0) = x_d$ and thereafter we have the plant state given by

$$x(t) = e^{(A+B_2K)t}x_d \tag{5.18}$$

and the controlled input given by

$$u_2(t) = Kx(t) \tag{5.19}$$

Thus using (5.18, 5.19) in the quadratic control cost, (5.17), yields

$$J_{QC} = x_d^* \left[\int_0^\infty e^{\bar{A}^*t} \bar{Q}_c e^{\bar{A}t} \mathrm{d}t \right] x_d \tag{5.20}$$

where

$$\bar{Q}_c = Q_c + K^* R_c K$$
$$\bar{A} = A + B_2 K$$

Now since we are assuming that K is stabilizing, i.e., $\bar{A}$ is stable, the integral, (5.20), converges to a real symmetric matrix P_{QC},

$$\int_0^\infty e^{\bar{A}^*t} \bar{Q}_c e^{\bar{A}t} \mathrm{d}t = P_{QC} \tag{5.21}$$

and the performance index, (5.20), can be written as

$$J_{QC} = x_d^* P_{QC} x_d \tag{5.22}$$

Recall from Chapter 4 that P_{QC}, (5.21), is the solution to the Lyapunov equation

$$\overline{A}^* P_{QC} + P_{QC}\overline{A} + \bar{Q}_c = \emptyset \tag{5.23}$$

Therefore, since $\overline{A}$, $\bar{Q}_c$, (5.20), depend on K, so does the solution, P_{QC}, to this Lyapunov equation and the performance index, J_{QC}, (5.22). We denote this fact by writing P_{QC} and J_{QC} as $P_{QC}(K)$ and $J_{QC}(K, x_d)$ respectively.

From the foregoing we conclude that our goal is to find K such that

(i) K is stabilizing, i.e., $A + B_2K$ is stable, and
(ii) the solution, $P_{QC}(K)$, to (5.23) minimizes $J_{QC}(K, x_d)$, (5.22), for all initial states, x_d.

In the following development we use K_o to denote the value of K which satisfies (i) and (ii). Thus K_o satisfies

$$J_{QC}(K, x_d) > J_{QC}(K_o, x_d) \qquad K \neq K_o \tag{5.24}$$

for all disturbed initial states, x_d.

Now we are going to use a matrix variational approach in connection with the Lyapunov equation, (5.23) to develop an equation for P_{QCo}, where P_{QCo} denotes the matrix P_{QC} which solves (5.23) when $K = K_o$. We begin by noting that any real and stabilizing K can be written as

$$K = K_o + \epsilon(\delta K) \tag{5.25}$$

where ϵ is a positive real scalar and δK is a real matrix having the same dimension as K with $\epsilon(\delta K)$ being referred to as a perturbation on K_o, the desired value for K.

Since P_{QC} depends on K through the Lyapunov equation, (5.23), the perturbation, $\epsilon(\delta K)$, on K_o produces a series expansion for P_{QC} about its optimal value P_{QCo},

$$P_{QC} = P_{QCo} + \epsilon(\delta_1 P_{QC}) + HOT \tag{5.26}$$

where HOT (higher order terms) stands for terms in the expansion involving powers of ϵ which are 2 or greater. The term $(\delta_1 P_{QC})$ is a symmetric matrix which depends on (δK) in a manner to be determined.

Next we substitute the expansion of P_{QC}, (5.26), into the expression for the cost J_{QC}, (5.22). This gives an expansion for J_{QC} about its minimum value J_{QCo}

$$J_{QC} = J_{QCo} + \epsilon(\delta_1 J_{QC}) + HOT \tag{5.27}$$

where

$$J_{QCo} = x_d^* P_{QCo} x_d \qquad (\delta_1 J_{QC}) = x_d^*(\delta_1 P_{QC}) x_d$$

and again HOT stands for terms in the expansion involving powers of ϵ which are 2 or greater.

Now since the terms indicated by HOT in both (5.26) and (5.27) depend on $\{\epsilon^i : i = 2, 3, \cdots\}$ we see that

$$\lim_{\epsilon \to 0} \left[\frac{HOT}{\epsilon} \right] = 0 \tag{5.28}$$

Therefore it follows that the scalar, $(\delta_1 J_{QC})$, in (5.27) and the matrix, $(\delta_1 P_{QC})$, in (5.26) can be expressed as

$$(\delta_1 J_{QC}) = \lim_{\epsilon \to 0} \left[\frac{J_{QC} - J_{QCo}}{\epsilon} \right] \tag{5.29}$$

$$(\delta_1 P_{QC}) = \lim_{\epsilon \to 0} \left[\frac{P_{QC} - P_{QCo}}{\epsilon} \right] \tag{5.30}$$

Notice that (5.29) is reminiscent of the derivative in calculus. This fact together with ideas we encounter in using calculus to obtain minima of a scalar function of a scalar variable enables us to visualize the present problem of minimizing a scalar function J_{QC} of a matrix K as follows.

Suppose, for simplicity we assume $u_2(t)$ is a scalar. Then K is a row vector with n scalar entries. Then we can think of J_{QC} together with the components of K as forming an $n+1$ dimensional space. Imagine making a plot of the scalar J_{QC} on the "vertical axis" vs. K with K being represented by a point in a "horizontal" hyperplane, or plane when $n = 2$. The resulting surface in this space would appear as a bowl with the bottom of the bowl as the "point" $\{J_{QCo}, K_o\}$, i.e., the optimal point. At this point $(\delta_1 J_{QC})$, (5.29), is zero both for all δK and for all initial disturbed states x_d. This is analogous to the derivative of a scalar function of a scalar variable being zero at points where the function has a local minimum.

Notice from the dependency of $(\delta_1 J_{QC})$ on $(\delta_1 P_{QC})$, (5.27), namely

$$(\delta_1 J_{QC}) = x_d^*(\delta_1 P_{QC})x_d$$

that $(\delta_1 J_{QC})$ is zero for all x_d if and only if $(\delta_1 P_{QC})$ is a null matrix. Therefore the optimal value for K, K_o, makes $(\delta_1 P_{QC})$ null for all (δK) which maintain K, (5.25), stabilizing. Therefore in order to determine K_o we need to determine an equation for the matrix $(\delta_1 P_{QC})$ which involves (δK). We can do this by substituting the expansions for K and for P_{QC} (5.25, 5.26), in the Lyapunov equation, (5.23). The results of doing this are determined as follows.

We begin by denoting the result of using the expansion of K, (5.25), in $\overline{A}$ and $\overline{Q_c}$, (5.20), as

$$\begin{aligned} \overline{A} &= \overline{A}_o + \epsilon B_2(\delta K) \\ \overline{Q_c} &= \overline{Q}_{co} + \epsilon[(\delta K)^* R_c K_o + K_o^* R_c(\delta K)] + HOT \end{aligned} \tag{5.31}$$

where

$$\overline{A}_o = A + B_2 K_o \qquad \overline{Q}_{co} = Q_c + K_o^* R_c K_o$$

Then substituting these expressions for $\overline{A}, \bar{Q}_c$ as well as the expansion of P_{QC}, (5.26), in the Lyapunov equation, (5.23), yields

$$\bar{A}_o^* P_{QCo} + P_{QCo}\bar{A}_o + \bar{Q}_{co} + \epsilon(M + M^*) + HOT = \emptyset \tag{5.32}$$

where

$$M = \bar{A}_o^*(\delta_1 P_{QC}) + (\delta K)^* M_1$$

$$M_1 = R_c K_o + B_2^* P_{QCo}$$

Notice that the first three terms in (5.32) sum to a null matrix since P_{QCo} satisfies the Lyapunov equation, (5.23), when $K = K_o$. Thus (5.32) becomes

$$\epsilon(M + M^*) + HOT = \emptyset \tag{5.33}$$

Then dividing (5.33) through by ϵ, letting ϵ go to zero, and using the limiting property,

(5.28), yields

$$M + M^* = \emptyset$$

Finally, substituting for M from (5.32) yields the following Lyapunov equation relating $(\delta_1 P_{QC})$ to (δK)

$$\bar{A}_o^*(\delta_1 P_{QC}) + (\delta_1 P_{QC})\bar{A}_o + (\delta K)^* M_1 + M_1^* \delta K = \emptyset \tag{5.34}$$

Recall, from discussion earlier in this subsection, that $K = K_o$ when $(\delta_1 P_{QC})$ is a null matrix for all δK which maintain K, (5.25) stabilizing. Therefore when $(\delta_1 P_{QC})$ is a null matrix, (5.34) becomes

$$(\delta K)^* M_1 + M_1^* \delta K = \emptyset \tag{5.35}$$

which is satisfied for all stabilizing perturbations, δK, only if M_1 is a null matrix. Therefore we see from the definition of M_1, (5.32) that

$$K_o = -R_c^{-1} B_2^* P_{QCo} \tag{5.36}$$

where we have used the assumption $R_c > 0$ to insure that R_c is invertible.

Notice, at this stage of the development, that the state feedback matrix K_o is readily determined from (5.36) once we have P_{QCo}. It turns out that P_{QCo} is a solution to a nonlinear matrix equation known as the quadratic control algebraic Riccati equation (QCARE). This equation can be developed by setting $K = K_o$, (5.36), in the Lyapunov equation (5.23). After carrying this out, we get the QCARE as

$$A^* P_{QCo} + P_{QCo} A - P_{QCo} \bar{R}_{QC} P_{QCo} + Q_c = \emptyset \tag{5.37}$$

where

$$\bar{R}_{QC} = B_2 R_c^{-1} B_2^*$$

Having determined the QCARE it remains to solve it for P_{QCo} such that $A + B_2 K_o$ is stable when P_{QCo} is used to determine K_o from (5.36).

We can recap this section as follows. We began by introducing a matrix variational approach and using it to determine an equation for the optimal feedback matrix, K_o, (5.36). This equation involves P_{QCo}, the matrix which solves the Lyapunov equation, (5.23), when $K = K_o$. Then we saw that by eliminating K_o from the Lyapunov equation by using (5.36) we obtained a nonlinear matrix equation in P_{QCo} referred to as the QCARE, (5.37).

In the next section we will be concerned with conditions on the plant state model parameters and the parameters of the performance index, R_c and Q_c, which ensure the existences of the so-called stabilizing solution, P_{QCo}, to the QCARE, i.e., the solution which makes $A - B_2 R_c^{-1} B^* P_{QCo}$ stable.

5.4 Solving the QCARE

We have just seen that the determination of the state feedback matrix K_o, (5.36), which minimizes the performance index J_{QC}, (5.17), requires the determination of a stabilizing solution to the QCARE, (5.37). The main goal of the present section is to establish conditions on the plant state model parameters and weighting matrices Q_c and R_c which ensure the existence of a stabilizing solution to the QCARE. We will be aided in this task by resorting to a $2n \times 2n$ matrix called the quadratic control Hamiltonian matrix. Before doing this we consider aspects of the closed loop stability which are brought out using the fact the P_{QCo} is the solution to the Lyapunov equation, (5.23), when $K = K_o$.

5.4.1 Stabilizing solutions

Consider the following theorem and its proof.

Theorem 5.1 $\bar{A}_o$ is stable if

(i) (A, B_2) is a stabilizable pair and
(ii) P_{QCo} satisfies the QCARE with $P_{QCo} \geq 0$ and
(iii) (A, Q_c) is a detectable pair.

where $\bar{A}_o = A + B_2K_o$ with K_o given by (5.36) and P_{QCo} satisfying (5.37).

Proof The necessity of (i) is obvious since we have, from Chapter 2, that a pair (A, B) is stabilizable if we can find K such the $A + BK$ is stable. Therefore assuming (i) holds, we will show that (ii) and (iii) together are sufficient for $\bar{A}_o$ to be stable.

Notice from (5.36, 5.37) that

$$-P_{QCo}\bar{R}_{QC}P_{QCo} = P_{QCo}B_2K_o = K_o^*B_2^*P_{QCo}$$

Therefore adding and subtracting $P_{QCo}\bar{R}_{QC}P_{QCo}$ on the left side of the QCARE, (5.37), gives the Lyapunov equation, (5.23), with $K = K_o$,

$$\bar{A}_o^*P_{QCo} + P_{QCo}\bar{A}_o + \bar{Q}_{Co=}\emptyset \tag{5.38}$$

where

$$\bar{A}_o = A + B_2K_o \qquad \bar{Q}_{Co} = K_o^*R_cK_o + Q_c$$

We proceed now to use the ideas from Chapter 4 concerning stability in connection with the solution of Lyapunov equations.

First we rewrite (5.38) by expanding $\bar{Q}_{Co}$, (5.31), using K_o, (5.36). This enables (5.38) to be rewritten as

$$\bar{A}_o^*P_{QCo} + P_{QCo}\bar{A}_o - -P_{QCo}B_2R_c^{-1}B_2^*P_{QCo} - Q_c \tag{5.39}$$

Then pre and post-multiply this equation by v^{i*} and v^i respectively where $\bar{A}_ov^i = \lambda v^i$ yields

$$2\text{Re}[\lambda_i](v^{i*}P_{QCo}v^i) = -v^{i*}(P_{QCo}B_2R_c^{-1}B_2^*P_{QCo})v^i - v^{i*}Q_cv^i \tag{5.40}$$

Next recalling that $R_c > 0$ and that $Q_c \geq 0$ we have

$$P_{QCo}B_2R_c^{-1}B_2^*P_{QCo} \geq 0 \tag{5.41}$$

and the right side of (5.40) is either negative or zero, i.e., either

$$\text{(a)} \qquad -v^{i*}(P_{QCo}B_2R_c^{-1}B_2^*P_{QCo})v^i - v^{i*}Q_cv^i < 0$$

or

$$\text{(b)} \qquad -v^{i*}(P_{QCo}B_2R_c^{-1}B_2^*P_{QCo})v^i - v^{i*}Q_cv^i = 0$$

Suppose (a) holds. Then we see from (5.40) that

$$2\text{Re}[\lambda_i](v^{i*}P_{QCo}v^i) < 0 \tag{5.42}$$

and neither $\text{Re}[\lambda_i]$ nor $v^{i*}P_{QCo}v^i$ can be zero. Therefore we see from condition (ii) in the theorem that $v^{i*}P_{QCo}v^i > 0$. Thus (5.42) is satisfied only if $\text{Re}[\lambda_i] < 0$.

Alternatively, suppose (b) holds. Then we see from (5.41) that both

$$v^{i*}(P_{QCo}B_2R_c^{-1}B_2^*P_{QCo})v^i = 0 \qquad \text{and} \qquad v^{i*}Q_cv^i = 0$$

This implies that

$$B_2^*P_{QCo}v^i = \emptyset \qquad \text{and} \qquad Q_cv^i = \emptyset \tag{5.43}$$

and we see that

$$\lambda_iv^i = \bar{A}_ov^i = (A - B_2R_c^{-1}B_2^*P_{QCo})v^i = Av^i \tag{5.44}$$

Thus (λ_i, v^i) is an eigenvalue, right-eigenvector pair for A satisfying $Q_cv^i = \emptyset$ implying that λ_i is an unobservable eigenvalue for the pair (A, Q_c). However (A, Q_c) is detectable, condition (iii). Therefore if (b) holds we have $\text{Re}[\lambda_i] < 0$.

To recap, since the right side of (5.40) is either negative or zero and we have shown that in either case $\text{Re}[\lambda] < 0$ when the conditions in the theorem hold. Thus we have shown that the conditions in the theorem are sufficient for $\bar{A}_o$ to be stable. ■

We will show in the next subsection that condition (iii) in the statement of the foregoing theorem is not necessary for $\bar{A}_o$ to be stable. Instead, we will show that only condition (i) in Theorem 5.1 and the condition that any imaginary axis eigenvalues of A be observable eigenvalues for the pair (A, Q_c) are needed to ensure the existence of a stabilizing solution to the QCARE. Some appreciation of this weakening of condition (iii) can be obtained by reconsidering the foregoing proof as follows.

First, notice that when case (b) in the proof holds, (5.40) implies that

$$2\text{Re}[\lambda_i](v^{i*}P_{QCo}v^i) = 0$$

and either

$$(\text{b1}) : \text{Re}[\lambda_i] = 0$$

or

$$(\text{b2}) : P_{QCo} v^i = \emptyset$$

However, recall from the proof that when case (b) holds the eigenvalue λ_i of $\bar{A}_o$ is also an unobservable eigenvalue for the pair (A, Q_c). Therefore if we impose the condition that (A, Q_c) has no unobservable imaginary axis eigenvalues, then case (b1) is impossible.

Second, though not apparent at this stage in the discussion, we will see in the next subsection that if both condition (i) in Theorem 5.1 and the condition that (A, Q_c) have no unobservable imaginary axis eigenvalues are satisfied, then eigenvectors of A can only satisfy the condition for case (b2) to apply if the corresponding eigenvalue of A is stable.

Finally, notice in the proof of Theorem 5.1, that case (b) is impossible when the pair (A, Q_c) is observable and thus $P_{QCo} > 0$ when (A, Q_c) is observable. Again, though not obvious now, we will show that we can replace this condition for $P_{QCo} > 0$ in case (b) by the condition that (A, Q_c) have no stable unobservable eigenvalues. Therefore this condition together with condition (i) in Theorem 5.1 and the condition that (A, Q_c) have no imaginary axis eigenvalues are the necessary and sufficient conditions for $P_{QCo} > 0$.

The uniqueness of the stabilizing solution is shown in the following theorem.

Theorem 5.2 The stabilizing solution to the QCARE is unique, when it exists.

Proof Suppose P_{QC1} and P_{QC2} are two stabilizing solutions to the QCARE. This enables us to write

$$A^* P_{QC1} + P_{QC1} A - P_{QC1} \overline{R}_{QC} P_{QC1} + Q_c = 0$$

$$A^* P_{QC2} + P_{QC2} A - P_{QC2} \overline{R}_{QC} P_{QC2} + Q_c = 0$$

where $\bar{A}_i = A - \bar{R}_{QC} P_{QCi} : i = 1, 2$ are each stable.

Next taking the difference between these two equations yields

$$A^* (\Delta P_{QC}) + (\Delta P_{QC}) A - P_{QC1} \overline{R}_{QC} P_{QC1} + P_{QC2} \overline{R}_{QC} P_{QC2} = 0$$

where $(\Delta P_{QC}) = P_{QC1} - P_{QC2}$.

Then adding and subtracting $P_{QC1} \overline{R}_{QC} P_{QC2}$ enables this equation to be rewritten as

$$\bar{A}_1^* (\Delta P_{QC}) + (\Delta P_{QC}) \overline{A}_2 = 0 \tag{5.45}$$

Now proceeding in the same manner as was done to prove Theorem 4.5, we can show that when $\bar{A}_1, \bar{A}_2$ are stable the matrix equation

$$\bar{A}_1^* (\wedge P_{QC}) + (\Delta P_{QC}) \overline{A}_2 = M$$

has solution ΔP_{QC} which can be written as the integral

$$\Delta P_{QC} = \int_0^\infty e^{\bar{A}_1^* t} M e^{\bar{A}_2 t} \mathrm{d}t$$

However since the right side in (5.45) is null, M is null and the integrand of the foregoing integral is null for all time. Therefore we have $\Delta P_{QC} = \emptyset$ which implies that $P_{QC1} = P_{QC2}$ and the stabilizing solution to the QCARE is unique. ■

Having examined certain properties of stabilizing solutions to the QCARE, we turn next to questions concerning the existence and calculation of such solutions. An important tool for answering these questions is the Hamiltonian matrix.

5.4.2 The Hamiltonian matrix for the QCARE

In this section we will show the relation between the solutions of the QCARE and properties of a $2n \times 2n$ matrix made up from the parameters of the QCARE, (5.37). This matrix, called the quadratic control Hamiltonian matrix, is defined as

$$H_{QC} = \begin{bmatrix} A & \bar{R}_{QC} \\ Q_c & -A^* \end{bmatrix} \tag{5.46}$$

where $\bar{R}_{QC} = B_2 R_c^{-1} B_2^*$.

Notice that the blocks of H_{QC} depend on all the matrices in the QCARE, (5.37), except its solutions P_{QC}. Recall that we are interested in the stabilizing solution denoted P_{QCo}. In the following theorem, we give a property of H_{QC} which is important in establishing conditions for the existence of a stabilizing solution to the QCARE.

Theorem 5.3 It is not possible to find a stabilizing solution to the QCARE, (5.37), if H_{QC}, (5.46), has imaginary axis eigenvalues.

Proof Define the $2n \times 2n$ matrix T as

$$T = \begin{bmatrix} I & 0 \\ -P_{QC} & I \end{bmatrix} \tag{5.47}$$

where P_{QC} is any Hermitian matrix. Notice that T is invertible for all Hermitian matrices, P_{QC}, having inverse given by

$$T^{-1} = \begin{bmatrix} I & 0 \\ P_{QC} & I \end{bmatrix}$$

Next let $\bar{H}_{QC}$ be related to H_{QC} as

$$\overline{H}_{QC} = T^{-1} H_{QC} T = \begin{bmatrix} \bar{A} & \bar{R}_{QC} \\ Z & -\bar{A}^* \end{bmatrix} \tag{5.48}$$

where

$$\bar{A} = A - \bar{R}_{QC} P_{QC}$$

$$Z = A^* P_{QC} + P_{QC} A - P_{QC} \bar{R}_{QC} P_{QC} + Q_c$$

Notice that

$$\lambda[\overline{H}_{QC}] = \lambda[H_{QC}]$$

independent of P_{QC}. Therefore if we choose P_{QC} so that Z is null, $\overline{H}_{QC}$ is block upper-triangular and

$$\lambda[H_{QC}] = \lambda[\bar{A}] \cup \lambda[-\bar{A}^*] \tag{5.49}$$

However since complex eigenvalues of $\bar{A}$ occur in conjugate pairs, we have $\lambda[\bar{A}^*] = \lambda[\bar{A}]$. This together with the fact that $\lambda[-\bar{A}] = -\lambda[\bar{A}]$ allow us to rewrite (5.49) as

$$\lambda[H_{QC}] = \lambda[\bar{A}] \cup -\lambda[\bar{A}] \tag{5.50}$$

This shows that the eigenvalues of H_{QC} are mirror images of each other across the imaginary axis, i.e., if $\eta \in \lambda[H_{QC}]$ then $-\eta \in \lambda[H_{QC}]$.

Now when we choose P_{QC} so that Z is null, P_{QC} is a solution to the QCARE, (5.37). Moreover, recalling the expression for the feedback matrix K_o, (5.36), we see that $\bar{A}$, (5.48) can be rewritten as

$$\bar{A} = A - \bar{R}_{QC}P_{QC} = A + B_2K_o$$

when $P_{QC} = P_{QCo}$, the stabilizing solution to the QCARE. However, if H_{QC} has an imaginary axis eigenvalue, (5.50) shows that this eigenvalue must also be an eigenvalue for $\bar{A}$ for all Hermitian matrices P_{QC} which satisfy the QCARE. Therefore it is impossible to choose P_{QC} to satisfy the QCARE, i.e., to make Z null, so that $\bar{A}$ is stable. ■

To recap, the $2n$ eigenvalues of H_{QC} split into two sets of n eigenvalues each such that each eigenvalue in one set has a corresponding eigenvalue in the other set which is its mirror image across the imaginary axis. Thus if P_{QC} is a stabilizing solution to the QCARE, the eigenvalues of $\bar{A}$, (5.48), equal n of the eigenvalues of H_{QC} which are in the open left half plane. However, if H_{QC} has imaginary axis eigenvalues, H_{QC} does not have n eigenvalues in the open left half plane. Therefore, in this situation, a stabilizing solution to QCARE does not exist since the eigenvalues of $\bar{A}$ must include at least one imaginary axis eigenvalue of H_{QC} for all solutions, P_{QC}, to the QCARE.

We need now to establish conditions which ensure that H_{QC} does not have imaginary axis eigenvalues. This is done in the following theorem.

Theorem 5.4 The quadratic control Hamiltonian matrix H_{QC}, (5.46), is devoid of imaginary axis eigenvalues if

(i) $(A, \bar{R}_{QC})$ has no uncontrollable eigenvalues on the imaginary axis and
(ii) (A, Q_c) has no unobservable eigenvalues on the imaginary axis

Proof Suppose H_{QC} has an eigenvalue at $j\omega$. Then we have

$$H_{QC}v = j\omega v \tag{5.51}$$

$$\begin{bmatrix} A & \bar{R}_{QC} \\ Q_c & -A^* \end{bmatrix} \begin{bmatrix} v_1 \\ v_2 \end{bmatrix} = j\omega \begin{bmatrix} v_1 \\ v_2 \end{bmatrix}$$

or

$$(j\omega I - A)v_1 = \bar{R}_{QC}v_2 \tag{5.52}$$

$$(j\omega I + A^*)v_2 = Q_c v_1 \tag{5.53}$$

Next, we premultiply (5.52) by v_2^* and (5.53) by v_1^* to obtain

$$v_2^*(j\omega I - A)v_1 = v_2^*\bar{R}_{QC}v_2 \tag{5.54}$$

$$v_1^*(j\omega I + A^*)v_2 = v_1^*Q_cv_1 \tag{5.55}$$

Then taking the conjugate transpose on either side of (5.55) and recalling that Q_c is Hermitian enables us to rewrite (5.55) as

$$v_2^*(j\omega I - A)v_1 = -v_1^*Q_cv_1 \tag{5.56}$$

Now we see, by comparing (5.54) with (5.56), that

$$v_2^*\bar{R}_{QC}v_2 = -v_1^*Q_cv_1 \tag{5.57}$$

However Q_c and $\bar{R}_{QC}$ are each non-negative. Therefore the only way (5.57) can be satisfied is for each side to be zero. Thus we have

$$\bar{R}_{QC}v_2 = \emptyset \tag{5.58}$$

$$Q_cv_1 = \emptyset \tag{5.59}$$

and (5.52, 5.53) become

$$Av_1 = j\omega v_1 \tag{5.60}$$

$$A^*v_2 = -j\omega v_2 \tag{5.61}$$

This shows that $j\omega$ is an eigenvalue of A and that v_1, v_2 are the corresponding right and left-eigenvectors of A.

However if $v_2 \neq \emptyset$, (5.58, 5.61) imply that $(A, \bar{R}_{QC})$ has an uncontrollable eigenvalue on the imaginary axis and condition (i) is not satisfied. Therefore if (i) is satisfied, we must have $v_2 = \emptyset$.

Alternatively, if $v_1 \neq \emptyset$, (5.59, 5.60) imply that (A, Q_c) has an unobservable eigenvalue on the imaginary axis and condition (ii) is not satisfied. Therefore if (ii) is satisfied, we must have $v_1 = \emptyset$.

From the foregoing we see that if conditions (i) and (ii) are satisfied, (5.51) holds only if $v = \emptyset$. Thus, contrary to the assumption we made at the beginning of the proof, H_{QC} has no imaginary axis eigenvalues when the conditions of the theorem are satisfied. ■

Notice that since R_c is non-singular and $\bar{R}_{QC} = B_2R_c^{-1}B_2^*$, we have

$$v_2^*\bar{R}_{QC} = \emptyset \qquad \text{only if} \qquad v_2^*B_2 = \emptyset$$

Therefore condition (i) in the foregoing theorem is equivalent to

(i) (A, B_2) has no uncontrollable eigenvalues on the imaginary axis.

We will show in the next subsection that, though sufficient for H_{QC} to have no imaginary axis eigenvalues, the conditions in the foregoing theorem are not sufficient to ensure that the QCARE has a stabilizing solution.

5.4.3 Finding the stabilizing solution

The proof of Theorem 5.3 suggests a way of determining the stabilizing solution to the QCARE. More specifically, we saw that when the stabilizing solution to the QCARE exists, we can find T, (5.47), such that $\bar{H}_{QC}$, (5.48), is block upper-triangular with the top diagonal block being stable. This suggests that the stabilizing solution to the QCARE can be determined from the eigenvectors of H_{QC} corresponding to the stable eigenvalues. However, it turns out that we can avoid having to compute eigenvectors. Instead we can use Schur decomposition on H_{QC} to determine the stabilizing solution to the QCARE. In order to see how to do this we need the following basic result form linear algebra.

Theorem 5.5 If

$$MU = UW \tag{5.62}$$

where M and W are square with W being invertible and U having independent columns, then:

(i) each eigenvalue of W is an eigenvalue of M, i.e., $\lambda[W] \subset \lambda[M]$
(ii) range$[U]$ is an M-invariant subspace corresponding to $\lambda[W]$, i.e., range$[U]$ is spanned by the eigenvectors of M corresponding to the eigenvalues of W.

Proof Let (λ_i, v^i) be any eigenvalue–eigenvector pair for W. Then premultiplying $Wv^i = \lambda_i v^i$ by U and using (5.62) we obtain

$$Mw = \lambda_i w$$

where $w = Uv^i$ and we see that λ_i is also an eigenvalue of M. This completes the proof of (i).

Next assuming the eigenvectors of W are complete, we can express any vector q of the same dimension as these eigenvectors as

$$q = \sum_{i=1}^{n_1} \alpha_i v^i$$

so that post-multiplying (5.62) by q gives

$$Ms = r$$

where

$$r = Up \qquad p = \sum_{i=1}^{n_1} (\lambda_i \alpha_i) v^i$$

$$s = Uq$$

and we see that M maps range$[U]$ into itself, i.e., range$[U]$ is M-invariant.

Finally, notice that if V_1 is a matrix whose columns are the eigenvectors of M corresponding to the eigenvalues of M which are also eigenvalues of W, then

$$MV_1 = V_1\Lambda_1$$

where Λ_1 is diagonal with the eigenvalues of W along its diagonal. Then post-multiplying this equation by the non-singular matrix Θ such that $U = V_1\Theta$ and inserting $\Theta\Theta^{-1}$ between V_1 and Λ_1 gives

$$MU = UW$$

where

$$U = V_1\Theta \qquad\qquad W = \Theta^{-1}\Lambda_1\Theta$$

Thus $\lambda[W] = \lambda[\Lambda_1]$. Moreover since Θ is non-singular, we have

$$\text{range}[U] = \text{range}[V_1\Theta] = \text{range}[V_1]$$

This shows that range$[U]$ is spanned by eigenvectors of M corresponding to the eigenvalues of W. ■

In order to develop a relation between the quadratic control Hamiltonian matrix and the QCARE which we can use as a basis for solving the QCARE for its stabilizing solution, we proceed in the following manner.

Suppose we have the stabilizing solution, P_{QCo}. Then the QCARE, (5.37), can be written as

$$A^*P_{QCo} + Q_c = -P_{QCo}(A - \bar{R}_{QC}P_{QCo})$$

and we see that

$$H_{QC}\begin{bmatrix} I \\ -P_{QCo} \end{bmatrix} = \begin{bmatrix} I \\ -P_{QCo} \end{bmatrix}(A - \bar{R}_{QC}P_{QCo}) \tag{5.63}$$

where

$$H_{QC} = \begin{bmatrix} A & \bar{R}_{QC} \\ Q_c & -A^* \end{bmatrix}$$

Now since $A - \bar{R}_{QC}P_{QCo}$ is stable, we see from Theorem 5.5 that (5.63) implies

$$\text{range}\begin{bmatrix} I \\ -P_{QCo} \end{bmatrix}$$

is the H_{QC}-invariant subspace corresponding to the stable eigenvalues of H_{QC}.

The standard way of computing this subspace, when it exists, is to use Schur decomposition. This numerically reliable method computes an orthogonal (unitary when H_{QC} is complex) matrix T such that the transformed quadratic control Hamiltonian matrix, $\bar{H}_{QC}$, (5.48), is block upper-triangular with first diagonal block, H_{11}, being stable, i.e.,

$$T^* H_{QC} T = \bar{H}_{QC} = \begin{bmatrix} H_{11} & H_{12} \\ \emptyset & H_{22} \end{bmatrix} \tag{5.64}$$

Notice from (5.48) that $H_{11} = A - \bar{R}_{QC} P_{QCo} = \bar{A}_o$, the stable system matrix for the state feedback control system.

Now since T is unitary, $T^* = T^{-1}$, we see that premultiplying (5.64) by T and equating the first n columns on either side of the resulting equation gives

$$H_{QC} T_1 = T_1 H_{11} \tag{5.65}$$

where T is partitioned into $n \times n$ blocks, T_{ij}, with T_1 denoting the first n columns of T, i.e.,

$$T_1 = \begin{bmatrix} T_{11} \\ T_{21} \end{bmatrix} \qquad T = [T_1 \quad T_2] = \begin{bmatrix} T_{11} & T_{12} \\ T_{21} & T_{22} \end{bmatrix}$$

Then we see from Theorem 5.5 and (5.65) that

(i) $\lambda[H_{11}] \subset \lambda[H_{QC}]$
(ii) range$[T_1]$ is an H_{QC}-invariant subspace corresponding to $\lambda[H_{11}]$.

Therefore since H_{11} is stable, the columns of T_1 span the stable eigenspace of H_{QC}, i.e., the columns of T_1 can be expressed in terms of the eigenvectors of H_{QC} whose corresponding eigenvalues lie in the open left half plane.

Recall that range$[T_1]$ =range$[T_1\Theta]$ for any non-singular matrix Θ of appropriate dimension. Hence if T_{11} is non-singular the columns of $T_1 T_{11}^{-1}$ also span the stable eigenspace of H_{QC}. Therefore assuming T_{11} is invertible, we postmultiply (5.65) by T_{11}^{-1} and insert $T_{11}^{-1} T_{11}$ between T_1 and H_{11} on the right side of (5.65) to obtain

$$H_{QC} \begin{bmatrix} I \\ T_{21} T_{11}^{-1} \end{bmatrix} = \begin{bmatrix} I \\ T_{21} T_{11}^{-1} \end{bmatrix} (T_{11} H_{11} T_{11}^{-1})$$

Then comparing this equation with (5.63) we see that P_{QCo} is given by

$$P_{QCo} = -T_{21} T_{11}^{-1} \tag{5.66}$$

and we have the stabilizing solution to the QCARE.

Notice that in order to use the foregoing approach to solve the QCARE for its stabilizing solution, not only must H_{QC} have no imaginary axis eigenvalues, (Theorem 5.3), but in addition T_{11} must be invertible. Notice also that when T_{11} is invertible we have $P_{QCo} > 0$ only if T_{21} is also invertible. The following theorem establishes conditions on the parameters of the plant and control cost which ensure that T_{11} and T_{21} are invertible.

Theorem 5.6 Suppose we can find a unitary matrix T so that H_{11} is stable and

$$T^* H_{QC} T = \bar{H}_{QC} = \begin{bmatrix} H_{11} & H_{12} \\ 0 & H_{22} \end{bmatrix}$$

where

$$H_{QC} = \begin{bmatrix} A & \bar{R}_{QC} \\ Q_c & -A^* \end{bmatrix} \qquad T = [T_1 \quad T_2] = \begin{bmatrix} T_{11} & T_{12} \\ T_{21} & T_{22} \end{bmatrix}$$

Then we have the following results:

(i) T_{11} is invertible if (A, B_2) is stabilizable;
(ii) if T_{11} is invertible, then T_{21} is invertible if (A, Q_c) has no stable unobservable eigenvalues.

Proof (i) From (5.65) and the blocks in H_{QC} and H_{11} we have

$$AT_{11} + \bar{R}_{QC}T_{21} = T_{11}H_{11} \tag{5.67}$$

$$Q_cT_{11} - A^*T_{21} = T_{21}H_{11} \tag{5.68}$$

Then letting let (λ, v) be any eigenvalue-right-eigenvector pair for H_{11} and post-multiplying (5.67, 5.68) by v gives

$$AT_{11}v + \bar{R}_{QC}T_{21}v = \lambda T_{11}v \tag{5.69}$$

$$Q_cT_{11}v - A^*T_{21}v = \lambda T_{21}v \tag{5.70}$$

Now suppose $T_{11}v = \emptyset$. Then T_{11} is not invertible and (5.69, 5.70) become

$$\bar{R}_{QC}w = \emptyset \tag{5.71}$$

$$A^*w = -\lambda w \tag{5.72}$$

where $w = T_{21}v$. Then if $w = \emptyset$ we would have

$$\begin{bmatrix} T_{11} \\ T_{21} \end{bmatrix} v = \emptyset$$

implying that T has dependent columns and is therefore not invertible. This is impossible since we are assuming that T is unitary. Therefore $w \neq \emptyset$.

Now since H_{11} is stable, (5.71, 5.72) imply that $(A, \bar{R}_{QC})$ is not stabilizable, and since $\bar{R}_{QC} = B_2R_c^{-1}B_2^*$ with R_c^{-1} non-singular, we have (A, B_2) is not stabilizable. This contradicts the condition given in (i). Therefore no eigenvector v, of H_{11} satisfies $T_{11}v = \emptyset$ when (A, B_2) is stabilizable. Thus assuming the eigenvectors of H_{11} are complete, the foregoing condition shows that T_{11} is invertible when (A, B_2) is stabilizable.

Proof (ii) Suppose T_{11} is invertible and $T_{21}v = \emptyset$ for an eigenvector v of H_{11}. Then (5.69, 5.70) become

$$As = \lambda s \tag{5.73}$$

$$Q_cs = \emptyset \tag{5.74}$$

where $T_{11}v = s$.

Therefore since H_{11} is stable and λ is an eigenvalue of H_{11}, we have that (A, Q_c) has a stable unobservable eigenvalue. This contradicts the condition given in (ii). Therefore no eigenvector v of H_{11} satisfies $T_{21}v \neq \emptyset$ if (A, Q_c) has no stable unobservable eigenvalues. Finally assuming H_{11} has a complete set of eigenvectors, the foregoing condition shows that T_{21} is non-singular when (A, Q_c) has no stable unobservable eigenvalues. ■

Although more involved, it can be shown that the foregoing theorem holds even when H_{11} does not have a complete set of eigenvectors. This is done using generalized eigenvectors.

In summary, in this section, we have considered the problem of solving the QCARE for its stabilizing solution. The so-called Hamiltonian matrix was introduced for this purpose. By using this matrix it was shown that the computationally useful technique of Schur decomposition could be employed to obtain the stabilizing solution to the QCARE when it exists. In addition, the Hamiltonian matrix enabled the development of important conditions on the plant and performance index parameters which are necessary and sufficient for the existence of a stabilizing solution to the QCARE.

5.5 Quadratic State Estimation

So far we have solved the quadratic control problem under the assumption that the state of the plant is known. In this case the controlled input is

$$u_{2S}(t) = K_o x(t)$$

However, when the plant state is unknown, we must resort to the use of an observer in an observer based feedback control scheme as discussed at the beginning of this chapter. Doing this and using the optimal quadratic state feedback matrix K_o, developed in previous sections, gives the controlled input as

$$u_2(t) = K_o \hat{x}(t) = u_{2S}(t) - u_{2E}(t) \tag{5.75}$$

where

$$u_{2E}(t) = K_o \tilde{x}(t) \qquad \tilde{x}(t) = x(t) - \hat{x}(t)$$

Since $u_{2S}(t) = K_o x(t)$ is the controlled input which minimizes the quadratic state feedback control cost, J_{QC}, (5.17), the departure of the controlled input, $u_{2E}(t)$, from its optimal value, $u_{2S}(t)$, causes the cost J_{QC} to increase. Our goal in this section is to design an observer to minimize this increase in the quadratic control performance index. We approach this goal by solving the optimization problem of choosing the observer's L matrix so as to minimize the energy in the state estimation error caused by both an impulse disturbance in the measurement of the plant output and an initial plant state estimation error.

5.5.1 Problem formulation

Recall, from Section 3.3, that ideally the observer estimates the state of the plant based on the true plant output, $y(t) = Cx(t)$, where for simplicity, we assume, as in the quadratic state feedback control problem, that the plant is strictly proper. However, in practice, only a measured plant output is available and this output differs from the true plant output by an additive measurement noise signal, $w(t)$. Thus if $y_m(t)$ denotes the measured output we have

$$y_m(t) = y(t) + w(t)$$

It should be emphasized that, unlike the quadratic state feedback control problem where the disturbance is at the plant input, the disturbance here is at the input to the observer.

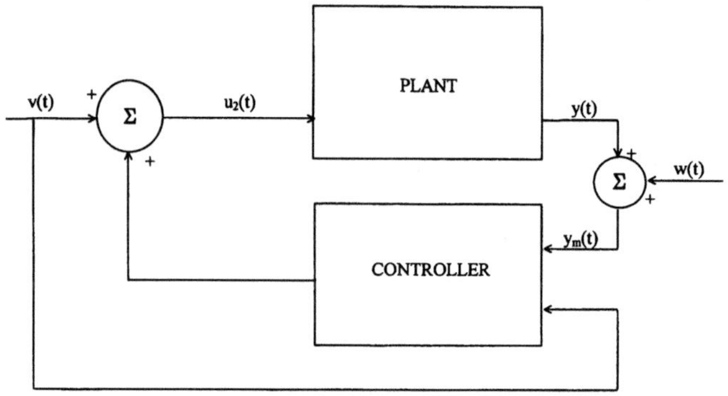

Figure 5.3 Setup for output measurement noise attenuation

$w(t)$	=	output measurement noise	$y(t)$	=	control system output
$u_2(t)$	=	controlled input	$v(t)$	=	control system command input

Now applying $y_m(t)$ to the input of an observer, we obtain an estimate of the plant state from

$$\dot{\hat{x}} = A\hat{x}(t) + Bu(t) + L[\hat{y}(t) - y_m(t)]$$
$$\hat{y}(t) = C\hat{x}(t)$$

and we see that the plant state estimation error, $\tilde{x}(t) = x(t) - \hat{x}(t)$, is governed by the differential equation

$$\dot{\tilde{x}}(t) = \tilde{A}\tilde{x}(t) + Lw(t) \qquad \tilde{A} = A + LC \tag{5.76}$$

As in the state feedback control problem, we assume the unknown disturbance is a large-amplitude, short-duration signal which we model as an impulse, $w(t) = \delta(t - t_0)$.

Suppose the control system has been in operation with a constant command input and no measurement noise, for a time which is great enough so that the plant state estimate has converged to the steady-state value for the plant state, i.e.,

$$\hat{x}(t) = x_s \qquad \tilde{x}(t) = \emptyset$$

Then shifting the time origin to t_0, the time where the output measurement noise impulse occurs, we have $w(t) = \delta(t)$, $\tilde{x}(0) = \emptyset$ and we see from (5.76) that the state estimation error becomes

$$\tilde{x}(t) = e^{\tilde{A}t}L \tag{5.77}$$

Now we want to return the state estimation error to its null value as rapidly as possible following the occurrence of the measurement noise impulse. Therefore (5.77) implies that $L = \emptyset$ is the best choice for this purpose since then $\tilde{x}(t) = \emptyset$ and the disturbance would have no effect on the state estimate. However, this approach to setting up the quadratic state estimation problem does not take into account effects on the state estimation error which arise from the initial plant state being unknown. For instance, if the plant is unstable, we have seen in Chapter 3 that choosing L null causes any estimation error present at start-up to grow without bound.

To overcome this problem, we reformulate the foregoing optimization problem by assuming that the initial state estimation error has an additional component not caused by the output measurement noise, $w(t)$, i.e., we take $\tilde{x}(0)$ as

$$\tilde{x}(0) = \tilde{x}_d$$

Then we see from (5.76) that the plant state estimation error, $\tilde{x}(t)$, caused by both $\tilde{x}(0) = \tilde{x}_d$ and $w(t) = \delta(t)$ is given by

$$\tilde{x}(t) = e^{\tilde{A}t}\tilde{x}_d + e^{\tilde{A}t}L \tag{5.78}$$

Now we want to choose L to decrease the effect on the plant state estimation error of having both a non-null initial state estimation error, and an impulsive output measurement noise. One way of doing this is to solve the problem of choosing L so as to minimize the energy in the state estimation error, i.e., find L such that

$$\min_L J_{QE} = J_{QE0}$$

where

$$J_{QE} = \int_0^\infty \tilde{x}^*(t)\tilde{x}(t)\mathrm{d}t$$

However recalling the use we made of the trace of a matrix in connection with the development of the controllability Gramian in Chapter 4, we see here that J_{QE} can be rewritten as

$$J_{QE} = \text{trace}\left[\int_0^\infty \tilde{x}(t)\tilde{x}^*(t)\mathrm{d}t\right]$$

which using (5.78) becomes

$$J_{QE} = \text{trace}\left[\int_0^\infty e^{\tilde{A}t}Me^{\tilde{A}^*t}\mathrm{d}t\right] \tag{5.79}$$

where

$$M = LL^* + \tilde{x}_d L^* + L\tilde{x}_d^* + \tilde{x}_d\tilde{x}_d^* \qquad \tilde{A} = A + LC$$

Now in order to make the optimization problem more tractable we replace M by $\tilde{Q}_e$,

$$\tilde{Q}_e = Q_e + LR_eL^*$$

where R_e, Q_e are real symmetric matrices with $Q_e \geq 0$, $R_e > 0$. Then our problem is to choose L to minimize J_{QE} such that $\tilde{A}$ is stable, where

$$J_{QE} = \text{trace}\left[\int_0^\infty e^{\tilde{A}t}\tilde{Q}_e e^{\tilde{A}^*t}\text{d}t\right] \tag{5.80}$$

with

$$\tilde{A} = A + LC \qquad \tilde{Q}_e = Q_e + LR_eL^*$$

The development of equations for doing this proceeds along lines similar to those used to develop the equations for K_o in Section 5.3.

5.5.2 Problem solution

We begin the development by noting that the required stability of the observer ensures that the integral, (5.80), converges to a real symmetric matrix P_{QE} so that

$$J_{QE} = \text{trace}[P_{QE}] \tag{5.81}$$

where

$$P_{QE} = \int_0^\infty e^{\tilde{A}t}\tilde{Q}_e e^{\tilde{A}^*t}\text{d}t \tag{5.82}$$

Recall, from Chapter 4, that P_{QE}, (5.82), solves the Lyapunov equation

$$\tilde{A}P_{QE} + P_{QE}\tilde{A}^* + \tilde{Q}_e = 0 \tag{5.83}$$

Notice that since $\tilde{A}$ and $\tilde{Q}_e$, (5.80), depend on L so does P_{QE} and J_{QE}. We use the notation $P_{QE}(L)$ and $J_{QE}(L)$ to denote this fact. Therefore our goal is to find L so that

(i) $\tilde{A}(L)$ is stable, and
(ii) the solution, $P_{QE}(L)$, to (5.83) minimizes $J_{QE}(L)$, (5.81).

We denote the value of L that satisfies (i) and (ii) as L_o. Thus L_o, the optimal value of L, satisfies

$$J_{QE}(L) > J_{QE}(L_o) \qquad L \neq L_o$$

Now we employ the matrix variational technique used in Section 5.3 to develop L_o. Thus suppose L is perturbed away from L_o as

$$L = L_o + \epsilon(\delta L) \tag{5.84}$$

where ϵ is a small positive real scalar and (δL) is any matrix having the same dimensions as L such that $\tilde{A}$ is stable.

Then we can write the expansion of P_{QE} about P_{QEo} as

$$P_{QE} = P_{QEo} + \epsilon\left(\delta_1 P_{QE}\right) + HOT \tag{5.85}$$

where HOT (higher order terms) stands for terms in the series involving powers of ϵ which are 2 or greater. Next we obtain the expansion of J_{QE} about its minimum value, J_{QE0}, by substituting (5.85) in (5.81). Doing this and using the fact that the trace of sum of matrices is the sum of the trace of each matrix gives

$$\begin{aligned} J_{QE} &= \text{trace}\left[P_{QE}\right] = \text{trace}[P_{QEo}] + \epsilon\,\text{trace}[(\delta_1 P_{QE})] + HOT \\ &= J_{QE0} + \epsilon\left(\delta_1 J_{QE}\right) + HOT \end{aligned} \tag{5.86}$$

Notice that

$$\left(\delta_1 J_{QE}\right) = \text{trace}[\delta_1 P_{QE}] \tag{5.87}$$

Now comparing the present development of equations for L_o with the development used to obtain equations for K_o, Section 5.3, we see that the optimal value for L, L_o, makes $(\delta_1 J_{QE}) = 0$ for all δL. Therefore we see from (5.87) that an equation for $\left(\delta_1 P_{QE}\right)$ is needed.

We begin the determination of an equation for $\left(\delta_1 P_{QE}\right)$ by using (5.84) to express $\tilde{A}$ and $\tilde{Q}_e$, (5.80), as

$$\tilde{A} = \tilde{A}_o + \epsilon(\delta L)C$$

$$\tilde{Q}_e = \tilde{Q}_o + \epsilon[L_o R_e(\delta L)^* + (\delta L)R_e L_0^*] + HOT$$

where

$$\tilde{A}_o = A + L_o C \qquad \tilde{Q}_o = Q_e + L_o R_e L_0^*$$

Then substituting these expressions for $\tilde{A}, \tilde{Q}$ and the expansion for P_{QE}, (5.85), in the Lyapunov equation, (5.83), yields

$$\tilde{A}_o P_{QEo} + P_{QEo}\tilde{A}_o^* + \tilde{Q}_o + \epsilon[M + M^*] + HOT = \emptyset \tag{5.88}$$

where

$$M = \tilde{A}_o(\delta_1 P_{QE}) + M_1(\delta L)^* \qquad M_1 = L_o R_e + P_{QEo} C^*$$

Now the first three terms on the left side of (5.88) sum to a null matrix since P_{QEo} satisfies the Lyapunov equation, (5.83), when $L = L_o$. Thus using this fact and recalling from Section 5.3 that

$$\lim_{\epsilon \to 0} \left[\frac{HOT}{\epsilon} \right] = 0$$

we see that dividing (5.88) by ϵ and letting ϵ go to zero gives

$$(\delta_1 P_{QE})\tilde{A}_o^* + \tilde{A}_o(\delta_1 P_{QE}) + (\delta L)M_1^* + M_1(\delta L)^* = \emptyset \tag{5.89}$$

Notice that (5.89) is a Lyapunov equation in $(\delta_1 P_{QE})$. Therefore, since $\tilde{A}_o$ is required to be stable we must have $(\delta_1 P_{QE}) \geq 0$. Thus no eigenvalue, σ_i, of $\delta_1 P_{QE}$ is negative, i.e., $\{\sigma_i \geq 0 : i = 1, 2 \cdots, n\}$. Moreover, referring to the development in Section 5.3, we see that J_{QE} is minimized by choosing L so that $\delta_1 J_{QE} = 0$ for all δL. Thus after rewriting (5.87) as

$$\delta_1 J_{QE} = \text{trace}[(\delta_1 P_{QE})] = \sum_{i=1}^{n} \sigma_i$$

we see that $\delta_1 J_{QE} = 0$ for all δL only if all the σ_is are zero for all δL which implies that $(\delta_1 P_{QE})$ is null for all δL.

Now referring to (5.89) we see that if $(\delta_1 P_{QE})$ is null then

$$(\delta L)M_1^* + M_1(\delta L)^* = \emptyset$$

which holds for all δL only if M_1 is null. Therefore L_o must satisfy

$$M_1 = L_o R_e + P_{QEo} C^* = \emptyset$$

which since $R_e > 0$ can be solved for L_o as

$$L_o = -P_{QEo} C^* R_e^{-1} \tag{5.90}$$

Notice that in the same way that K_o depends on P_{QCo}, the stabilizing solution to the QCARE, so L_o depends on P_{QEo}, the stabilizing solution to an algebraic Riccati equation commonly referred to as the quadratic filtering algebraic Riccati equation (QFARE). This equation is obtained by substituting L_o, (5.90) in the Lyapunov equation, (5.83). Thus after some algebra we obtain the QFARE as

$$AP_{QEo} + P_{QEo}A^* - P_{QEo}\bar{R}_{QE}P_{QEo} + Q_e = 0 \tag{5.91}$$

where $\bar{R}_{QE} = C^* R_e^{-1} C$. This completes the development of the equations for the optimal value of L.

In summary, the same variational approach which was used in Section 5.3 to determine equations for the optimal state feedback matrix K_o was used to determine equations for the optimal observer matrix L_o. In the next section we will find P_{QEo} so that the QFARE,

(5.91), is satisfied and $A - P_{QEo}\bar{R}_{QE}$ is stable by resorting to a Hamiltonian matrix approach similar to the approach used to solve the QCARE for its stabilizing solution.

5.6 Solving the QFARE

In this section we exploit the dual nature of the quadratic state feedback and state estimation problems. We begin by recalling the QCARE, (5.37) and QFARE, (5.91)

$$\text{QCARE}: \quad A^* P_{QCo} + P_{QCo}A - P_{QCo}\bar{R}_{QC}P_{QCo} + Q_c = 0$$

$$\text{QFARE}: \quad AP_{QEo} + P_{QEo}A^* - P_{QEo}\bar{R}_{QE}P_{QEo} + Q_e = 0$$

where

$$\bar{R}_{QC} = B_2 R_c^{-1} B_2^* \qquad \bar{R}_{QE} = C^* R_e^{-1} C$$

Then comparing these equations we obtain th correspondences given in Table 5.1.

Recall that if (A^*, C^*) is stabilizable then (A, C) is detectable. Therefore the foregoing correspondences together with Theorem 5.1 in Section 5.4.1, imply the following sufficient conditions for the existence of a stabilizing solution to the QFARE.

Theorem 5.7 $\tilde{A}_o$ is stable if

(i) (A, C) is a detectable pair and
(ii) P_{QEo} satisfies the QFARE with $P_{QEo} \geq 0$ and
(iii) (A, Q_e) is a stabilizable pair.

where $\tilde{A}_o = A + L_o C$ with L_o given by (5.90) and P_{QEo} satisfying (5.91).

Next recalling the quadratic control Hamiltonian matrix, H_{QC}, (5.46), we see from the foregoing correspondences that the quadratic estimation Hamiltonian matrix, H_{QE}, needed to solve the QFARE for its stabilizing solution should be defined as

$$H_{QE} = \begin{bmatrix} A^* & \bar{R}_{QE} \\ Q_e & -A \end{bmatrix} \tag{5.92}$$

where $\bar{R}_{QE} = C^* R_e^{-1} C$

Then substituting P_{QE} in place of P_{QC} in the similarity transformation matrix T, (5.47)

Table 5.1 QCARE-QFARE Symbol Correspondences

Estimation	**Control**
A	A^*
C^*	B_2
R_e	R_c
Q_e	Q_c
P_{QEo}	P_{QCo}

gives the transformed quadratic estimation Hamiltonian matrix as

$$T^{-1}H_{QE}T = \bar{H}_{QE} = \begin{bmatrix} \tilde{A}^* & \bar{R}_{QE} \\ Y & -\tilde{A} \end{bmatrix}$$

where

$$\tilde{A} = A - P_{QE}\bar{R}_{QE}$$

$$Y = P_{QE}A^* + AP_{QE} - P_{QE}\bar{R}_{QE}P_{QE} + Q_e$$

Notice that H_{QE} and $\bar{H}_{QE}$ have the same eigenvalues for all T. Therefore we can see, by choosing T so that Y is null, that H_{QE} has eigenvalues which are mirror images of each other across the imaginary axis. This means that any imaginary axis eigenvalues of H_{QE} are also eigenvalues of $\tilde{A}$ for all P_{QE} that make Y null, i.e., that satisfy the QFARE. Thus when H_{QE} has imaginary axis eigenvalues the QFARE does not have a stabilizing solution, cf., Theorem 5.3. Conditions which ensure that H_{QE} has no imaginary axis eigenvalues are determined by analogy with Theorem 5.4.

Theorem 5.8 The quadratic estimation Hamiltonian matrix, H_{QE}, (5.92), is devoid of imaginary axis eigenvalues if

(i) (A, C) has no unobservable eigenvalues on the imaginary axis and
(ii) (A, Q_e) has no uncontrollable eigenvalues on the imaginary axis

Assuming a stabilizing solution to the QFARE exists, we can use the Schur decomposition of H_{QE} to determine this solution as was done in Section 5.4.3 to determine the stabilizing solution to the QCARE. Therefore referring to Section 5.4.3, we use the first n columns of T, the unitary or orthogonal matrix which transforms H_{QE} to a block upper triangular matrix with the first diagonal block stable. Then the stabilizing solution to the QFARE is given as

$$P_{QEo} = -T_{21}T_{11}^{-1}$$

where

$$T = [T_1 \quad T_2] = \begin{bmatrix} T_{11} & T_{12} \\ T_{21} & T_{22} \end{bmatrix}$$

Conditions required for the existence of the stabilizing solution to the QFARE can be obtained by analogy with Theorem 5.6.

Theorem 5.9 Suppose we can find a unitary matrix T so that H_{11} is stable and

$$T^*H_{QE}T = \bar{H}_{QE} = \begin{bmatrix} H_{11} & H_{12} \\ 0 & H_{22} \end{bmatrix}$$

where

$$H_{QE} = \begin{bmatrix} A^* & \bar{R}_{QC} \\ Q_e & -A \end{bmatrix} \qquad\qquad T = [T_1 \quad T_2] = \begin{bmatrix} T_{11} & T_{12} \\ T_{21} & T_{22} \end{bmatrix}$$

Then we have the following results:

(i) T_{11} is invertible if (A, C) is detectable;
(ii) if T_{11} is invertible, then T_{21} is invertible if (A, Q_e) has no stable uncontrollable eigenvalues.

Finally we can show that the stabilizing solution to the QFARE is unique, when it exists, by proceeding in the same fashion as was used to prove Theorem 5.2. This completes the discussion of the stabilizing solution to the QFARE.

5.7 Summary

We began this chapter with a description of observer based state feedback and its properties. This was followed by the derivation of design equations for both an optimal quadratic state feedback controller and for an optimal quadratic state estimator (observer). In each case the design equations were nonlinear matrix equations referred to as algebraic Riccati equations, i.e., the QCARE and QFARE. The requirement for the state feedback system and observer to be stable restricted the use of the solutions of each of these equations to their so-called stabilizing solution. Conditions for the existence of these stabilizing solutions were developed, in each case, in terms of the parameters of the plant and performance index. This was done through the use of a related four-block matrix known as a Hamiltonian matrix. We saw, in Theorems 5.4 and 5.6, that the QCARE has a stabilizing solution if and only if (A, B_2) is stabilizable and (A, Q_c) has no unobservable imaginary axis eigenvalues. Alternatively we saw in Theorems 5.8 and 5.9 that the QFARE has a stabilizing solution if and only if (A, C) is detectable and (A, Q_e) has no uncontrollable imaginary axis eigenvalues.

5.8 Notes and References

The algebraic Riccati equation has played a role of great importance in the evolution of control theory. An extensive view of this subject is given in [3]. The use of the Schur decomposition of the Hamiltonian matrix to solve the quadratic state feedback control problem, is implemented in the MATLAB Control System Toolbox under the command lqr. More detailed information on the Schur decomposition is given in [17].

The matrix variational technique used here to develop the design equations for the optimal K and L matrices was used previously to reduce the effect of inaccurate modeling of the plant on the performance of a quadratic control system [28]. For more recent uses of this approach see [13] and the references therein.

Since the stabilizing solutions, P, to either of the algebraic Riccati equations encountered in this chapter are uniquely determined from the corresponding Hamiltonian matrix, H, we have $H \mapsto P$ is a function. This fact has lead to the use of $H \in$dom(Ric) to denote those H that admit a stabilizing P and $P =$Ric(H) to denote the stabilizing solution P determined from H. This notation has become standard in the control literature. For a more detailed discussion see Chapter 13 of [47].

When either the plant is time-varying, or when the control cost is evaluated over a finite interval of time, the optimal controller is time-varying and is determined by solving two Riccati matrix differential equations. The development of these conditions is beyond the scope of the present text. The reader is referred to [1] and [10]for more information on this problem.

6
LQG Control

6.1 Introduction

This chapter continues the development, begun in Chapter 5, of methods for choosing the K and L matrices in an observer based feedback control system. As in Chapter 5, our goal here is to combat the effect that unknown disturbances have on the control system's output. Thus, as discussed in Section 5.3.1, we can assume, without loss of generality, that the control system's command input is null so that the control system output is caused solely by the disturbance input. The assumed sporadic, short duration character of the disturbances used in Chapter 5, is replaced here by disturbances characterized as being persistent random signals which when applied as an input to a linear system produce an output with bounded average power, where $y(t)$ has bounded average power if

$$\lim_{T\to\infty}\frac{1}{T}\int_0^T y^*(t)y(t)\mathrm{d}t < \infty$$

Notice that the persistence of the disturbances makes the output energy unbounded. Therefore unlike the quadratic control problem, we can no longer use the output energy caused by the disturbance input as a measure of the control system's ability to attenuate disturbances. Instead the steady-state average power output is used as a performance measure.

We will assume throughout that the disturbances $w_i(t)$ are zero mean Gaussian random vectors with covariance given as

$$E[w(t_1)w^*(t_2)] = \begin{bmatrix} W_{00} & W_{01} & W_{02} \\ W_{10} & W_{11} & W_{12} \\ W_{20} & W_{21} & W_{22} \end{bmatrix} \delta(t_1 - t_2) \tag{6.1}$$

where

$$w(t) = \begin{bmatrix} w_0(t) \\ w_1(t) \\ w_2(t) \end{bmatrix}$$

and $E[\cdot]$ is the expectation operator from probability theory. Moreover we will assume that the output caused by these random disturbances is ergodic so that the steady-state average power at the output is given as

$$\lim_{T\to\infty}\frac{1}{T}\int_0^T y^*(t)y(t)\mathrm{d}t = \lim_{t\to\infty} E[y^*(t)y(t)] \tag{6.2}$$

Now we will begin by assuming that the plant to be controlled has state model specified as

$$\dot{x}(t) = Ax(t) + w_0(t) + B_2u_2(t) \tag{6.3}$$

$$y_1(t) = C_1x(t) + w_1(t) + D_{12}u_2(t) \tag{6.4}$$

$$y_2(t) = C_2x(t) + w_2(t) + D_{22}u_2(t) \tag{6.5}$$

where

$w_i(t)$	=	disturbances	$y_1(t)$	=	controlled output
$u_2(t)$	=	controlled input	$y_2(t)$	=	measured output

and $u_2(t), y_1(t), y_2(t)$ have dimensions m_2, p_1, p_2 respectively.

Our goal here is to choose K, L in an observer based feedback control system so that J_{GC} is minimized where

$$J_{GC} = \lim_{t\to\infty} E[y_1^*(t)y_1(t)] \tag{6.6}$$

Since the disturbance signals are assumed to be zero mean Gaussian random vectors, this type of optimal control is known by the acronym LQG (Linear Quadratic Gaussian). More recently this optimal control problem was interpreted as a minimum norm problem and the term $\mathcal{H}_2$ optimal control was also used. More will be said about this in Chapter 8 following the development of ideas involving signal and system spaces.

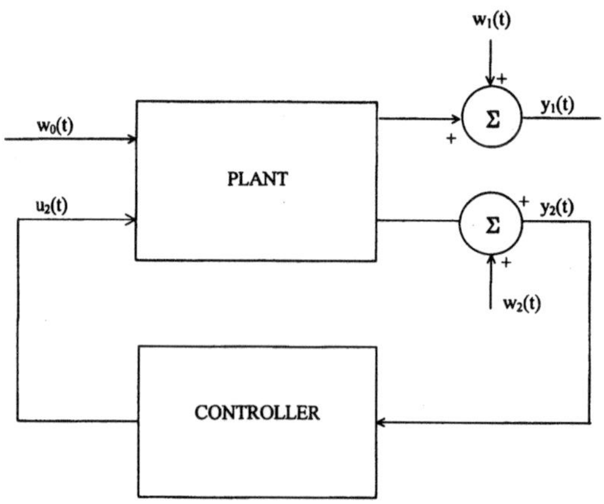

Figure 6.1 LQG control configuration.

As in the quadratic control problem, the separation principle allows us to obtain the solution to the LQG control problem by combining the solutions to two subproblems:

(i) the LQG state feedback control problem
(ii) the LQG state estimation problem.

The LQG control problem is also referred to as

(iii) the LQG measured output feedback control problem.

In what follows we assume that all matrices are real so that we can write $(\cdot)^T$ in place of $(\cdot)^*$

6.2 LQG State Feedback Control Problem

In this section we develop the design equations for the optimal state feedback control matrix K which minimizes J_{GC}, (6.6). We will do this using the variational approach introduced in the previous chapter. We will see that the determination of the optimal K requires the solution of an algebraic Riccati equation.

6.2.1 *Problem formulation*

Suppose the plant state, $x(t)$, is known. Then we want to determine the state feedback matrix, K, involved in generating the controlled input, $u_2(t) = Kx(t)$, so that J_{GC}, (6.6), is minimized. Notice from (6.3, 6.4) that under state feedback the system relating the disturbances to the controlled output is given as

$$\dot{x}(t) = \bar{A}x(t) + w_0(t) \tag{6.7}$$

$$y_1(t) = \bar{C}x(t) + w_1(t) \tag{6.8}$$

where

$$\bar{A} = A + B_2K \qquad \bar{C} = C_1 + D_{12}K$$

and $w_0(t)$ and $w_1(t)$ are unknown zero mean Gaussian random vectors having covariances given by (6.1).

Now in order to proceed we must have $w_1(t) = \emptyset$. This requirement can be seen by noting from (6.1) that when $w_1(t) \neq \emptyset$ the performance index, J_{GC}, (6.6), is unbounded for all K because of the impulsive nature of the covariance of $w_1(t)$. Therefore in order for the LQG state feedback control problem to be properly posed we must have $w_1(t) = \emptyset$. Therefore under this assumption the statement of the LQG state feedback control problem becomes:

$$\min_K J_{GC}$$

given that

$$J_{GC} = \lim_{t\to\infty} E[y_1^T(t)y_1(t)] \tag{6.9}$$

$$\dot{x}(t) = \bar{A}x(t) + w_0(t) \tag{6.10}$$

$$y_1(t) = \bar{C}x(t) \tag{6.11}$$

with $\bar{A}$ stable where $(\bar{A}, \bar{C})$ are given in (6.7, 6.8) and $w_0(t)$ is characterized from (6.1) as

$$E\left[w_0(t_1)w_0^T(t_2)\right] = W_{00}\ \delta(t_1 - t_2)$$

The variational method used in the previous chapter is used now to develop equations for K to solve this problem.

6.2.2 Development of a solution

Since $\bar{A}$ is stable and J_{GC}, (6.9), involves the steady-state behavior, the performance is unaffected by the initial state of the plant. Therefore only the zero state controlled output response is needed. Thus referring to (6.10, 6.11) we see that this response is given by

$$y_1(t) = \bar{C}\int_o^t e^{\bar{A}(t-\tau)}w_0(\tau)\mathrm{d}\tau \tag{6.12}$$

Then using (6.12) we see that the expectation needed in the cost, (6.9), can be written as

$$E[y_1(t)^T y_1(t)] = E\left[\int_0^t w_0^T(\tau_1)e^{\bar{A}^T(t-\tau_1)}\bar{C}^T\mathrm{d}\tau_1 \int_0^t \bar{C}e^{\bar{A}(t-\tau_2)}w_0(\tau_2)\mathrm{d}\tau_2\right] \tag{6.13}$$

Now we can express the right side of this equation by employing the trace relation used in the development of the controllability Gramian in Chapter 4. This relation is restated here as

$$(v_1^T\Theta)v_2 = \mathrm{trace}[v_2 v_1^T\Theta] \tag{6.14}$$

for any compatible vectors v_i and matrix Θ.

Therefore identifying $w_0(\tau_1)$ with v_1 and $w_0(\tau_2)$ with v_2 enables (6.13) to be rewritten as

$$E[y_1^T(t)y_1(t)] = E\left[\mathrm{trace}\left[\int_0^t\left(\int_0^t w_0(\tau_2)w_0^T(\tau_1)e^{\bar{A}^T(t-\tau_1)}\bar{C}^T\mathrm{d}\tau_1\right)\bar{C}e^{\bar{A}(t-\tau_2)}\mathrm{d}\tau_2\right]\right] \tag{6.15}$$

Next evaluating the expectation using (6.1) reduces (6.15) to

$$E[y_1^T(t)y_1(t)] = \mathrm{trace}\left[\int_0^t\left(\int_0^t W_{00}\delta(\tau_2-\tau_1)e^{\bar{A}^T(t-\tau_1)}\bar{C}^T\mathrm{d}\tau_1\right)\bar{C}e^{\bar{A}(t-\tau_2)}\mathrm{d}\tau_2\right] \tag{6.16}$$

Finally, use of the sifting property of the impulse gives

$$E[y_1^T(t)y_1(t)] = \text{trace}\left[\int_0^t W_{00}e^{\bar{A}^T(t-\tau_2)}\bar{C}^T\bar{C}e^{\bar{A}(t-\tau_2)}\text{d}\tau_2\right] \tag{6.17}$$

and substituting $\tau = t - \tau_2$ and allowing $t \to \infty$ enables us to express the performance index, J_{GC}, (6.9) as

$$J_{GC} = \text{trace}[W_{00}P_{GC}] \tag{6.18}$$

where

$$P_{GC} = \int_0^\infty e^{\bar{A}^T\tau}\bar{C}^T\bar{C}e^{\bar{A}\tau}\text{d}\tau$$

Now recalling Theorem 4.5 in Chapter 4, we see that P_{GC} satisfies the Lyapunov equation

$$\bar{A}^T P_{GC} + P_{GC}\bar{A} + \bar{C}^T\bar{C} = \emptyset \tag{6.19}$$

and is the observability Gramian for the pair $(\bar{A}, \bar{C})$. Thus since $(\bar{A}, \bar{C})$ depends on K so does P_{GC}.

Now the equations which determine K so that J_{GC}, (6.9), is minimized can be developed by following the variational approach used in the previous chapter. Thus if K_o denotes the optimal value for K we have

$$J_{GC}(K) > J_{GC}(K_o) = J_{GCo}$$

and expanding J_{GC} and P_{GC} about K_o yields

$$\begin{aligned} J_{GC} &= J_{GCo} + \epsilon(\delta_1 J_{GC}) + HOT \\ &= \text{trace}[W_{00}P_{GCo}] + \epsilon\text{trace}[W_{00}(\delta_1 P_{GC})] + HOT \end{aligned} \tag{6.20}$$

where

$$P_{GC} = P_{GCo} + \epsilon(\delta_1 P_{GC}) + HOT \qquad K = K_o + \epsilon(\delta K)$$

and ϵ is a small positive real scalar with δK being any real matrix having the same dimension as K such that $\bar{A}$ is stable. In addition, recall that HOT denotes higher order terms in ϵ and therefore satisfies

$$\lim_{\epsilon \to 0} \frac{HOT}{\epsilon} = 0 \tag{6.21}$$

Thus we see from (6.20, 6.21) that

$$(\delta_1 J_{GC}) = \lim_{\epsilon \to 0}\left[\frac{J_{GC} - J_{GC0}}{\epsilon}\right] = \text{trace}[W_{00}(\delta_1 P_{GC})] \tag{6.22}$$

Recall from the previous chapter that J_{GC} is minimum when $(\delta_1 J_{GC}) = 0$ for all δK. Thus we see from (6.22) that we need a relation between $(\delta_1 P_{GC})$ and δK. This relation is developed by substituting the expansions for K and P_{GC} as given in (6.20) into the Lyapunov equation (6.19). This gives

$$P_{GCo}\bar{A}_o + \bar{A}_o^T P_{GCo} + \bar{C}_o^T \bar{C}_o + \epsilon(E_1 + F_1) + HOT = 0 \tag{6.23}$$

where

$$\begin{aligned}
\bar{A}_o &= A + B_2 K_o \\
\bar{C}_o &= C_1 + D_{12} K_o \\
E_1 &= (\delta_1 P_{GC})\bar{A}_o + \bar{A}_0^T(\delta_1 P_{GC}) + P_{GCo} B_2(\delta K) + (\delta K)^T B_2^T P_{GCo} \\
F_1 &= \bar{C}_o^T D_{12}(\delta K) + (\delta K)^T D_{12}^T \bar{C}_o
\end{aligned}$$

Notice that the first three terms on the left side of (6.23) sum to a null matrix since P_{GCo} satisfies (6.19) when $K = K_o$. Therefore dividing (6.23) through by ϵ, letting ϵ go to zero, and using (6.21) yields

$$\bar{A}_o^T(\delta_1 P_{GC}) + (\delta_1 P_{GC})\bar{A}_o + (\delta K)^T M_1 + M_1^T \delta K = \emptyset \tag{6.24}$$

where

$$M_1 = B_2^T P_{GCo} + D_{12}^T C_1 + D_{12}^T D_{12} K_o$$

and we see that $(\delta_1 P_{GC})$ is governed by a Lyapunov equation.

Now in order for $(\delta_1 J_{QC})$, (6.22), to be zero for all δK, we must have $W_{00}(\delta_1 P_{GC})$ null for all δK. However W_{00} is independent of δK. Therefore we can have $W_{00}(\delta_1 P_{GC})$ null for all δK only if $(\delta_1 P_{GC})$ is null for all δK.

Thus we see from (6.24) that if $(\delta_1 P_{GC}) = \emptyset$ for all δK then we must have $M_1 = \emptyset$. Therefore assuming that the columns of D_{12} are independent so that $D_{12}^T D_{12}$ invertible, we can solve $M_1 = \emptyset$ for K_o as

$$K_o = -(D_{12}^T D_{12})^{-1} Q_0 \tag{6.25}$$

where

$$Q_0 = B_2^T P_{GCo} + D_{12}^T C_1$$

As in the case of the optimal quadratic control problem, the matrix P_{GCo} is obtained as the solution to an algebraic Riccati equation. We develop this equation by substituting, K_o, (6.25), in the Lyapunov equation, (6.19). Thus after substituting for $\bar{A}, \bar{C}$ from (6.7, 6.8) in (6.19) we obtain

$$A^T P_{GCo} + P_{GCo} A + K_o^T Q_0 + Q_0^T K_o + C_1^T C_1 + K_o^T D_{12}^T D_{12} K_o = \emptyset \tag{6.26}$$

which after substituting for K_o from (6.25) becomes

$$A^T P_{GCo} + P_{GCo} A - 2Q_0^T(D_{12}^T D_{12})^{-1} Q_0 + C_1^T C_1 + Q_0^T(D_{12}^T D_{12})^{-1} Q_0 = \emptyset$$

Finally substituting for Q_0, (6.25), gives

$$[A^T - C_1^T D_{12}(D_{12}^T D_{12})^{-1} B_2^T] P_{GCo} + P_{GCo}[A - B_2(D_{12}^T D_{12})^{-1} D_{12}^T C_1]$$
$$-P_{GCo} B_2(D_{12}^T D_{12})^{-1} B_2 P_{GCo} + C_1^T C_1 - C_1^T D_{12}(D_{12}^T D_{12})^{-1} D_{12}^T C_1 = \emptyset$$

which we can write more compactly as

$$A_1^T P_{GCo} + P_{GCo} A_1 - P_{GCo} R_1 P_{GCo} + Q_1 = \emptyset \tag{6.27}$$

where

$$A_1 = A - B_2(D_{12}^T D_{12})^{-1} D_{12}^T C_1$$
$$R_1 = B_2(D_{12}^T D_{12})^{-1} B_2^T$$
$$Q_1 = C_1^T [I - D_{12}(D_{12}^T D_{12})^{-1} D_{12}^T] C_1$$

Notice that

$$A_1 - R_1 P_{GCo} = A + B_2 K_o$$

Now the algebraic Riccati equation (6.27) is referred to as the Gaussian control algebraic Riccati equation, (GCARE). Thus we obtain the optimal K, K_o, by solving the GCARE for its stabilizing solution, i.e., the solution P_{GCo} which makes $A + B_2 K_o$ stable. As in the previous chapter, we will show that the GCARE can be solved for its stabilizing solution by using an appropriate Hamiltonian matrix. This matter is considered further after taking up the state estimation problem.

Before leaving this section, notice that the requirement that $D_{12}^T D_{12}$ be invertible is similar to the requirement we encountered in Section 5.3.2 that R_c be invertible. Both R_c and $D_{12}^T D_{12}$ play the role of introducing a penalty on the use of the controlled input in minimizing the control costs J_{QC} and J_{GC} respectively.

6.3 LQG State Estimation Problem

In this problem we assume that the plant state $x(t)$ is unknown and that we want to design an observer to estimate it from the controlled input, $u_2(t)$ and measured output, $y_2(t)$, (6.3, 6.5). Unlike the problem just treated, the controlled output $y_1(t)$, (6.4) is not used. Our goal now is to choose the observer matrix L so that J_{GE}, the steady-state average power in the state estimation error, $\tilde{x}(t)$, is minimized where

$$J_{GE} = \lim_{t\to\infty} E[\tilde{x}^T(t)\tilde{x}(t)] = \lim_{t\to\infty} \text{trace}\left[E\left[\tilde{x}(t)\tilde{x}^T(t)\right]\right] \tag{6.28}$$

Notice that this performance index involves the steady-state, state estimation error covariance matrix P_{GE} where

$$P_{GE} = \lim_{t\to\infty} E\left[\tilde{x}(t)\tilde{x}^T(t)\right]$$

In order to proceed we need to review the role played by L in the evolution of the estimation error. This is done as follows.

6.3.1 Problem formulation

Recall, from the beginning of this chapter, that the state model for the plant having the measured output, $y_2(t)$, as its only output is given as

$$\dot{x}(t) = Ax(t) + w_0(t) + B_2u_2(t) \tag{6.29}$$

$$y_2(t) = C_2x(t) + w_2(t) + D_{22}u_2(t) \tag{6.30}$$

where the unknown disturbances $w_0(t)$ and $w_2(t)$ are zero mean Gaussian random vectors with covariance matrices as specified by (6.1). Then referring to Chapter 3, we see that an observer for this state model can be cast as

$$\dot{\hat{x}}(t) = A\hat{x}(t) + B_2u_2(t) + L[\hat{y}_2(t) - y_2(t)] \tag{6.31}$$

$$\hat{y}_2(t) = C_2\hat{x}(t) + D_{22}u_2(t) \tag{6.32}$$

where we have replaced the unknown disturbances by their expected values.

Again as in Chapter 3, we can obtain a differential equation for the state estimation error by subtracting (6.31) from (6.29) and using (6.30, 6.32). This gives

$$\dot{\tilde{x}}(t) = \tilde{A}\tilde{x}(t) + \tilde{B}\tilde{w}(t) \tag{6.33}$$

where

$$\tilde{w}(t) = \begin{bmatrix} w_0(t) \\ w_2(t) \end{bmatrix} \qquad \tilde{x}(t) = x(t) - \hat{x}(t)$$

$$\tilde{A} = A + LC_2 \qquad \tilde{B} = [I \quad L]$$

Notice from (6.1) that the composite zero mean Gaussian random disturbance vector, $\tilde{w}(t)$, has covariance

$$E[\tilde{w}(t_1), \tilde{w}^T(t_2)] = \tilde{W}\delta(t_1 - t_2) \tag{6.34}$$

where

$$\tilde{W} = \begin{bmatrix} W_{00} & W_{02} \\ W_{20} & W_{22} \end{bmatrix}$$

Recall from Chapter 3, that in the absence of disturbances, we chose L so that $\tilde{A}$ is stable. This ensures that the observer generates a state estimate which approaches the plant state for any initial plant state estimation error. Unfortunately this beneficial asymptotic behavior of the state estimator is not possible in the presence of persistent disturbances like those being considered here. However the expected value of the state estimation error will be null in the steady-state when $\tilde{A}$ is stable and the expected values of the disturbances are null. Estimators which have this desirable property are said to be unbiased. Therefore since we are assuming zero mean disturbances, we obtain unbiased estimates of the plant state provided our choice of L to minimize J_{GE}, (6.28), is constrained by the requirement that $\tilde{A}$ be stable. This plus the fact that the performance index, J_{GE}, (6.28), concerns the steady-state behavior of the estimator, implies that we can ignore the state estimation error response due to any initial state estimation error. Therefore using (6.33) we obtain

$$\tilde{x}(t) = \int_0^t e^{\tilde{A}(t-\tau)} \tilde{B}\tilde{w}(\tau) \mathrm{d}\tau \tag{6.35}$$

Now this relation can be used to develop an expression for the state estimation error covariance matrix P_{GE}. We do this by following the steps we used to develop P_{GC} in the previous section. Doing this yields

$$\begin{aligned} E[\tilde{x}(t)\tilde{x}^T(t)] &= E\left[\int_0^t \mathrm{d}\tau_2 \int_0^t e^{\tilde{A}(t-\tau_1)} \tilde{B}\tilde{w}(\tau_1)\tilde{w}^T(\tau_2)\tilde{B}^T e^{\tilde{A}^T(t-\tau_2)} \mathrm{d}\tau_1\right] \\ &= \int_0^t \mathrm{d}\tau_2 \int_0^t e^{\tilde{A}(t-\tau_1)} \tilde{B}\tilde{W}\delta(\tau_1-\tau_2)\tilde{B}^T e^{\tilde{A}^T(t-\tau_2)} \mathrm{d}\tau_1 \\ &= \int_0^t \mathrm{d}\tau_2 e^{\tilde{A}(t-\tau_2)} \tilde{B}\tilde{W}\tilde{B}^T e^{\tilde{A}^T(t-\tau_2)} \end{aligned} \tag{6.36}$$

Finally, we can obtain the steady-state error covariance matrix, P_{GE}, by setting $\tau = t - \tau_2$ and taking the limit as t tends to infinity. This yields

$$\lim_{t\to\infty} E[\tilde{x}(t)\tilde{x}^T(t)] = \int_0^\infty e^{\tilde{A}\tau} \tilde{B}\tilde{W}\tilde{B}^T e^{\tilde{A}^T\tau} \mathrm{d}\tau = P_{GE} \tag{6.37}$$

Notice from Chapter 4, (Theorem 4.5), that P_{GE}, (6.37), is also the controllability Gramian for $(\tilde{A}, \tilde{B})$ and satisfies a Lyapunov equation

$$\tilde{A}P_{GE} + P_{GE}\tilde{A}^T + \tilde{B}\tilde{W}\tilde{B}^T = 0 \tag{6.38}$$

where P_{GE} depends on the observer matrix L since both $\tilde{A}$ and $\tilde{B}$ depend on L. Thus we see from (6.28) that we need to adjust L so that $\tilde{A}$ is stable and so that the solution to (6.38), P_{GE}, has minimum trace. A solution to this problem is obtained now by using the duality between the present problem and the LQG state feedback control problem which we solved in the previous section.

6.3.2 Problem solution

We begin by using the dependency of $\bar{A}$ and $\bar{C}$, on K, (6.7, 6.8) to expand the Lyapunov equation (6.19) and by using the dependency of $\tilde{A}$ and $\tilde{B}$ on L, (6.33) to expand the

Lyapunov equation (6.38). This gives

$$\begin{aligned} &A^T P_{GC} + P_{GC} A + K^T B_2^T P_{GC} + P_{GC} B_2 K \\ &+ C_1^T C_1 + C_1^T D_{12} K + K^T D_{12}^T C_1 + K^T D_{12}^T D_{12} K = \emptyset \end{aligned} \tag{6.39}$$

$$\begin{aligned} &A P_{GE} + P_{GE} A^T + L C_2 P_{GE} + P_{GE} C_2^T L^T \\ &\quad + W_{00} + L W_{20} + W_{02} L^T + L W_{22} L^T = \emptyset \end{aligned} \tag{6.40}$$

In order to compare the symbols in these Lyapunov equations we make use of the fact that covariance matrices are symmetric. Therefore we can factorize $\tilde{W}$, (6.34), as

$$\tilde{W} = \begin{bmatrix} B_1 \\ D_{21} \end{bmatrix} \begin{bmatrix} B_1^T & D_{21}^T \end{bmatrix} = \begin{bmatrix} W_{00} & W_{02} \\ W_{20} & W_{22} \end{bmatrix} \tag{6.41}$$

so that

$$W_{00} = B_1 B_1^T \qquad W_{02} = W_{20}^T = B_1 D_{21}^T \qquad W_{22} = D_{21} D_{21}^T$$

Notice that these factorizations allow us to relate $w_0(t)$ and $w_2(t)$ to a zero mean Gaussian random vector $u_1(t)$ as

$$w_0(t) = B_1 u_1(t) \qquad w_2(t) = D_{21} u_1(t) \tag{6.42}$$

where

$$E\left[u_1(t_1) u_1^T(t_2)\right] = I \delta(t_1 - t_2)$$

Now we can use the foregoing factorizations, (6.41), to rewrite the Lyapunov equation (6.40) as

$$\begin{aligned} &A P_{GE} + P_{GE} A^T + L C_2 P_{GE} + P_{GE} C_2^T L^T \\ &+ B_1 B_1^T + L D_{21} B_1^T + B_1 D_{21}^T L^T + L D_{21} D_{21}^T L^T = \emptyset \end{aligned} \tag{6.43}$$

Then comparing terms in (6.43) with terms in (6.39) we obtain the symbol correspondences for the LQG state feedback control problem and for the LQG state estimation problem given in Table 6.1.

Thus we see that the formulas for the optimal L and P_{GE}, i.e., L_o and P_{GEo}, can be obtained by substituting from these correspondences in the formulas for K_o, P_{GCo} given in (6.25, 6.27)). Doing this yields

$$L_o = -\left[P_{GEo} C_2^T + B_1 D_{21}^T\right] (D_{21} D_{21}^T)^{-1} \tag{6.44}$$

$$A_2 P_{GEo} + P_{GEo} A_2^T - P_{GEo} R_2 P_{GEo} + Q_2 = 0 \tag{6.45}$$

Table 6.1 GCARE-GFARE Symbol Correspondences

Estimation	Control
P_{GE}	P_{GC}
A	A^T
L	K^T
C_2	B_2^T
B_1	C_1^T
D_{21}	D_{12}^T

where

$$A_2 = A - B_1 D_{21}^T (D_{21} D_{21}^T)^{-1} C_2$$

$$R_2 = C_2^T (D_{21} D_{21}^T)^{-1} C_2$$

$$Q_2 = B_1 [I - D_{21}^T (D_{21} D_{21}^T)^{-1} D_{21}] B_1^T$$

Notice that

$$A_2 - P_{GEo} R_2 = A + L_o C_2$$

Now the algebraic Riccati equation (6.45) is referred to as the Gaussian filtering algebraic Riccati equation, (GFARE). Thus L_o is obtained by substituting P_{GEo} in (6.44) where P_{GEo} is called the stabilizing solution to the GFARE since the resulting L_o makes $A + L_o C_2$ stable. As in the previous chapter, the GFARE can be solved for its stabilizing solution by using an appropriate Hamiltonian matrix. We will show this in Section 6.5.

Before going on, notice that the foregoing solution to the LQG state estimation problem requires that $D_{21} D_{21}^T$ be invertible. This implies that D_{21} must have independent rows. Thus when this condition is satisfied the elements in the random vector $w_2(t)$ are deterministically independent of each other in the sense that a subset of the elements in $w_2(t)$ can not be used to determine another element in $w_2(t)$ exactly. When this condition is not satisfied, the estimation problem is referred to as being singular.

6.4 LQG Measured Output Feedback Problem

Having solved the LQG state feedback control problem and the LQG state estimation problem, the separation principle allows us to determine the LQG measured output feedback controller by combining these solutions as follows.

First, replacing the plant state by the observer's estimate of the plant state in the expression for the controlled input gives

$$u_2(t) = K_o \hat{x}(t)$$

where K_o is obtained from (6.25, 6.27) and $\hat{x}(t)$ is the state of the observer. Then substituting this expression for $u_2(t)$ in the observer's state differential equation, (6.31,

6.32), yields the state model for the controller as

$$\dot{\hat{x}}(t) = A_{ce}\hat{x}(t) - L_o y_2(t) \tag{6.46}$$

$$u_2(t) = K_o\hat{x}(t) \tag{6.47}$$

where

$$A_{ce} = A + L_o(C_2 + D_{22}K_o) + B_2K_o$$

and L_o is obtained from (6.44, 6.45).

This completes the formulation of the controller which solves the LQG control problem. It remains now to develop conditions on the plant state model parameters which ensure the existence of stabilizing solutions to the GCARE and the GFARE.

6.5 Stabilizing Solution

Since stabilizing solutions to both the GCARE and GFARE are needed to implement an LQG controller, sufficient conditions for the existence of these solutions, expressed as conditions on the plant parameters, are of considerable importance to designers of these controllers. As in the quadratic control problem, the stabilizing solutions to the GCARE, (6.27), and to the GFARE, (6.45), can be computed, when they exist, by using appropriate Hamiltonian matrices. We will use this fact, in the same manner as was done in the previous chapter, to develop conditions on the parameters of the plant which ensure the existence of stabilizing solutions for the GCARE and GFARE. The duality of the GCARE to the GFARE allows us to obtain these results for both by treating the GCARE in depth and extending the results obtained to the GFARE through the use of duality.

6.5.1 *The Hamiltonian matrix for the GCARE*

Referring to Section 5.4 we see that the GCARE has a stabilizing solution if and only if the associated Hamiltonian matrix H_{GC} has

(i) no eigenvalues on the imaginary axis,
(ii) T_{11} invertible where

$$T_1 = \begin{bmatrix} T_{11} \\ T_{21} \end{bmatrix}$$

is any $2n \times n$ matrix whose columns span the invariant subspace of H_{GC} corresponding to the stable eigenvalues of H_{GC}.

where H_{GC} is composed from the GCARE, (6.27), in the same manner as H_{QC} is composed from the QCARE in Section 5.4.2. Thus we see that

$$H_{GC} = \begin{bmatrix} A_1 & R_1 \\ Q_1 & -A_1^T \end{bmatrix} \tag{6.48}$$

with

$$A_1 = A - B_2(D_{12}^T D_{12})^{-1} D_{12}^T C_1$$
$$R_1 = B_2(D_{12}^T D_{12})^{-1} B_2^T$$
$$Q_1 = C_1^T [I - D_{12}(D_{12}^T D_{12})^{-1} D_{12}^T] C_1$$

Also recall, from the previous chapter, that condition (i) is necessary and sufficient for the existence of a nonsingular $2n \times 2n$ matrix T, (which can be orthogonal), so that H_{GC} can be transformed to block upper triangular form such that the $n \times n$ matrix H_{11} is stable where

$$T^{-1} H_{GC} T = \begin{bmatrix} H_{11} & H_{12} \\ \emptyset & H_{22} \end{bmatrix} \qquad T = \begin{bmatrix} T_{11} & T_{12} \\ T_{21} & T_{22} \end{bmatrix} \tag{6.49}$$

Then condition (ii) is required to construct the stabilizing solution to the GCARE from

$$P_{GCo} = -T_{21} T_{11}^{-1}$$

and we see that $P_{GCo} > 0$ requires that T_{21} be invertible.

In what follows we obtain necessary and sufficient conditions on the plant parameters so that (i) and (ii) hold. We do this by modifying theorems obtained in the previous chapter taking into account the differences in the dependency of the Hamiltonian matrices H_{QC} and H_{GC} on the plant parameters. Conditions on the plant parameters which ensure that $P_{GCo} > 0$ will also be developed.

6.5.2 Prohibition of imaginary eigenvalues

Recall that in the quadratic state feedback control problem we obtained conditions on the plant and performance index which ensured that H_{QC} has no imaginary axis eigenvalues. These conditions were given in Theorem 5.4. Therefore using Theorem 5.4 and comparing H_{QC} with H_{GC}, we see that H_{GC}, (6.48), has no imaginary axis eigenvalues if:

(a) the pair (A_1, R_1) has no uncontrollable imaginary axis eigenvalues and
(b) the pair (A_1, Q_1) has no unobservable imaginary axis eigenvalues.

Notice that these conditions are not immediately applicable to the given data of a plant state model and performance index since they do not explicitly state how these parameters are involved in contributing to the presence of imaginary axis poles for H_{GC}. This deficiency is overcome in the following theorem.

Theorem 6.1 H_{GC}, (6.48), has no imaginary axis eigenvalues if

(i) (A, B_2) has no uncontrollable imaginary axis eigenvalues,
(ii)

$$\operatorname{rank}\left[\begin{bmatrix} A - j\omega I & B_2 \\ C_1 & D_{12} \end{bmatrix}\right] = n + m_2 \qquad \omega \in (-\infty, \infty)$$

where B_2 is $n \times m_2$

Proof (i) Referring to the proof of Theorem 5.4, we see that since $D_{12}^T D_{12}$ is nonsingular we have

$$q^T R_1 = \emptyset \qquad \text{if and only if} \qquad q^T B_2 = \emptyset$$

and therefore λ is an uncontrollable eigenvalue of (A_1, R_1) if and only if λ is an uncontrollable eigenvalue of (A, B_2). Thus condition (i) here is equivalent to condition (i) in Theorem 5.4.

Proof (ii) We begin by considering the pair $(A_1, \tilde{Q}_1)$ where

$$A_1 = A - B_2(D_{12}^T D_{12})^{-1} D_{12}^T C_1$$

$$\tilde{Q}_1 = \left[I - D_{12}(D_{12}^T D_{12})^{-1} D_{12}^T\right] C_1$$

Thus if $j\omega$ is an unobservable eigenvalue of $(A_1, \tilde{Q}_1)$ we have

$$A_1 v = j\omega v \qquad \tilde{Q}_1 v = \emptyset$$

which can be rewritten as

$$(A - j\omega I)v - B_2(D_{12}^T D_{12})^{-1} D_{12}^T C_1 v = \emptyset$$

$$\left(I - D_{12}(D_{12}^T D_{12})^{-1} D_{12}^T\right) C_1 v = \emptyset$$

Now these equations can be written as

$$\Theta\Delta \begin{bmatrix} v \\ \emptyset \end{bmatrix} = \begin{bmatrix} \emptyset \\ \emptyset \end{bmatrix} \tag{6.50}$$

where

$$\Theta = \begin{bmatrix} A - j\omega I & B_2 \\ C_1 & D_{12} \end{bmatrix} \qquad \Delta = \begin{bmatrix} I & \emptyset \\ -(D_{12}^T D_{12})^{-1} D_{12}^T C_1 & I \end{bmatrix}$$

Since Δ is invertible, $\Theta\Delta$ has independent columns only if Θ has independent columns. However (6.50) implies that $\Theta\Delta$ has dependent columns. Therefore Θ must have dependent columns and

$$\text{rank}[\Theta] < n + m_2$$

This shows that condition (ii) is sufficient for $(A_1, \tilde{Q}_1)$ to have no imaginary axis eigenvalues. Now we want to use this fact to show that condition (ii) is sufficient for (A_1, Q_1) to have no imaginary axis eigenvalues.

In order to proceed we need to be able to factor D_c as

$$D_c = \left(D_c^{\frac{1}{2}}\right)^T D_c^{\frac{1}{2}} \tag{6.51}$$

where here

$$D_c = I_{p_1} - D_{12}(D_{12}^T D_{12})^{-1} D_{12}^T$$

However, we can only factorize D_c in the foregoing manner if D_c is symmetric and non-negative. The symmetry of D_c is obvious. We can show $D_c \geq 0$ as follows.

Since D_{12} has independent columns, we can solve the matrix equation

$$D_{12}u = y_r \qquad y_r \in \text{range}[D_{12}]$$

as

$$u = (D_{12}^T D_{12})^{-1} D_{12}^T y_r$$

Therefore it follows that

$$y_r = D_{12}u = \left[D_{12}(D_{12}^T D_{12})^{-1} D_{12}^T\right] y_r$$

and

$$\left[I_{p_1} - D_{12}(D_{12}^T D_{12})^{-1} D_{12}^T\right] y_r = D_c y_r = \emptyset \tag{6.52}$$

However since any p_1 dimensional real vectors y can be written uniquely as

$$y = y_r + y_\perp \tag{6.53}$$

where

$$y_r \in \text{range}[D_{12}] \qquad y_\perp \in \text{null}\left[D_{12}^T\right]$$

we see from (6.52) that

$$D_c y = y_\perp \tag{6.54}$$

and D_c is said to be a projector.

Finally, since in general $y_r^T y_\perp = 0$, we see from (6.54, 6.53) that

$$y^T D_c y = y_\perp^T y_\perp \tag{6.55}$$

for all real p_1 dimensional vectors, y. Therefore since the right side of the foregoing equation can not be negative we have shown that $D_c \geq 0$ and therefore we can factorize D_c as (6.51) and we can proceed to show that when condition (ii) is satisfied (A_1, Q_1) has no imaginary axis eigenvalues.

Suppose condition (ii) is satisfied. Then we showed earlier in the proof that any eigenvector-imaginary axis eigenvalue pair $(v, j\omega)$ of A_1 satisfies

$$\tilde{Q}_1 v \neq \emptyset \tag{6.56}$$

where

$$\tilde{Q}_1 = D_c C_1 = \left(D_c^{\frac{1}{2}}\right)^T D_c^{\frac{1}{2}} C_1$$

Thus we must have

$$D_c^{\frac{1}{2}} C_1 v \neq \emptyset$$

so that

$$\left(D_c^{\frac{1}{2}} C_1 v\right)^T D_c^{\frac{1}{2}} C_1 v \neq 0$$

or

$$v^T C_1^T \left(D_c^{\frac{1}{2}}\right)^T D_c^{\frac{1}{2}} C_1 v = v^T Q_1 v \neq 0$$

This shows that

$$Q_1 v \neq \emptyset$$

for any eigenvector v of A_1 associated with an imaginary axis eigenvalue which implies that the pair (A_1, Q_1) has no unobservable imaginary axis eigenvalues. ■

The foregoing proof reveals an important property of D_c which we will make use of in Chapter 9 and 10 in connection with the H_∞ feedback control problem. Not only is D_c square and non-negative but in addition is a contraction since it has the property that

$$u^T u \leq y^T y$$

for any p_1 dimensional vectors u, y which are related through D_c as $u = D_c y$. This can be seen by noting from (6.53–6.55) that

$$y^T y = y_r^T y_r + y_\perp^T y_\perp \qquad\qquad u^T u = y_\perp^T y_\perp$$

A matrix or operator having this property is referred to as a contraction.

In addition, notice that matrices which are contractions have eigenvalues which are bounded above by one, i.e.,

$$\lambda_{\max}[D_c] \leq 1$$

This property can be seen by refering back to Chapter 4 and the discussion of the significance of the largest eigenvalue of a non-negative matrix (Section 4.6.3).

Finally, notice that this constraint on the eigenvalues of D_c is also a consequence of D_c being a projector since D_c projects from a higher to a lower dimenional space.

6.5.3 Invertability of T_{11} and T_{21}

Having established conditions on the plant state model which ensure that H_{GC} has no eigenvalues on the imaginary axis, it remains to determine conditions on the plant which ensure that T_{11}, (6.49), is non-singular. We can do this by appropriately modifying Theorem 5.6 as follows.

Theorem 6.2 If H_{GC}, (6.48), has no imaginary axis eigenvalues so that we can find T, (6.49), so that H_{11} is stable where

$$T^{-1} H_{GC} T = \begin{bmatrix} H_{11} & H_{12} \\ \emptyset & H_{22} \end{bmatrix} \qquad T = [\,T_1 \quad T_2\,] \qquad T_1 = \begin{bmatrix} T_{11} \\ T_{21} \end{bmatrix} \tag{6.57}$$

then we have

(i) T_{11} invertible if (A, B_2) is stabilizable;
(ii) if T_{11} is invertible, then T_{21} is invertible if $(A_1, D_c^{\frac{1}{2}}C_1)$ has no stable unobservable eigenvalues where

$$A_1 = A - B_2(D_{12}^T D_{12})^{-1} D_{12}^T C_1$$

$$D_c = I - D_{12}(D_{12}^T D_{12})^{-1} D_{12}^T = \left(D_c^{\frac{1}{2}}\right)^T D_c^{\frac{1}{2}}$$

Proof (i) We begin by recognizing that (6.57) implies that

$$H_{GC}T_1 = T_1 H_{11}$$

which we can expand as

$$A_1 T_{11} + R_1 T_{21} = T_{11} H_{11} \tag{6.58}$$

$$Q_1 T_{11} - A_1^T T_{21} = T_{21} H_{11} \tag{6.59}$$

Now suppose H_{11} has an eigenvalue–eigenvector pair (λ, v) such that

$$T_{11} v = \emptyset$$

Then post-multiplying (6.58, 6.59) by v we see that

$$R_1 w = \emptyset \tag{6.60}$$

$$-A_1^T w = \lambda w \tag{6.61}$$

where $w = T_{21} v$.

Now if $w = \emptyset$ we would have

$$\begin{bmatrix} T_{11} \\ T_{21} \end{bmatrix} v = \emptyset$$

implying that T_1 and hence T has dependent columns. This is impossible since T is invertible. Therefore $w \neq \emptyset$.

Finally since H_{11} is stable we have $\mathrm{Re}[\lambda] < 0$ and therefore (6.60, 6.61) imply that (A_1, R_1) is not stabilizable. This is equivalent to (A, B_2) not being stabilizable since $R_1 = B_2(D_{12}^T D_{12})^{-1} B_2^T$. Therefore no eigenvector, v, of H_{11} satisfies $T_{11} v = \emptyset$ when (A, B_2) is stabilizable. Thus assuming the eigenvectors of H_{11} are complete we have shown that T_{11} must be invertible when (A, B_2) is stabilizable. If H_{11} lacks a complete set of eigenvectors, the proof is completed by resorting to generalized eigenvectors.

Proof (ii) Suppose $T_{21} v = \emptyset$ with $s = T_{11} v \neq \emptyset$ for some eigenvector, v, of H_{11}. Then post-multiplying (6.58, 6.59) by v we obtain

$$A_1 s = \lambda s \tag{6.62}$$

$$Q_1 s = M^T M s = C_1^T \left(D_c^{\frac{1}{2}}\right)^T D_c^{\frac{1}{2}} C_1 s = \emptyset \tag{6.63}$$

where

$$M = D_c^{\frac{1}{2}} C_1$$

However since, in general, we have null$\left[M^T\right] \perp$ range $[M]$, we see that (6.63) is satisfied only if $s \in$null$[M]$. Therefore if (6.63) is satisfied then

$$D_c^{\frac{1}{2}} C_1 s = \emptyset \tag{6.64}$$

Moreover since H_{11} is stable, (6.62, 6.64) imply that $(A_1, D_c^{\frac{1}{2}} C_1)$ has a stable unobservable eigenvalue. Therefore no eigenvector, v, of H_{11} satisfies $T_{21} v = \emptyset$ when $(A_1, D_c^{\frac{1}{2}} C_1)$ has no stable unobservable eigenvalues. Thus assuming the eigenvectors of H_{11} are complete we have shown that T_{21} is invertible when $(A_1, D_c^{\frac{1}{2}} C_1)$ has no stable unobservable eigenvalue. Again as in the proof of (i), if H_{11} lacks a complete set of eigenvectors, the proof is completed by resorting to generalized eigenvectors. ■

Notice that when (A, C_1) is not observable neither is $(A_1, D_c^{\frac{1}{2}} C_1)$. Therefore when the GCARE has a stabilizing solution, a necessary condition for T_{21} to be non-singular so that $P_{GCo} > 0$ is that (A, C_1) have no stable unobservable eigenvalues. However this condition is not sufficient for $P_{GCo} > 0$. This can be seen from the following considerations.

First, reviewing (6.51–6.56) we see that

$$\text{rank}[D_c] = \text{rank}\left[D_c^{\frac{1}{2}}\right] = p_1 - m_2$$

where null$\left[D_c^{\frac{1}{2}}\right]$ is of dimension m_2 and it may turn out that $C_1 v \neq \emptyset$ but $D_c^{\frac{1}{2}} C_1 v = \emptyset$.

Second, since $D_{12}^T D_{12} = R_D$ is symmetric and invertible it can be factored and its square root $R_D^{\frac{1}{2}}$ is invertible. Hence we can write (6.51) as

$$\left(D_c^{\frac{1}{2}}\right)^T D_c^{\frac{1}{2}} = I - D_{12}\left(R_D^{\frac{1}{2}}\right)^{-1}\left(R_D^{\frac{1}{2}}\right)^{-T} D_{12}^T$$

which implies that

$$\begin{bmatrix} D_c^{\frac{1}{2}} \\ \left(R_D^{\frac{1}{2}}\right)^{-T} D_{12}^T \end{bmatrix}$$

has orthonormal columns. Therefore it is impossible to have a non-null vector q such that

$$D_c^{\frac{1}{2}} q = \emptyset \qquad \text{and} \qquad \left(R_D^{\frac{1}{2}}\right)^{-T} D_{12}^T\, g = \emptyset$$

Thus if v is an eigenvector of A which satisfies $C_1 v \neq \emptyset$ but $D_c^{\frac{1}{2}} C_1 v = \emptyset$, then we have $(R_D^{\frac{1}{2}})^{-T} D_{12}^T C_1 v \neq \emptyset$ so that

$$A_1 v = \left[A - B_2 \left(R_D^{\frac{1}{2}} \right)^{-1} \left(R_D^{\frac{1}{2}} \right)^{-T} D_{12}^T C_1 \right] v \neq \lambda v$$

and v is an eigenvector of A but not of A_1. Notice in this case that unlike the eigenvectors of A, the eigenvectors of A_1 depend on B_2 implying that the positive definiteness of the stabilizing solution depends on B_2.

In summary, the GCARE has a stabilizing solution, P_{GCo}, if

(1c) (A, B_2) is stabilizable;

(2c)

$$\operatorname{rank} \left[\begin{bmatrix} A - j\omega I & B_2 \\ C_1 & D_{12} \end{bmatrix} \right] = n + m_2 \qquad \omega \in (-\infty, \infty)$$

Finally if a stabilizing solution, P_{GCo}, exists then

(3c) $P_{GCo} > 0$ if $(A_1, D_c^{\frac{1}{2}} C_1)$ has no stable unobservable eigenvalues.

6.5.4 Conditions for solving the GFARE

In order to obtain conditions on the plant parameters which ensure that the GFARE has a stabilizing solution, we can use the correspondence between matrices in the GCARE and the GFARE given in Table 6.1. Therefore substituting these correspondences in (1c–3c) we see that the GFARE has a stabilizing solution, P_{GEo} if

(1e) (A^T, C_2^T) is stabilizable;

(2e)

$$\operatorname{rank} \left[\begin{bmatrix} A^T - j\omega I & C_2^T \\ B_1^T & D_{21}^T \end{bmatrix} \right] = n + p_2 \qquad \omega \in (-\infty, \infty)$$

Moreover, if a stabilizing solution, P_{GEo}, exists then

(3e) $P_{GEo} > 0$ if $(A_2^T, D_e^{\frac{1}{2}} B_1^T)$ has no stable unobservable eigenvalues, where

$$\left(D_e^{\frac{1}{2}} \right)^T D_e^{\frac{1}{2}} = I_{m_1} - D_{21}^T \left(D_{21} D_{21}^T \right)^{-1} D_{21}$$

However these conditions can be restated in a more convenient form by using the facts that the stabilizability of any pair (A, B) is equivalent to the detectability of the pair (A^T, B^T), and that the rank of a matrix equals the rank of its transpose. Thus the GFARE has a stabilizing solution, P_{GEo}, if

(1e) (A, C_2) is detectable;

(2e)

$$\operatorname{rank} \left[\begin{bmatrix} A - j\omega I & B_1 \\ C_2 & D_{21} \end{bmatrix} \right] = n + p_2 \qquad \omega \in (-\infty, \infty)$$

Moreover, if a stabilizing solution, P_{GEo}, exists then

(3e) $P_{GEo} > 0$ if $\left(A_2,\ B_1 D_e^{\frac{T}{2}}\right)$ has no stable uncontrollable eigenvalues.

6.6 Summary

In this chapter we have given a derivation of the design equations for the LQG optimal feedback controller for a linear time-invariant continuous-time plant. The performance criterion or cost function which is minimized by the optimal LQG controller is the steady-state expected or average value of a quadratic form in the output vector to be controlled (the controlled output) when the disturbance input is a zero mean Gaussian random vector. Necessary and sufficient conditions were given for being able to design the LQG controller.

6.7 Notes and References

Again, as in the quadratic control problem treated in the previous chapter, we have seen that the algebraic Riccati equation plays a central role in the design of optimal controllers. For other instances where the algebraic Riccati equation arises in control theory see Chapter 13 of [47]. The LQG optimal control problem was recently referred to as the H_2 optimal control problem. The reason for this will be given in Chapter 7. Assuming the appropriate existence conditions are satisfied, the LQG controller can be calculated using the command h2lqg in the MATLAB Robust Control Toolbox.

The observer obtained by solving the LQG state estimation problem was originally given by Kalman as an alternative to the Wiener filter for extracting a desired signal from an additive combination of the desired signal and noise. This aspect of the LQG problem has had a great impact on a wide variety of industrial problems. There are many books which deal with this subject. For example, informative treatments are given in [7, 10, 25].

7

Signal and System Spaces

7.1 Introduction

In the previous two chapters we were concerned with the problem of designing controllers to attenuate the effects of disturbance inputs on the output of a feedback control system. We assumed that the disturbance input was either an impulse (Chapter 5) or a random signal with an impulsive covariance (Chapter 6). In the next two chapters we develop ideas needed to solve the disturbance attenuation problem for a broader class of disturbance input. Signals in this class have finite energy and are denoted by $\mathcal{L}_2[0,\infty)$ where $\mathcal{L}$ is used in recognition of the mathematician H. L. Lebesgue, pronounced "Lebeg", the subscript 2 is used in recognition of the quadratic nature of energy, and the bracketed quantities indicate the time interval over which the signals are not always zero. Therefore $f(t) \in \mathcal{L}_2[0,\infty)$ if

(i) $f(t)$ has bounded L_2 norm (finite energy),

$$\left[\int_0^\infty f^*(t)f(t)\mathrm{d}t\right]^{\frac{1}{2}} = \|f(t)\|_2 < \infty \tag{7.1}$$

where $[\cdot]^{\frac{1}{2}}$ denotes the positive square root and $f^*(t)$ denotes the conjugate transpose of $f(t)$.

(ii) $f(t)$ is null for negative time

$$f(t) = \emptyset \qquad \text{for all } t \in (-\infty, 0) \tag{7.2}$$

In this chapter we will use the foregoing idea of the L_2 norm for signals to develop system norms defined in both the time and frequency domains. This is made possible by Parseval's theorem which relates the L_2 norm in the time domain to the L_2 norm in the frequency domain.

7.2 Time Domain Spaces

In classical control theory there was no apparent need to consider signals defined for negative time since control problems usually have a well defined start time. However the need for more sophisticated mathematics to deal with new approaches to control problems requires that we enlarge the domain of definition of signals to include the negative time axis and to consider the operation of systems in both positive and negative time.

7.2.1 Hilbert Spaces for Signals

Consider a signal vector, $f(t)$, which unlike signals in $\mathcal{L}_2[0,\infty)$, is not zero for all negative time but still has finite energy over the time interval $(-\infty,\infty)$, i.e.,

$$\left[\int_{-\infty}^{\infty} f^*(t)f(t)\mathrm{d}t\right]^{\frac{1}{2}} = \|f(t)\|_2 < 0 \tag{7.3}$$

Any $f(t)$ which satisfies (7.3) belongs to the normed space $\mathcal{L}_2(-\infty,\infty)$.

Alternatively, another related normed space, denoted $\mathcal{L}_2(-\infty,0]$, consists of finite energy signals which are null for all positive time. Notice that $\mathcal{L}_2[0,\infty)$ and $\mathcal{L}_2(-\infty,0]$ are each subspaces of $\mathcal{L}_2(-\infty,\infty)$, i.e.,

$$\mathcal{L}_2(-\infty,0] \subset \mathcal{L}_2(-\infty,\infty) \qquad \mathcal{L}_2[0,\infty) \subset \mathcal{L}_2(-\infty,\infty)$$

Now from the foregoing we see that if $f(t) \in \mathcal{L}_2(-\infty,\infty)$ it is always possible to write

$$f(t) = f_+(t) + f_-(t) \tag{7.4}$$

where $f_+(t) \in \mathcal{L}_2[0,\infty)$ and $f_-(t) \in \mathcal{L}_2(-\infty,0]$ with

$$\begin{aligned} f_+(t) &= f(t) \quad t > 0 \qquad f_+(0) = \kappa f(0) \\ f_-(t) &= f(t) \quad t < 0 \qquad f_-(0) = (1-\kappa)f(0) \end{aligned}$$

where $\kappa \in [0,1]$.

Notice that once κ is specified, $f_+(t), f_-(t)$ are uniquely dependent on $f(t)$. Moreover since $f_+(t)$ and $f_-(t)$ are nonzero on disjoint intervals, we see that

$$\int_{-\infty}^{\infty} f_+^*(t)f_-(t)\mathrm{d}t = 0 \tag{7.5}$$

Now (7.4) together with (7.5) enables the L_2 norm of any $f(t) \in \mathcal{L}_2(-\infty,\infty)$ to be decomposed in the following fashion

$$\begin{aligned} \|f(t)\|_2 &= \left[\int_{-\infty}^{\infty} f^*(t)f(t)\mathrm{d}t\right]^{\frac{1}{2}} \\ &= \left[\int_0^{\infty}[f_+^*(t)f_+(t)]\mathrm{d}t + \int_{-\infty}^{0}[f_-^*(t)f_-(t)]\mathrm{d}t\right]^{\frac{1}{2}} \\ &= [\|f_+(t)\|_2^2 + \|f_-(t)\|_2^2]^{\frac{1}{2}} \end{aligned} \tag{7.6}$$

which is reminiscent of the Pythagorean theorem for right-angled triangles.

In the foregoing signal analysis we encountered an integral involving two signals in $\mathcal{L}_2(-\infty,\infty)$. In general the quantity

$$\int_a^b \alpha^*(t)\beta(t)\mathrm{d}t \tag{7.7}$$

is referred to as the inner product of $\alpha(t)$ and $\beta(t)$ for any equal length vectors $\alpha(t),\beta(t) \in \mathcal{L}_2(a,b)$ and is denoted as $<\alpha(t),\beta(t)>$, i.e.,

$$<\alpha(t),\beta(t)> = \int_a^b \alpha^*(t)\beta(t)\mathrm{d}t \tag{7.8}$$

Notice that, unlike the norm which involves only one signal and is always a real non-negative scalar, the inner product involves two signals and is not restricted to be positive or real. Notice that the norm and the inner product are related as

$$\|\alpha(t)\|_2 = [<\alpha,\alpha>]^{\frac{1}{2}} \tag{7.9}$$

where again $[\cdot]^{\frac{1}{2}}$ denotes the positive square root.

A signal space with an inner product is referred to as inner product space. Under additional technical constraints (completion), an inner product space is referred to as a Hilbert space. Hilbert space has found widespread use in applied mathematics, physics and engineering. In addition to being normed spaces, $\mathcal{L}_2[0,\infty)$, $\mathcal{L}_2(-\infty,\infty)$, and $\mathcal{L}_2(-\infty,0]$ are each Hilbert spaces with inner product defined by (7.7).

An important consequence of inner products is the property of orthogonality. Two signals are said to be orthogonal or form an orthogonal pair if they have an inner product which is zero. Thus we see from (7.5) that $f_+(t), f_-(t) \in \mathcal{L}_2(-\infty,\infty)$ are orthogonal.

The classical example of signals which are orthogonal arises in connection with the Fourier series decomposition of periodic signals. This decomposition exploits the periodicity and orthogonality of the following signal pairs: $\{\cos(\omega t),\ \sin(\omega t)\}$; $\{\cos(n\omega t),\ \cos(m\omega t)\}$; $\{\sin(n\omega t),\ \sin(m\omega t)\}$; where n,m are unequal integers and the interval of integration used in the inner product is the period $\frac{2\pi}{\omega}$ of the periodic signal being decomposed.

Two spaces S_1, S_2 are said to be orthogonal, denoted $S_1 \perp S_2$ if

$$<\alpha(t),\beta(t)> = 0 \qquad \text{for all } \alpha(t) \in S_1 \quad \text{and all } \beta(t) \in S_2$$

Thus we see that $\mathcal{L}_2(-\infty,0] \perp \mathcal{L}_2[0,\infty)$. More important, $\mathcal{L}_2(-\infty,0] \cap \mathcal{L}_2[0,\infty) = \emptyset$ so that $\mathcal{L}_2(-\infty,0]$ and $\mathcal{L}_2[0,\infty)$ are orthogonal complements of each other, denoted by

$$\mathcal{L}_2(-\infty,0] = \mathcal{L}_2^{\perp}[0,\infty) \qquad \mathcal{L}_2[0,\infty) = \mathcal{L}_2^{\perp}(-\infty,0]$$

and the decomposition of any signal $f(t) \in \mathcal{L}_2(-\infty,\infty)$ given by (7.4) is captured by the following relation between the involved Hilbert spaces

$$\mathcal{L}_2(-\infty,\infty) = \mathcal{L}_2(-\infty,0] \oplus \mathcal{L}_2[0,\infty) \tag{7.10}$$

where $\oplus$ is referred to as the direct sum of the spaces involved.

7.2.2 The L_2 Norm of the Weighting Matrix

Recall, from Chapter 1, that the zero state response of a single-input, single-output system can be calculated from its state model as

$$y(t) = \int_0^t Ce^{A(t-\tau)}Bu(\tau)\mathrm{d}\tau \tag{7.11}$$

with impulse response denoted $y_I(t)$ being given as

$$y_I(t) = Ce^{At}B \tag{7.12}$$

In addition recall from Section 1.8.3 that the causality constraint requires

$$y_I(t) = 0 \qquad \text{for all } t < 0$$

Therefore the square of the L_2 norm of $y_I(t)$ for stable systems can be determined by recalling the observability and controllability Gramians, W_o, W_c from Chapter 4. Thus we have

$$\begin{aligned} \| y_I(t)\|_2^2 &= \int_0^\infty y_I^*(t)y_I(t)\mathrm{d}t = \int_0^\infty B^*e^{A^*\tau}C^*Ce^{A\tau}B\mathrm{d}\tau \\ &= B^*W_oB \end{aligned} \tag{7.13}$$

Alternatively, since $y_I(t)$ is a scalar, we can determine the square of the L_2 norm of $y_I(t)$ as

$$\begin{aligned} \| y_I(t)\|_2^2 &= \int_0^\infty y_I(t)y_I^*(t)\mathrm{d}t = \int_0^\infty Ce^{At}BB^*e^{A^*t}C^*\mathrm{d}t \\ &= CW_cC^* \end{aligned} \tag{7.14}$$

Thus we see from (7.13) or (7.14) that the impulse response of a stable system has finite L_2 norm and hence $y_I(t) \in \mathcal{L}_2[0,\infty)$.

The extension of the foregoing result to multi-input, multi-output systems is more involved since, in this case, $Ce^{At}B$ is a matrix referred to as the system's "weighting matrix". This terminology is used in recognition of the fact that $Ce^{At}B$ applies differing weights to the components in the system's input vector in the calculation of the zero state response

$$y(t) = \int_0^t Ce^{A(t-\tau)}Bu(\tau)\mathrm{d}\tau$$

Notice that the Laplace transform of a system's weighting matrix is the system's transfer function. Notice also that the weighting matrix is the impulse response when the system is single-input, single-output. In what follows we will use the sum of the energies in each entry in the weighting matrix to generalize the L_2 norm of the impulse response to the multi-input, multi-output case.

Suppose $u(t)$ is a vector of length m and $y(t)$ is a vector of length p , i.e., the system has m scalar inputs and p scalar outputs and the weighting matrix, $Ce^{At}B$, is $p \times m$. Then if the input is

$$u(t) = u_I^i(t) = I_m^i \delta(t)$$

where I_m^i is the ith column of the $m \times m$ identity matrix, then the output is

$$y(t) = y_I^i(t) = Ce^{At}B^i$$

where B^i is the ith column of the $n \times m$ matrix B.

Now we want to calculate the sum of the energy in each of the outputs in the set of outputs $\{y_I^i(t) : i = 1, 2, \cdots m\}$ resulting from the application of each input in the set $\{u_I^i : i = 1, 2, \cdots m\}$ respectively. Clearly, we can write the desired energy as

$$\begin{aligned} \|\bar{y}_I(t)\|_2 &= \left(\sum_{i=1}^{m} \left(\int_0^\infty [y_I^i(t)]^* y_I^i(t) \mathrm{d}t\right)\right)^{\frac{1}{2}} \\ &= \left(\int_0^\infty \sum_{i=1}^{m} (B^i)^* e^{A^* t} C^* C e^{At} B^i \mathrm{d}t\right)^{\frac{1}{2}} \end{aligned} \tag{7.15}$$

where $\bar{y}_I(t)$ is the pm dimensional vector

$$\bar{y}_I(t) = \begin{bmatrix} y_I^1(t) \\ y_I^2(t) \\ \vdots \\ y_I^m(t) \end{bmatrix}$$

Now we can rewrite (7.15) as

$$\|\bar{y}_I(t)\|_2 = \left(\text{trace}\left[\int_0^\infty (Ce^{At}B)^* Ce^{At} B \mathrm{d}t\right]\right)^{\frac{1}{2}} \tag{7.16}$$

$$= (\text{trace}[B^* W_o B])^{\frac{1}{2}} \tag{7.17}$$

Alternatively, recall (Theorem 4.2) that for any matrix M we have

$$\text{trace}[M^* M] = \text{trace}[MM^*]$$

Therefore setting $M = Ce^{At}B$, we see that (7.16) can be rewritten as

$$\|\bar{y}_I(t)\|_2 = (\text{trace}[CW_c C^*])^{\frac{1}{2}} \tag{7.18}$$

Finally, since $\|\bar{y}_I(t)\|_2$, (7.16), is the sum of the energy in each scalar signal in the $p \times m$

matrix $Ce^{At}B$, we define the L_2 norm of the $p \times m$ matrix $Ce^{At}B$ as

$$\|Ce^{At}B\|_2 = \left(\text{trace}\left[\int_0^\infty (Ce^{At}B)^* Ce^{At}B\text{d}t\right]\right)^{\frac{1}{2}} \tag{7.19}$$

which we see from (7.17, 7.18) can be written in either of the following two forms

$$\|Ce^{At}B\|_2 = (\text{trace}[B^* W_o B])^{\frac{1}{2}} \tag{7.20}$$

$$= (\text{trace}[CW_c C^*])^{\frac{1}{2}} \tag{7.21}$$

We will see in Section 7.3.6 that the foregoing time domain L_2 norm of the weighting matrix equals an appropriately defined frequency domain L_2 norm of the system's transfer function which is referred to as the H_2 norm.

7.2.3 Anticausal and antistable systems

We begin by considering the single-input, single-output case. Suppose $u(t) \in \mathcal{L}_2(-\infty, 0]$. Then substituting negative t in (7.11) and considering t decreasing, i.e., reversing the direction of integration in (7.11), we see that the output for negative time is given as

$$y(-|t|) = -\int_0^{-|t|} Ce^{A(-|t|-\tau)}Bu(\tau)\text{d}\tau \qquad t \in [0, \infty)$$

or

$$y(t) = -\int_0^{t} Ce^{A(t-\tau)}Bu(\tau)\text{d}\tau \qquad t \in (-\infty, 0] \tag{7.22}$$

Notice that this system operates in negative time, i.e., runs backward in time. Thus an impulse input at the time origin, $u(t) = \delta(t)$, affects the output at earlier (negative) times. This effect is contrary to nature where dynamic processes have the general behavioral property that the output now is independent of the future input and is solely caused by past initial conditions and the cumulative effect of the input over the past interval of time. Natural systems having this behavior, where the cause precedes the effect, are referred to as causal systems. Artificial systems having the predictive behavior exhibited in (7.22), where the effect precedes the cause, are referred to as anticausal systems.

Now if we let $u(t) = \delta(t)$ then we see from (7.22) that the impulse response of an anticausal system is given by

$$\begin{aligned} y_I(t) &= -Ce^{At}B \qquad && t \in (-\infty, 0] \\ y_I(t) &= 0 && \text{for all } t > 0 \end{aligned} \tag{7.23}$$

Notice that $y_I(-\infty)$ is zero only if $-A$ is stable, i.e. only if A has all its eigenvalues in the open right half-plane or equivalently, only if no eigenvalue of A is in the closed left half-plane. Systems that have this property are referred to as being antistable. Recall that a system is unstable if at least one eigenvalue of its A matrix lies in the closed right half-plane. Thus we see that antistable systems are a special class of unstable systems.

The introduction of this class of system allows us to express the transfer function of an unstable system, provided it has no imaginary axis poles, as the sum of two transfer functions, one for a stable system and one for an antistable system. We can obtain this sum decomposition by expanding the given transfer function in partial fractions.

The foregoing readily extends to the multi-input, multi-output case as follows. If an anticausal system is antistable the system's weighting matrix satisfies

$$\lim_{t \to -\infty} W(t) = \emptyset$$

and the system's A matrix has no eigenvalues in the closed left half-plane. Moreover, the system's zero state response is given by

$$y(t) = -\int_0^t W(t-\tau)u(\tau)\mathrm{d}\tau \qquad t \in (-\infty, 0]$$

$$W(t) = -Ce^{At}B \qquad t \in (-\infty, 0]$$

so that

$$\| W(t) \|_2 = \left(\operatorname{trace}\left[\int_{-\infty}^0 W^*(t)W(t)\mathrm{d}t \right] \right)^{\frac{1}{2}}$$

Finally, in summary, we have shown that the weighting matrices for causal systems which are stable satisfy

$$W(t) \in \mathcal{L}_2[0, \infty)$$

whereas the weighting matrices for anticausal systems which are antistable satisfy

$$W(t) \in \mathcal{L}_2(-\infty, 0]$$

7.3 Frequency Domain Hilbert Spaces

We begin this section by relating the time domain Hilbert space $\mathcal{L}_2(-\infty, \infty)$ to the frequency domain Hilbert space $\mathcal{L}_2$. This relation is made possible by Parseval's theorem in the general theory of the Fourier transform. Following this we introduce two frequency domain Hilbert spaces, $\mathcal{H}_2$ and $\mathcal{H}_2^\perp$ which constitute an orthogonal decomposition of $\mathcal{L}_2$ and are known as Hardy spaces.

7.3.1 *The Fourier transform*

Recall that the Fourier transform and the inverse Fourier transform are defined as

$$\mathcal{F}[f(t)] = F(j\omega) = \int_{-\infty}^{\infty} f(t)e^{-j\omega t}\mathrm{d}t \tag{7.24}$$

$$f(t) = \frac{1}{2\pi}\int_{-\infty}^{\infty} F(j\omega)e^{j\omega t}\mathrm{d}\omega \tag{7.25}$$

Also recall the following well known sufficient condition for the convergence of the integral defining the Fourier transform of $f(t)$, (7.24)

$$\int_{-\infty}^{\infty}[f^*(t)f(t)]^{\frac{1}{2}}\mathrm{d}t < \infty \tag{7.26}$$

where again $[\cdot]^{\frac{1}{2}}$ is to be interpreted as the positive square root. Notice that when $f(t)$ is a scalar, (7.26) reduces to $f(t)$ being absolute value integrable, i.e.,

$$\int_{-\infty}^{\infty} | f(t) | \,\mathrm{d}t < \infty \tag{7.27}$$

Signals satisfying (7.26), or (7.27) if appropriate, are said to be in the Lebesgue space $\mathcal{L}_1(-\infty,\infty)$.

Now we are interested in signals $f(t) \in \mathcal{L}_2(-\infty,\infty)$. Therefore we need to be concerned about the convergence of the Fourier integral, (7.24) when $f(t) \in \mathcal{L}_2(-\infty,\infty), f(t) \notin \mathcal{L}_1(-\infty,\infty)$. It turns out that the Fourier integral of signals in this latter subspace converges for allmost all ω except for, what is referred to as, a set of measure zero. This means that Parseval's theorem

$$\int_{-\infty}^{\infty} f^*(t)f(t)\mathrm{d}t = \frac{1}{2\pi}\int_{-\infty}^{\infty} F^*(j\omega)F(j\omega)\mathrm{d}\omega \tag{7.28}$$

applies to all $f(t) \in \mathcal{L}_2(-\infty,\infty)$ independent of the values assigned to $F(j\omega)$ at points ω where the Fourier integral does not converge. This fact is discussed further in the next subsection.

The important conclusion to be drawn from the foregoing is that the L_2 norm of a signal in the time domain, $\|f(t)\|_2$, equals an appropriately defined L_2 norm in the frequency domain, $\|F(j\omega)\|_2$,

$$\|f(t)\|_2 = \|F(j\omega)\|_2 \tag{7.29}$$

where

$$\|f(t)\|_2 = \left[\int_{-\infty}^{\infty} f^*(t)f(t)\mathrm{d}t\right]^{\frac{1}{2}} \tag{7.30}$$

$$\|F(j\omega)\|_2 = \left[\frac{1}{2\pi}\int_{-\infty}^{\infty} F^*(j\omega)F(j\omega)d\omega\right]^{\frac{1}{2}} \tag{7.31}$$

Notice that the definition of the frequency domain L_2 norm, (7.31) includes the scaling factor $(2\pi)^{-1}$ which is missing from the definition of the time domain L_2 norm, (7.31).

Any $F(j\omega)$ satisfying

$$\|F(j\omega)\|_2 < \infty$$

is said to belong to the L_2 frequency domain space denoted $\mathcal{L}_2$. Notice that the frequency and time domain L_2 norms are denoted by $\mathcal{L}_2$ and $\mathcal{L}_2(-\infty,\infty)$ respectively.

Now Parseval's theorem, (7.28), allows us to calculate the time domain inner product of $f_1(t), f_2(t) \in \mathcal{L}_2(-\infty, \infty)$ in the frequency domain as

$$< f_1(t), f_2(t) > = < F_1(j\omega), F_2(j\omega) > \tag{7.32}$$

where $F_i(j\omega) = \mathcal{F}[f_i(t)]$ and

$$< f_1(t), f_2(t) >= \int_{-\infty}^{\infty} f_1^*(t) f_2(t) \mathrm{d}t \tag{7.33}$$

$$< F_1(j\omega), F_2(j\omega) >= \frac{1}{2\pi} \int_{-\infty}^{\infty} F_1^*(j\omega) F_2(j\omega) \mathrm{d}\omega \tag{7.34}$$

Thus $\mathcal{L}_2$ is an inner product space which can be shown to be complete. Therefore $\mathcal{L}_2$ is a Hilbert space with inner product defined by (7.34).

The reason which make it possible for us to extend Parseval's theorem from $\mathcal{L}_1(-\infty, \infty)$ to $\mathcal{L}_2(-\infty, \infty)$ arises from the theory of Lebesgue integration which is beyound the scope of this book. However the following subsection is included in an attempt to provide some insight into this matter.

7.3.2 Convergence of the Fourier integral

Suppose $f(t)$ is specified as

$$\begin{aligned} f(t) &= (t+1)^{-1} \qquad & t \geq 0 \\ &= 0 & t < 0 \end{aligned} \tag{7.35}$$

Then

$$f(t) \in \mathcal{L}_2(-\infty, \infty) \qquad f(t) \notin \mathcal{L}_1(-\infty, \infty)$$

or more specifically

$$\int_0^\infty \left(\frac{1}{t+1}\right)^2 \mathrm{d}t < \infty \quad \text{and} \quad \int_0^\infty \frac{1}{t+1} \mathrm{d}t = \infty$$

However we can determine the Fourier transform of (7.35) at all frequencies except $\omega = 0$. To see this we need to show that

$$\int_0^\infty \frac{1}{1+t} \sin(\omega t) \mathrm{d}t < \infty \qquad \text{and} \qquad \int_0^\infty \frac{1}{1+t} \cos(\omega t) \mathrm{d}t < \infty \qquad \omega \neq 0$$

We can do this by noting that

$$\int_0^\infty \frac{1}{1+t} \sin(\omega t) \mathrm{d}t = \sum_{k=0}^{\infty} c_k \tag{7.36}$$

whcrc for $k - 0, 1, 2 \cdots$ wc havc

$$c_k = \int_{t_k}^{t_{k+1}} \frac{1}{1+t} \sin(\omega t) \mathrm{d}t$$

$$t_k = \frac{k\pi}{\omega}$$

Then since $f(t)$ is a positive monotonically decreasing function, the foregoing series is an alternating series satisfying

$$|c_k| > |c_{k+1}| \qquad \lim_{k\to\infty} c_k = 0 \qquad k = 0, 1, 2\cdots$$

which is known to converge. In a similar manner we can show that the integral involving $\cos\omega t$ converges. Therefore in this example (7.24) defines $F(j\omega)$ unambiguously for every ω except $\omega = 0$.

Now since $F(j0)$ is undefined in this example, suppose we set it to $\mathbf{\Omega}$. Then we would find that the L_2 norm $\|F(j\omega\|_2^2$ is independent of $\mathbf{\Omega}$. This fact is stated mathematically, in this case, by saying that the Fourier transform of $f(t)$ specified as (7.35) is uniquely determined for all $\omega \in (-\infty, \infty)$ except $\omega = 0$ which is a point of measure zero.

The foregoing example demonstrates the fact that, in general, when $f(t) \notin \mathcal{L}_1(-\infty, \infty)$ and $f(t) \in \mathcal{L}_2(-\infty, \infty)$ the Fourier transform $F(j\omega)$ is defined uniquely as an element of the Hilbert space $\mathcal{L}_2$ but is only determined "almost everywhere" as a point function of $j\omega$, i.e., is uniquely determined except on a set of measure zero. Therefore the Fourier transform is a unitary operator between Hilbert spaces $\mathcal{L}_2(-\infty, \infty)$ and $\mathcal{L}_2$ since for any $f_1(t), f_2(t) \in \mathcal{L}_2(-\infty, \infty)$ we have

$$< \mathcal{F}[f_1(t)], \mathcal{F}[f_2(t)] > = < f_1(t), f_2(t) > \tag{7.37}$$

where the inner product on the left is defined by (7.34) and the inner product on the right is defined by (7.33). Hilbert spaces which are related in this way are said to be isomorphic and the unitary operator between the Hilbert spaces is called an isomorphism. In the present case the isomorphism is the Fourier transform and $\mathcal{L}_2(-\infty, \infty)$ is isomorphic to $\mathcal{L}_2$.

7.3.3 The Laplace transform

In order to characterize the properties of the Fourier transforms of causal and anticausal signals in $\mathcal{L}_2[0, \infty)$ and $\mathcal{L}_2(-\infty, 0]$, respectively, we need to define two Laplace transforms, the usual one denoted $\mathcal{L}_+[f(t)]$ for causal signals and one denoted $\mathcal{L}_-[f(t)]$ for anticausal signals. These transforms are defined as

$$\mathcal{L}_+[f(t)] = \lim_{a\to 0} \int_{-a}^{\infty} f(t)e^{-st}\mathrm{d}t \quad : \ \mathrm{Re}[s] > \alpha_+ \tag{7.38}$$

$$\mathcal{L}_-[f(t)] = \lim_{a\to 0} \int_{-\infty}^{a} f(t)e^{-st}\mathrm{d}t \quad : \ \mathrm{Re}[s] < \alpha_- \tag{7.39}$$

where a, α_+, and α_- are real scalars with a being positive.

Recall that the existence of the Laplace transform, $\mathcal{L}_+[f(t)]$, requires that $f(t)$ have a lower bounded exponential order, i.e., there must exist a finite real scalar α_+, called the abscissa of convergence, such that

$$\begin{aligned} \lim_{t\to\infty} f(t)e^{-\alpha t} &= 0 \qquad \alpha > \alpha_+ \\ &= \infty \qquad \alpha < \alpha_+ \end{aligned} \tag{7.40}$$

This implies that $\mathcal{L}_+[f(t)]$ has no poles to the right of a line called the axis of convergence which is parallel to the imaginary axis and which cuts the real axis at the abscissa of convergence, $s = \alpha_+$. Now it turns out that $f(t) \in \mathcal{L}_2[0, \infty)$ then $f(t)$ satisfies (7.40) with $\alpha_+ = 0$. Therefore the integral (7.38) converges for $\text{Re}[s] > 0$ when $f(t) \in \mathcal{L}_2[0, \infty)$. This implies that $\mathcal{L}_+[f(t)]$ is devoid of poles in the open right half-plane when $f_+(t) \in \mathcal{L}_2[0, \infty)$.

We can use a similar argument to show that $\mathcal{L}_-[f(t)]$ is devoid of poles in the open left half-plane when $f(t) \in \mathcal{L}_2(-\infty, 0]$

Now suppose we are given a causal signal $f_+(t) \in \mathcal{L}_2[0, \infty)$ and an anticausal signal $f_-(t) \in \mathcal{L}_2(-\infty, 0]$. Then we have

$$f_+(t)e^{-\alpha_+ t} \in \mathcal{L}_1(-\infty, \infty) \qquad \alpha_+ \in (0, \infty)$$

$$f_-(t)e^{-\alpha_- t} \in \mathcal{L}_1(-\infty, \infty) \qquad \alpha_- \in (-\infty, 0)$$

and

$$\mathcal{F}[f_+(t)e^{-\alpha_+ t}] = \int_{-\infty}^{\infty} f_+(t)e^{-\alpha_+ t}e^{-j\omega t}\mathrm{d}t = F_+(\alpha_+ + j\omega)$$

$$\mathcal{F}[f_-(t)e^{-\alpha_- t}] = \int_{-\infty}^{\infty} f_-(t)e^{-\alpha_- t}e^{-j\omega t}\mathrm{d}t = F_-(\alpha_- + j\omega)$$

for any real scalars $\alpha_+ > 0$, $\alpha_- < 0$.

Recall that the Fourier integral, (7.24), converges for almost all ω when $f(t) \in \mathcal{L}_2(-\infty, \infty)$. Therefore if we take α_+ and α_- zero we have the Fourier transforms for $f_+(t)$ and $f_-(t)$ as

$$\begin{aligned} F_+(j\omega) &= \mathcal{L}_+[f_+(t)]|_{s=j\omega} \\ F_-(j\omega) &= \mathcal{L}_-[f_-(t)]|_{s=j\omega} \end{aligned} \tag{7.41}$$

with $\{F_+(j\omega) : \omega \in S_+\}$, $\{F_-(j\omega) : \omega \in S_-\}$ being sets of arbitrary finite elements where S_+, S_- are sets of measure zero. Then we see that $f(t) \in \mathcal{L}_2[0, \infty)$ ($f(t) \in \mathcal{L}_2(-\infty, 0]$) has a Fourier transform almost everywhere on the imaginary axis if and only if the Laplace transform $\mathcal{L}_+[f(t)]$ ($\mathcal{L}_-[f(t)]$) has no poles in the closed right (left) half-plane. Functions having these properties are said to belong to the Hardy space $\mathcal{H}_2$ ($\mathcal{H}_2^\perp$).

7.3.4 The Hardy spaces: $\mathcal{H}_2$ and $\mathcal{H}_2^\perp$

In this section we will introduce frequency domain spaces denoted $\mathcal{H}_2$ and $\mathcal{H}_2^\perp$ so that we can write any $F(j\omega) \in \mathcal{L}_2$ as

$$F(j\omega) = F_1(j\omega) + F_2(j\omega)$$

where

$$F_1(j\omega) \in \mathcal{H}_2 \qquad F_2(j\omega) \in \mathcal{H}_2^\perp$$

These spaces are called Hardy spaces after the mathematician G. H. Hardy who carried out extensive studies on these and other related spaces, e.g., $\mathcal{H}_\infty$ space.

The definition of the Hardy spaces $\mathcal{H}_2$ and $\mathcal{H}_2^\perp$ involve ideas from the theory of complex functions of a complex variable. A function of this sort is said to be analytic at a point, if it has a derivative at that point. Points where the derivative does not exist are referred to as singular points. The only singular points a rational function has are its poles.

$\mathcal{H}_2$ consists of all complex functions of a complex variable which are analytic at every point in the open right half-plane and have a finite L_2 norm, (7.31). Alternatively, the Hardy space $\mathcal{H}_2^\perp$ consists of all complex functions which are analytic in the open left half-plane and have a finite L_2 norm, (7.31). The analytic requirement in these definitions is needed to ensure that (7.31) is a well defined norm for these spaces. No function in either $\mathcal{H}_2$ or $\mathcal{H}_2^\perp$ can have imaginary axis poles.

In the rational case, $\mathcal{H}_2$ consists of all strictly proper functions which have no poles in the closed right half-plane, whereas $\mathcal{H}_2^\perp$ consists of all strictly proper functions which have no poles in the closed left half-plane. In summary, if $F(s)$ is real rational then

(i) $F(s) \in \mathcal{H}_2$ if and only if $F(s)$ is strictly proper and has no poles in the closed right half-plane.
(ii) $F(s) \in \mathcal{H}_2^\perp$ if and only if $F(s)$ is strictly proper and has no poles in the closed left half-plane.
(iii) $F(s) \in \mathcal{L}_2$ if and only if $F(s)$ is strictly proper and has no poles on the imaginary axis.

Notice that if $F(s)$ is the transfer function of some system, then in case (i) the system is stable, and in case (ii) the system is antistable.

7.3.5 Decomposing $\mathcal{L}_2$ space

Recall that the L_2 time domain Hilbert spaces decompose as

$$\mathcal{L}_2(-\infty,\infty) = \mathcal{L}_2(-\infty,0] \oplus \mathcal{L}_2[0,\infty)$$

Then since the Fourier transform maps $\mathcal{L}_2[0,\infty)$ onto $\mathcal{H}_2$ and $\mathcal{L}_2(-\infty,0]$ onto $\mathcal{H}_2^\perp$ and the same inner product is used for $\mathcal{L}_2, \mathcal{H}_2$ and $\mathcal{H}_2^\perp$, we have

$$\mathcal{L}_2 = \mathcal{H}_2 \oplus \mathcal{H}_2^\perp \tag{7.42}$$

and the Fourier transform is an isomorphism such that

$$\begin{array}{lll} \mathcal{L}_2(-\infty,\infty) & \text{is isomorphic to} & \mathcal{L}_2 \\ \mathcal{L}_2[0,\infty) & \text{is isomorphic to} & \mathcal{H}_2 \\ \mathcal{L}_2(-\infty,0] & \text{is isomorphic to} & \mathcal{H}_2^\perp \end{array}$$

As an illustration, suppose we are given the following signal

$$\begin{aligned} f(t) &= e^{-at} \qquad t \geq 0 \\ &= e^{bt} \qquad t \leq 0 \end{aligned}$$

where a, b are real and positive.

Then we can write $f(t)$ as

$$f(t) = f_+(t) + f_-(t)$$

where

$$\begin{aligned} f_+(t) &= e^{-at} \quad & f_-(t) &= 0 \quad & \text{for } t \geq 0 \\ f_+(t) &= 0 \quad & f_-(t) &= e^{bt} \quad & \text{for } t \leq 0 \end{aligned}$$

Since the Laplace transforms of these signals are given by

$$\mathcal{L}_+[f_+(t)] = \frac{1}{s+a} \qquad \text{Re}[s] > -a$$

$$\mathcal{L}_-[f_-(t)] = \frac{1}{s-b} \qquad \text{Re}[s] < b$$

we see that both $\mathcal{L}_+[f_+(t)]$ and $\mathcal{L}_-[f_-(t)]$ are free of poles on the imaginary axis. Therefore $f(t)$ has a Fourier transform which can be written as

$$\mathcal{F}[f(t)] = \frac{1}{j\omega - b} + \frac{1}{j\omega + a}$$

However $\mathcal{L}_+[f_+(t)]$ and $\mathcal{L}_-[f_-(t)]$ are each zero at $\omega = \infty$ with $\mathcal{L}_+[f_+(t)]$ being analytic in the closed right half-plane and $\mathcal{L}_-[f_-(t)]$ being analytic in the closed left half-plane. Therefore we have

$$\mathcal{L}_+[f_+(t)] \in \mathcal{H}_2 \qquad \mathcal{L}_-[f_-(t)] \in \mathcal{H}_2^\perp \qquad \mathcal{F}[f(t)] \in \mathcal{L}_2$$

The foregoing example demonstrate that, in general, systems with transfer function satisfying $G(j\omega) \in \mathcal{L}_2$ can be written as

$$G(s) = G_1(s) + G_2(s)$$

where

$$G_1(s) \in \mathcal{H}_2 \qquad G_2(s) \in \mathcal{H}_2^\perp$$

7.3.6 The H_2 system norm

Recall from Section 7.2.2 that when the state model for a given system is minimal, the time domain L_2 norm of the weighting matrix is finite only if A is stable and D is null. Therefore in this case the system's transfer function

$$G(s) = \mathcal{L}_+\left[Ce^{At}B\right] = C(sI - A)^{-1}B$$

is strictly proper, i.e., $G(\infty) = 0$ and has no poles in the closed right half-plane. Thus we have $G(s) \in \mathcal{H}_2$ and using Parseval's theorem we have

$$\|Ce^{At}B\|_2 = \left(\text{trace}\left[\int_0^\infty (Ce^{At}B)^*(Ce^{At}B)\mathrm{d}t\right]\right)^{\frac{1}{2}}$$

$$= \left(\text{trace}\left[\frac{1}{2\pi}\int_{-\infty}^{\infty} G^*(j\omega)G(j\omega)d\omega\right]\right)^{\frac{1}{2}} = \|G(j\omega\|_2 \tag{7.43}$$

Thus the time domain L_2 norm of the system's weighting matrix equals the frequency domain L_2 norm of the system's transfer function, $G(s)$. Moreover, since in this case we have $G(s) \in \mathcal{H}_2$, this norm is usually referred to as the system's H_2 norm.

Notice that if $G(s), U(s) \in \mathcal{H}_2$ with $G(s)$ rational, then $G(s)U(s) \in \mathcal{H}_2$ since each term in the product is analytic in the open right half-plane has no poles on the imaginary axis with the product being zero at infinity. Thus the zero state response from a stable causal system having a strictly proper transfer function satisfies $Y(s) \in \mathcal{H}_2$ or $y(t) \in \mathcal{L}_2[0,\infty)$ for any input $u(t) \in \mathcal{L}_2[0,\infty)$.

In order to extend this result to the case where the transfer function is only proper we need another Hardy space. In the next section we will see that transfer functions of stable systems which are only proper, i.e., $G(\infty) \neq 0$, lie in a Hardy space denoted $\mathcal{H}_\infty$. Thus if $U(s) \in \mathcal{H}_2$ and $G(s) \in \mathcal{H}_\infty$ then the product $G(s)U(s)$ is analytic in the open right half-plane and has no poles on the imaginary axis. Clearly, in the case when $U(s)$ is rational we have $G(\infty)U(\infty) = 0$ and $Y(s) \in \mathcal{H}_2$ or $y(t) \in \mathcal{L}_2[0,\infty)$ for any input $u(t) \in \mathcal{L}_2[0,\infty)$. It turns out that this result holds in the more general case when $U(s)$ is irrational. Thus we have

$$Y(s) = G(s)U(s) \in \mathcal{H}_2 \tag{7.44}$$

when

$$G(s) \in \mathcal{H}_\infty \qquad U(s) \in \mathcal{H}_2$$

Just as there is a frequency domain norm for functions in either $\mathcal{H}_2$ or $\mathcal{H}_2^\perp$, namely the L_2 norm, we will show in the next section that there is a norm for functions in $\mathcal{H}_\infty$, called the H_∞ norm. Thus $\mathcal{H}_\infty$ is a normed space. However, unlike $\mathcal{H}_2$ or $\mathcal{H}_2^\perp$ there is no inner product defined on $\mathcal{H}_\infty$. Therefore $\mathcal{H}_\infty$ is not a Hilbert space.

Now the space of rational transfer functions that have no poles in the closed left half-plane and are finite at infinity is denoted $\mathcal{H}_\infty^-$. Notice that we do not use $\mathcal{H}_\infty^\perp$ to denote this space since the concept of orthogonality is missing. Therefore in addition to (7.44) we have the dual result

$$Y(s) = G(s)U(s) \in \mathcal{H}_2^\perp \tag{7.45}$$

when

$$G(s) \in \mathcal{H}_\infty^- \qquad U(s) \in \mathcal{H}_2^\perp$$

We will see in the next three chapters that the Hardy spaces, $\mathcal{H}_\infty$, $\mathcal{H}_\infty^-$, $\mathcal{H}_2$ and $\mathcal{H}_2^\perp$, that we are examining in this chapter are essential to the solution of the $\mathcal{H}_\infty$ control problem. This solution consists of a feedback controller which stabilizes the plant and constrains the H_∞ norm of the plant's transfer function, from disturbance input to desired output, to be less than a specified scalar. The H_∞ norm is introduced now as follows.

7.4 The H_∞ Norm: SISO Systems

At the end of the previous section we introduced the H_2 norm of a system's transfer function, $\|G(s)\|_2$, by relating it to the time domain L_2 norm of the system's impulse response, $Ce^{At}B$. In this section we introduce another type of frequency domain system norm referred to as the "H infinity norm" denoted $\|G(s)\|_\infty$. We do this now by relating $\|G(s)\|_\infty$ to an equivalent time domain norm referred to as the "L_2 system gain".

Suppose we are given a stable single-input, single-output system having zero state response $y(t)$ when the input is $u(t)$. Then we define the L_2 system gain, γ_0, as

$$\gamma_0 = \sup_{\substack{u(t)\in\mathcal{L}_2[0,\infty) \\ u(t)\neq 0}} \frac{\|y(t)\|_2}{\|u(t)\|_2} \tag{7.46}$$

where "sup" is the abbreviation for supremium or smallest upper bound of a set, which in this case is the set of positive real numbers generated by the ratio of the L_2 norms in (7.46) as $u(t)$ is varied over $\mathcal{L}_2[0,\infty)$. Notice that "sup" is used instead of "max" in case the set over which the search is carried out does not contain a maximal element. In addition, notice that we do not consider the null input to avoid dividing by zero.

Now the use of a signal norm to define a system norm as in (7.46) is referred to by saying that the signal norm induces the system norm. Therefore the L_2 system gain, (7.46), is induced by the L_2 signal norm. Notice that, for a given system, this norm is, in general, not equal to the L_2 norm of the system's impulse response, $Ce^{At}B$.

Since we are assuming the system is stable, we saw in the previous section that $y(t)\in\mathcal{L}_2[0,\infty)$ when $u(t)\in\mathcal{L}_2[0,\infty)$. In addition we can use the fact that $\mathcal{L}_2[0,\infty)$ is isomorphic to $\mathcal{H}_2$ to rewrite the L_2 system gain, (7.46) in the frequency domain as

$$\gamma_0 = \sup_{\substack{U(s)\in\mathcal{H}_2 \\ U(s)\neq 0}} \frac{\|Y(s)\|_2}{\|U(s)\|_2} \tag{7.47}$$

Therefore comparing (7.46) and (7.47) we see that the L_2 system gain in the time domain is equivalent to the frequency domain system norm induced by the H_2 norm. This norm is referred to as the "H infinity norm", denoted as $\|G(s)\|_\infty$. Notice that this norm and the H_2 norm (Section 7.3.6), for a given system are, in general, not equal.

7.4.1 *Transfer function characterization of the H_∞ norm*

Suppose we are given a stable system and that we can choose the system input so that the ratio in the definition of the L_2 system gain, (7.46), is close to being maximized. More specifically suppose the zero state response, $y_{\text{opt}}(t)$, caused by the input $u_{\text{opt}}(t)$ satisfies

$$\gamma_0 = \frac{\|y_{\text{opt}}(t)\|_2}{\|u_{\text{opt}}(t)\|_2} + \epsilon \tag{7.48}$$

where ϵ is a small positive scalar in the sense that $\epsilon \ll \gamma_0$.

Then recalling that linear systems satisfy the principle of superposition, we have $\alpha y_{\text{opt}}(t)$ when $\alpha u_{\text{opt}}(t)$ for any constant α. Therefore (7.48) can be written as

$$\gamma_0 = \|y_{\text{nor}}(t)\|_2 + \epsilon$$

where

$$y_{\text{nor}}(t) = \alpha y_{\text{opt}}(t) \qquad \alpha = \frac{1}{\|u_{\text{opt}}(t)\|_2}$$

This shows that we can obtain the L_2 system gain, γ_0, (7.46), or H_∞ system norm, $\|G(s)\|_\infty$, (7.47), by restricting $u(t)$ or $U(s)$ to satisfy the following conditions:

$$\begin{array}{lll} \text{(i)} \;\; u(t) \in \mathcal{L}_2[0,\infty) & \text{or} & U(s) \in \mathcal{H}_2 \\ \text{(ii)} \;\; \|u(t)\|_2 = 1 & \text{or} & \|U(s)\|_2 = 1 \end{array}$$

Thus (7.46, 7.47) become

$$\begin{aligned} \gamma_0 &= \sup_{\substack{U(s)\in\mathcal{H}_2 \\ \|u(t)\|_2=1}} \|y(t)\|_2 = \sup_{\substack{U(s)\in\mathcal{H}_2 \\ \|U(s)\|_2=1}} \|Y(s)\|_2 \\ &= \sup_{\substack{U(s)\in\mathcal{H}_2 \\ \|U(s)\|_2=1}} \left(\frac{1}{2\pi}\int_{-\infty}^{\infty} |G(j\omega)U(j\omega)|^2 d\omega\right)^{\frac{1}{2}} \end{aligned} \tag{7.49}$$

Notice that

$$\frac{1}{2\pi}\int_{-\infty}^{\infty} |G(j\omega)U(j\omega)|^2 d\omega \le \frac{1}{2\pi}\int_{-\infty}^{\infty} \sigma_{\max}^2 |U(j\omega)|^2 d\omega \tag{7.50}$$

where

$$\sigma_{\max} = \sup_{\omega\in(-\infty,\infty)} |G(j\omega)| \tag{7.51}$$

and $\sigma_{\max}$ is referred to as the L_∞ norm of the function $G(j\omega)$.

However, since $U(s)$ is restricted in (7.49) as

$$\|U(s)\|_2 = \left(\frac{1}{2\pi}\int_{-\infty}^{\infty} |U(j\omega)|^2 d\omega\right)^{\frac{1}{2}} = 1 \tag{7.52}$$

we see that (7.50) becomes

$$\frac{1}{2\pi}\int_{-\infty}^{\infty} |G(j\omega)U(j\omega)|^2 d\omega \le \sigma_{\max}^2 \tag{7.53}$$

Now it turns out that we can choose $U(j\omega)$ in (7.53) subject to (7.52) so that the left side of (7.53) is arbitrarily close to $\sigma_{\max}^2$. This implies that the supremium in (7.49) is $\sigma_{\max}$ and we have

$$\begin{aligned}\gamma_0 &= \sigma_{\max}\\ &= \sup_{\omega\in(-\infty,\infty)} |G(j\omega)| = \|G(j\omega)\|_\infty\end{aligned} \tag{7.54}$$

Notice that

$$\sup_{\omega\in(-\infty,\infty)} |G(j\omega)| < \infty$$

provided $G(s)$ has no imaginary axis poles. However we have

$$\sup_{\substack{u(t)\in\mathcal{L}_2[0,\infty)\\ u(t)\neq 0}} \frac{\|y(t)\|_2}{\|u(t)\|_2} < \infty$$

only if $G(s)$ has no poles in the closed right half-plane, since $y(t)$ is unbounded for some $u(t) \in \mathcal{L}_2[0,\infty)$ otherwise. Therefore $\|G(j\omega)\|_\infty$ is referred to as:

(i) the L_∞ norm of $G(s)$ when the system is unstable and $G(s)$ has no poles on the imaginary axis including $s = \infty$;
(ii) the H_∞ norm of $G(s)$ when the system is stable and $G(\infty) < \infty$.

In addition, transfer functions that have bounded H_∞ (L_∞) norm form a normed space denoted $\mathcal{H}_\infty$ ($\mathcal{L}_\infty$). Notice that $\mathcal{H}_\infty \subset \mathcal{L}_\infty$. More is said about this in the next subsection.

Now there are two different ways of interpreting $\|G(j\omega)\|_\infty$. One arises from defining $\|G(j\omega)\|_\infty$ as

$$\|G(j\omega)\|_\infty = \sup_{\substack{U(j\omega)\in\mathcal{L}_2\\ U(j\omega)\neq 0}} \frac{\|Y(j\omega)\|_2}{\|U(j\omega)\|_2}$$

Then in this case we are considering $G(j\omega)$ to be an operator which maps $U(j\omega) \in \mathcal{L}_2$ to $Y(j\omega) \in \mathcal{L}_2$ and $\|G(j\omega)\|_\infty$ to be the operator norm induced by the frequency domain L_2 norm or, if the system is stable, by the time domain L_2 norm.

Alternatively, if we take (7.54) to be the definition of $\|G(j\omega)\|_\infty$ then we are considering $G(j\omega)$ to be a function of ω and $\|G(j\omega)\|_\infty$ is a function norm.

7.4.2 Transfer function spaces

Since any proper rational transfer function with no imaginary axis poles has a finite L_∞ norm, we can define three normed transfer function spaces depending on the location of the transfer function's poles. Therefore assuming the rational function $G(s)$ is proper we have

(i) $G(s) \in \mathcal{H}_\infty$ when all poles of $G(s)$ are in the open left half-plane
(ii) $G(s) \in \mathcal{H}_\infty^-$ when all poles of $G(s)$ are in the open right half-plane
(iii) $G(s) \in \mathcal{L}_\infty$ when $G(s)$ has no poles on the imaginary axis.

Notice that when $G(s) \in \mathcal{L}_\infty$ we can use the partial fraction expansion (after one cycle of division) to write $G(s)$ as

$$G(s) = G^+(s) + G^-(s)$$

where

$$G^+(s) \in \mathcal{H}_\infty \qquad G^-(s) \in \mathcal{H}_\infty^-$$

Then assuming some scheme for making $G^+(s), G^-(s)$ unique, e.g., we can insist that $G^-(s)$ be strictly proper, we can capture this spliting of a transfer function in $\mathcal{L}_\infty$ into the sum of stable and antistable transfer functions as a direct sum decomposition of the spaces involved

$$\mathcal{L}_\infty = \mathcal{H}_\infty \oplus \mathcal{H}_\infty^- \tag{7.55}$$

Alternatively, suppose two stable systems having transfer functions $G_1(s), G_2(s) \in \mathcal{H}_\infty$ form the components of a cascade connection. Then the transfer function of the composite system, $G_1(s)G_2(s)$, is stable since its poles are contained in the poles of the component transfer functions. In addition, the transfer function of the composite system is proper or strictly proper since the product of proper rational functions is proper independent of any pole-zero cancellations that might occur. Therefore $G_1(s)G_2(s) \in \mathcal{H}_\infty$ if $G_1(s), G_2(s) \in \mathcal{H}_\infty$.

Now we can readily show, from the definition of the H_∞ norm, (7.47), that the H_∞ norm of the transfer function of the cascade connected system is related to the H_∞ norms of the transfer functions of the component systems as

$$\|G_1(s)\|_\infty \|G_2(s)\|_\infty \geq \|G_1(s)G_2(s)\|_\infty \tag{7.56}$$

This result plays an important role in control theory. An example of its use is given in the following subsection.

7.4.3 The small gain theorem

Prior to the use of the H_∞ norm in control theory, the H_∞ norm appeared in a number of different guises in various engineering applications. For example, the resonant gain used to characterize the performance of an electric filter is the H_∞ norm of the transfer function relating the filter's input and output voltages. Another instance of this occurs when the Nyquist stability criterion is applied to determine the stability of a feedback system having transfer function $T(s)$

$$T(s) = \frac{G(s)}{1 + G(s)H(s)}$$

Then when $G(s), H(s) \in \mathcal{H}_\infty$ the distance from the origin to the point on the polar plot of $G(j\omega)H(j\omega)$ which is furthest from the origin is $\|G(s)H(s)\|_\infty$. This geometrical observation leads to the following result which is known as the small gain theorem.

Recall that when $G(s)H(s) \in \mathcal{H}_\infty$, the Nyquist criterion states that the feedback system is stable if and only if the polar plot of $G(j\omega)H(j\omega)$ does not encircle the point

$-1+j0$. Therefore since the foregoing geometrical interpretation of $\|G(s)H(s)\|_\infty$ implies that the polar plot of $G(j\omega)H(j\omega)$ cannot have any encirclements of $-1+j0$ if

$$\|G(s)H(s)\|_\infty < 1 \tag{7.57}$$

we have (7.57) as a sufficient condition for the feedback system to be stable when $G(s)H(s) \in \mathcal{H}_\infty$. This result, referred to as the small gain theorem, is important in connection with robust stabilization where a single fixed controller is required to stabilize any one plant in a set of plants. A simplified example of the application of the small gain theorem in this context is given as follows.

Suppose we are given a fixed controller having transfer function $H(s) \in \mathcal{H}_\infty$ which satisfies

$$\|H(s)\|_\infty \leq \alpha \tag{7.58}$$

Then using (7.56) we can show that (7.57) is satisfied if

$$\|G(s)\|_\infty < \frac{1}{\alpha} \tag{7.59}$$

Therefore any stable plant satisfying (7.59) is stabilized by the given fixed controller. Notice that by interchanging the role of the controller and plant, the foregoing development leads to a characterization of a set of stable controllers any one of which can be used to stabilize a given fixed stable plant.

This completes the development of the H_∞ (L_∞) norm for single-input, single-output systems. We need now to proceed to develop this norm for multi-input, multi-output systems.

7.5 The H_∞ Norm: MIMO Systems

We have just seen that the H_∞ norm of a single-input, single-output system is the H_2 induced norm of the system's transfer function, $G(s)$. In order to extend this idea to the multi-input, multi-output case, where $G(s)$ is a matrix, we need to introduce the induced 2-norm for constant matrices, denoted as $\|G(j\omega)\|$ (for fixed ω). Then we can obtain the H_∞ norm of the matrix $G(j\omega)$ as the largest induced 2-norm as ω varies over $(-\infty, \infty)$, i.e.,

$$\|G(j\omega)\|_\infty = \sup_{\omega \in (-\infty,\infty)} (\|G(j\omega)\|)$$

7.5.1 Singular value decomposition

Before we can develop the induced 2-norm for a constant matrix we need to introduce the singular value decomposition (SVD) of a constant matrix. This is done as follows.

Recall, from Chapter 4, that Hermitian matrices have real eigenvalues and orthogonal eigenvectors. Moreover notice that the product matrices M^*M and MM^* formed from any complex matrix M, are Hermitian and nonnegative. Therefore these product matrices have orthogonal eigenvectors and real nonnegative eigenvalues. This general observation leads to the matrix analysis technique known as SVD. The following theorem, which is proved in the appendix, defines the SVD of a matrix.

Theorem 7.1 Any $p \times m$ matrix of complex constants, M, which has rank r can be decomposed as

$$M = U\Sigma V^* \tag{7.60}$$

where U, and V are $p \times p$ and $m \times m$ unitary matrices with

$$U = [\,U_1 \quad U_2\,] \qquad \Sigma = \begin{bmatrix} \Sigma_0 & \emptyset \\ \emptyset & \emptyset \end{bmatrix} \qquad V = [\,V_1 \quad V_2\,]$$

$$U_1 \text{ is } p \times r \qquad U_2 \text{ is } p \times p - r$$

$$V_1 \text{ is } m \times r \qquad V_2 \text{ is } m \times m - r$$

and Σ_0 is an $r \times r$, real, positive definite, diagonal matrix denoted as

$$\Sigma_0 = \begin{bmatrix} \sigma_1 & 0 & \dots & 0 \\ 0 & \sigma_2 & \dots & 0 \\ \vdots & \vdots & \ddots & \vdots \\ 0 & 0 & \dots & \sigma_r \end{bmatrix} = \text{diag}[\sigma_1, \sigma_2, \cdots, \sigma_r]$$

with diagonal entries referred to as singular values and ordered so that

$$\sigma_i \geq \sigma_{i+1} \qquad i = 1, 2, \cdots, r-1$$

Proof See Appendix

The SVD of a matrix has become ever more useful since the establishment of practical methods for its computation in the early 1970s. The SVD is of practical use wherever we require the solution of problems involving matrices having small rank. One problem of this sort, which we encountered in Chapter 4, involves the determination of a lower order state model approximation of a state model having controllability and/or observability matrices which are almost rank deficient, i.e., the state model is almost uncontrollable and/or unobservable. We were able to obtain a solution to this problem without requiring the SVD by using the state model's Gramians. Since the Gramians are nonnegative Hermitian matrices, they enjoy the nice properties which make the SVD so useful, namely, real and nonnegative eigenvalues and mutually orthogonal eigenvectors.

7.5.2 Induced 2-norm for constant matrices

We now proceed with the development of the constant matrix norm induced by the 2-norm for constant vectors. This is done in the proof of the following theorem by employing the SVD and its properties.

Theorem 7.2 Any constant $p \times m$ matrix M has a norm induced by the 2-norm for constant vectors which equals σ_1, the largest singular value of M.

Proof The induced 2-norm for a matrix M, $\|M\|$, is defined as

$$\|M\| = \sup_{\substack{x \in \mathcal{C}^m \\ x \neq \emptyset}} \frac{\|Mx\|}{\|x\|} \tag{7.61}$$

where $\mathcal{C}^m$ is the vector space of all vectors of length m having constant complex entries and $\|x\|$ equals the positive square root of the sum of the squared absolute value of each entry in x, i.e.,

$$\|x\| = \left(\sum_{i=1}^{m} x_i^* x_i \right)^{\frac{1}{2}} = (x^* x)^{\frac{1}{2}} \tag{7.62}$$

Notice that unlike the L_2 norm for a time varying vector which we used earlier, the 2-norm for a constant vector has no subscript, i.e., $\|\cdot\|_2$ denotes the L_2 norm whereas $\|\cdot\|$ denotes the 2-norm.

Next since the transformation Mx is linear, multiplying x by a constant scalar changes both the numerator and the denominator in (7.61) by the same amount so that the ratio in (7.61) remains unchanged. Therefore we have the alternative definition of the induced 2-norm of M given by

$$\|M\| = \sup_{\substack{x \in \mathcal{C}^m \\ \|x\|=1}} \|Mx\| \tag{7.63}$$

Since the $m \times m$ matrix V in the SVD of M, (7.60), is invertible, we can use its columns $\{v^i : i = 1, 2, \cdots m\}$ as a basis for $\mathcal{C}^m$. Thus we can express any $x \in \mathcal{C}^m$ as

$$x = \sum_{i=0}^{m} \alpha_i v^i = V\alpha \tag{7.64}$$

where

$$\alpha^T = [\alpha_1 \quad \alpha_2 \quad \cdots \quad \alpha_m]$$

with $\alpha \in \mathcal{C}^m$.

Now, since V is unitary we have

$$\begin{aligned} \|x\|^2 &= \alpha^* V^* V\alpha \\ &= \alpha^* \alpha = \sum_{i=0}^{m} |\alpha_i|^2 = \|\alpha\|^2 \end{aligned} \tag{7.65}$$

and the 2-norms of x and of α are equal. In addition since V is invertible there is a unique α for each x and vice versa so that

$$\sup_{\substack{x \in \mathcal{C}^m \\ \|x\|=1}} \|Mx\| = \sup_{\substack{\alpha \in \mathcal{C}^m \\ \|\alpha\|=1}} \|MV\alpha\| \tag{7.66}$$

Then using the SVD of M, (7.60) we can write

$$\|MV\alpha\| = (\alpha^* V^* M^* M V \alpha)^{\frac{1}{2}}$$

$$= (\alpha^* \Sigma_m^2 \alpha)^{\frac{1}{2}} = \left(\sum_{i=1}^{r} \sigma_i^2 |\alpha_i|^2 \right) \tag{7.67}$$

where Σ_m^2 is the $m \times m$ diagonal matrix given by

$$\Sigma_m^2 = \begin{bmatrix} \Sigma_0^2 & \emptyset \\ \emptyset & \emptyset \end{bmatrix} \qquad \Sigma_0^2 = \text{diag}[\sigma_1^2, \sigma_2^2, \cdots, \sigma_r^2]$$

Therefore from (7.63), (7.66) and (7.67) we have

$$\|M\| = \left[\sup_{\substack{\alpha \in C^m \\ \|\alpha\|=1}} \sum_{i=1}^{r} \sigma_i^2 |\alpha_i|^2 \right]^{\frac{1}{2}} \tag{7.68}$$

Now the summation on right side of the foregoing equation satisfies the following inequality

$$\sum_{i=1}^{r} \sigma_i^2 |\alpha_i|^2 \leq \sigma_1^2 \sum_{i=1}^{r} |\alpha_i|^2 \tag{7.69}$$

where σ_1 is the largest singular value of M. However since α is restricted in (7.68) to have unit norm, we see from (7.69) that

$$\|M\| \leq \sigma_1 \tag{7.70}$$

Notice that if the components of α satisfy

$$\alpha_1 = 1 \qquad \alpha_i = 0 \qquad i \geq 2$$

then $\|\alpha\| = 1$ and

$$\sum_{i=1}^{r} \sigma_i^2 |\alpha_i|^2 = \sigma_1^2$$

which determines the supremium in (7.68) as

$$\|M\| = \sigma_1$$

■

Notice, in the foregoing proof, that the supremium is achieved for $\alpha = I^1$, the first column of the $m \times m$ identity matrix. Therefore, letting $\alpha = I^1$ in (7.64) gives the result that the unit 2-norm x that gives the supremium in (7.63) is the right singular vector of M, v^1, corresponding to the largest singular value of M, σ_1. Thus

$$\frac{\|Mx\|}{\|x\|} = \sigma_1$$

when $x = \beta v_1$ for any complex scalar, β. Further consideration of this fact leads us to the important conclusion that

$$\|Mx\| \leq \sigma_1 \|x\| \qquad \text{for all } x \in \mathcal{C}^m \tag{7.71}$$

with equality only when x is a scalar multiple of v_1. Finally, notice that $\|M\| = |M|$ when M is a scalar.

7.5.3 The L_∞, H_∞ norm for transfer function matrices

Having developed the idea of the induced 2-norm for constant matrices, we are now in a position to give the definition of the L_∞ norm of a transfer function matrix, $G(j\omega)$. This norm is defined as the supremium of the induced 2-norm of $G(j\omega)$ over all $\omega \in (-\infty, \infty)$

$$\|G(j\omega)\|_\infty = \sup_{\omega \in (-\infty,\infty)} \sigma_1(j\omega) \tag{7.72}$$

where $\sigma_1(j\omega)$ is the largest singular value of $G(j\omega)$.

As in the SISO case, when an MIMO system is stable its transfer function matrix $G(s)$ has finite L_∞ norm, which is referred to as the H_∞ norm and $G(s)$ is lies in the normed space denoted by $\mathcal{H}_\infty$. Alternatively, when the system is antistable, its transfer function matrix $G(s)$ has finite L_∞ norm, (7.72), and $G(s)$ lies in the normed space denoted $\mathcal{H}_\infty^-$. Finally notice that, as in the SISO case, all $G(s)$ with finite L_∞ norm are denoted by $G(s) \in \mathcal{L}_\infty$. Thus $\mathcal{H}_\infty$ and $\mathcal{H}_\infty^-$ are each subspaces of $\mathcal{L}_\infty$ which are disjoint and complete the $\mathcal{L}_\infty$ space so that (7.55) holds for MIMO systems.

In subsequent development we will need the following result which is an immediate consequence of the foregoing.

Theorem 7.3 A real rational transfer function $G(s) \in \mathcal{L}_\infty$ has an L_∞ norm less than a finite positive scalar γ, i.e.,

$$\|G(j\omega)\|_\infty < \gamma$$

if and only if $F(j\omega)$ is positive definite for all ω, i.e.,

$$F(j\omega) > 0 \qquad \forall \omega \in (-\infty, \infty)$$

where

$$F(j\omega) = \gamma^2 I - G^*(j\omega)G(j\omega)$$

Proof Let the SVD (Theorem 7.1) of the $p \times m$ matrix $G(j\omega)$ be given as

$$G(j\omega) = U(j\omega)\Sigma(j\omega)V^*(j\omega)$$

where $U(j\omega), V(j\omega)$ are $p \times p$ and $m \times m$ unitary matrices respectively and

$$\Sigma(j\omega) = \begin{bmatrix} \Sigma_0(j\omega) & \emptyset \\ \emptyset & \emptyset \end{bmatrix} \qquad \Sigma_0(j\omega) = \text{diag}[\sigma_1(j\omega), \sigma_2(j\omega), \cdots, \sigma_r(j\omega)]$$

with $r = \text{rank}[G(j\omega]$ and

$$\sigma_1(j\omega) \geq \sigma_i(j\omega) \qquad i > 1 \tag{7.73}$$

Then using the SVD of $G(j\omega)$ we can write $F(j\omega)$ as

$$F(j\omega) = \gamma^2 I_m - V(j\omega)\Sigma_m^2(j\omega)V^*(j\omega) \tag{7.74}$$

where $\Sigma_m^2(j\omega)$ is the m dimensional diagonal matrix

$$\Sigma_m^2(j\omega) = \text{diag}\left[\sigma_1^2(j\omega), \sigma_2^2(j\omega), \cdots, \sigma_r^2(j\omega), 0, \cdots, 0\right]$$

Next pre and post multiplying (7.74) by $V^*(j\omega)$ and $V(j\omega)$ respectively gives

$$V^*(j\omega)F(j\omega)V(j\omega) = \Sigma_F^2 \tag{7.75}$$

where

$$\Sigma_F^2(j\omega) = \text{diag}\left[(\gamma^2 - \sigma_1^2(j\omega)), (\gamma^2 - \sigma_2^2(j\omega)), \cdots, (\gamma^2 - \sigma_r^2(j\omega), \gamma^2, \cdots \gamma^2\right]$$

Now since $V(j\omega)$ is unitary and therefore invertible, we have

$$F(j\omega) > 0 \qquad \text{if and only if} \qquad V^*(j\omega)F(j\omega)V(j\omega) > 0$$

Thus we see from (7.75) that

$$F(j\omega) > 0 \qquad \text{if and only if} \qquad \left[\gamma^2 - \sigma_1^2(j\omega)\right] > 0 \tag{7.76}$$

However, since the L_∞, norm for $G(s)$ is defined as

$$\|G(j\omega)\|_\infty = \sup_{\omega \in (-\infty,\infty)} \sigma_1(j\omega)$$

we see from (7.76) that

$$F(j\omega) > 0 \quad \text{for all } \omega \in (-\infty, \infty) \qquad \text{if and only if} \qquad \|G(j\omega)\|_\infty < \gamma$$

and the theorem is proved. ■

In the next chapter we will use the foregoing theorem to obtain necessary and sufficient conditions on the state model parameters, $\{A, B, C, D\}$, of a system which ensures that the system's transfer function has L_∞ norm less than a given finite real scalar.

7.6 Summary

We began by formalizing the idea, used earlier in Chapters 5 and 6, of relating the size of a signal to the energy in the signal. This led to the time domain Hilbert space $\mathcal{L}_2(-\infty, \infty)$ and its orthogonal subspaces $\mathcal{L}_2(-\infty, 0]$ and $\mathcal{L}_2[0, \infty)$. Following this we used the Hilbert space isomorphism from $\mathcal{L}_2(-\infty, \infty)$ in the time domain to $\mathcal{L}_2$ in the frequency domain to introduce the frequency domain spaces $\mathcal{H}_2, \mathcal{H}_2^\perp$ known as Hardy spaces.

Next we introduced the notion of the size of a system. This was done by using the induced operator norm to characterize a system's ability to transfer energy from its input to its output. As with signal norms we saw that there is an equivalence between a time domain system norm known as the system's L_2 gain and a frequency domain norm known as the H_∞ norm. Unlike the Hardy spaces, $\mathcal{H}_2, \mathcal{H}_2^\perp$ for signals which are equipped with an inner product, the Hardy spaces, $\mathcal{H}_\infty, \mathcal{H}_\infty^-$ for systems are normed spaces only.

7.7 Notes and References

There are many texts on Hilbert space. A readily accessible treatment of this subject which includes material related to the control problems being considered here can be found in [46]. Unfortunately books on Hardy spaces require a considerable background in functional analysis and complex analysis. Some of these references are given in [14, p. 13]. An interesting treatment of complex analysis as it applies to Laplace and Fourier transforms is given in [26].

The example following (7.27) in Section 7.3.2 which is used to discuss the Fourier transform of signals in $\mathcal{L}_2$ that are not in $\mathcal{L}_1$ is taken from [45, p. 272]. The convergence of the series (7.36) is discussed in [36, p. 71]. The applicability of Parseval's theorem for functions in $\mathcal{L}_2(-\infty, \infty)$ is discussed in [37, p. 185]. An excellent reference for the SVD is [17].

The small gain theorem is used to solve a wide variety of robust control problems including both linear, [47] [18] and nonlinear, [42] [21] [33], systems. Finally a good basic introduction to the ideas in this and the next chapter can be found in [11].

8

System Algebra

8.1 Introduction

In the previous chapter we were interested in characterizing signals and systems in terms of normed spaces in both the time and frequency domains. This led to the introduction of the frequency domain spaces $\mathcal{H}_2, \mathcal{H}_2^{\perp}$ and $\mathcal{L}_2$ for signals and $\mathcal{H}_\infty, \mathcal{H}_\infty^-$ and $\mathcal{L}_\infty$ for systems.

In the present chapter we consider a number of operations involving systems in these spaces. We will be especially interested in obtaining state models for the systems which result from these operations. To facilitate this endeavour, we use the compact equivalence relation

$$G(s) \stackrel{s}{=} \begin{bmatrix} A & B \\ C & D \end{bmatrix} \tag{8.1}$$

to denote that

$$G(s) = C(sI - A)^{-1}B + D$$

We begin the discussion of these operations by considering the effect of connecting systems in parallel and in series.

8.1.1 Parallel connection

One of the simplest examples of operations with systems consists of connecting several systems in parallel so that all systems have the same input and the output from the connection is the sum of the outputs from each system. The connection of the first order component systems resulting from a partial fraction expansion of the system's transfer function is an example of the parallel connection of systems.

The input–output constraints which ensure that r component systems are connected in parallel are

$$u(t) = u_i(t) \qquad i = 1, 2, \cdots r \tag{8.2}$$

$$y(t) = \sum_{i=1}^{r} y_i(t) \tag{8.3}$$

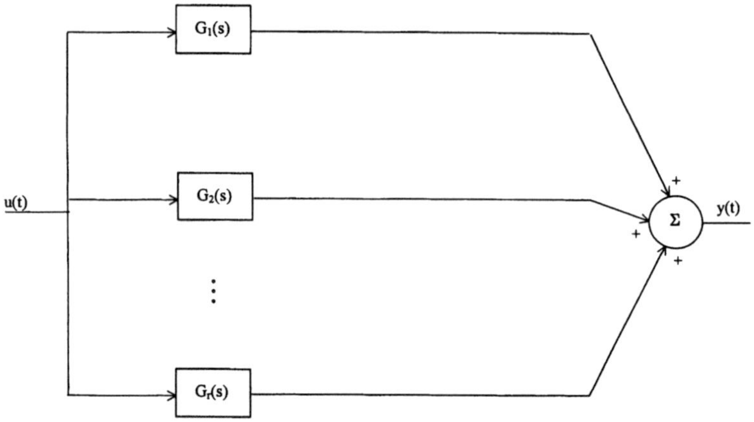

Figure 8.1 Parallel Connection

so that

$$Y(s) = G(s)U(s)$$

where

$$G(s) = \sum_{i=1}^{r} G_i(s)$$

with $G_i(s) : i = 1, 2 \cdots r$ being the transfer functions of the component systems.

Thus assuming we know state models for each of the component systems as

$$G_i(s) \stackrel{s}{=} \begin{bmatrix} A_i & B_i \\ C_i & D_i \end{bmatrix} \qquad i = 1, 2, \cdots r \tag{8.4}$$

we want to determine state model parameters $\{A, B, C, D\}$ for the composite system, i.e.,

$$G(s) \stackrel{s}{=} \begin{bmatrix} A & B \\ C & D \end{bmatrix}$$

We do this as follows.

First we write the state equations for each component system as

$$\begin{aligned} \dot{x}^i(t) &= A_i x^i(t) + B_i u_i(t) \\ y_i(t) &= C_i x^i(t) + D_i u_i(t) \end{aligned} \tag{8.5}$$

where $i = 1, 2, \cdots r$. Next we use (8.5) together with the constraints imposed by (8.2, 8.3) to give a state model for the composite system as

$$\dot{x}(t) = Ax(t) + Bu(t) \tag{8.6}$$
$$y(t) = Cx(t) + Du(t)$$

where

$$A = \begin{bmatrix} A_1 & \emptyset & \cdots & \emptyset \\ \emptyset & A_2 & \cdots & \emptyset \\ \vdots & \vdots & \ddots & \vdots \\ \emptyset & \emptyset & \cdots & A_r \end{bmatrix} \quad B = \begin{bmatrix} B_1 \\ B_2 \\ \vdots \\ B_r \end{bmatrix} \quad x(t) = \begin{bmatrix} x^1(t) \\ x^2(t) \\ \vdots \\ x^r(t) \end{bmatrix}$$

$$C = [\, C_1 \quad C_2 \quad \cdots \quad C_r \,] \quad D = \sum_{i=1}^{r} D_i$$

Notice that the dimension of the state model for the composite system is the sum of the dimensions of each component system. Moreover, MIMO systems can be connected in parallel if and only if each of the systems has: (i) same number of inputs, and (ii) the same number of outputs. Notice in the case of the partial fraction expansion of $G(s)$, the A_is are the poles of $G(s)$.

Finally notice that if we are given $G(s) \in \mathcal{L}_\infty$ then we can use partial fraction expansion to decompose $G(s)$ into the parallel connection of two systems such that $G(s) = G_1(s) + G_2(s)$ where $G_1(s) \in \mathcal{H}_\infty$ and $G_2(s) \in \mathcal{H}_\infty^-$ with the eigenvalues of A_1 (A_2) being in the open left (right) half plane. This fact was given in the previous chapter as the direct sum decomposition of the spaces involved, i.e., $\mathcal{L}_\infty = \mathcal{H}_\infty \oplus \mathcal{H}_\infty^-$.

8.1.2 Series connection

Next consider a series connection of two systems, often referred to as system composition. The constraints imposed by this connection are

$$u(t) = u_2(t) \qquad y_2(t) = u_1(t) \qquad y_1(t) = y(t) \tag{8.7}$$

so that

$$Y(s) = G(s)U(s) \tag{8.8}$$

where

$$G(s) = G_1(s)G_2(s)$$

Now we can develop the state model for the composed system, by imposing the constraints (8.7) on the state models for the component systems, (8.4), with $r = 2$. The result is as follows.

$$\begin{aligned} \dot{x}^1(t) &= A_1 x^1(t) + B_1 C_2 x^2(t) + B_1 D_2 u(t) \\ \dot{x}^2(t) &= A_2 x^2(t) + B_2 u(t) \\ y(t) &= C_1 x^1(t) + D_1 C_2 x^2(t) + D_1 D_2 u(t) \end{aligned}$$

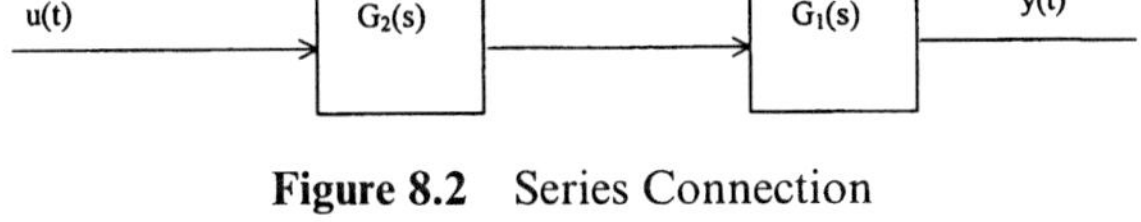

Figure 8.2 Series Connection

which can be rewritten as

$$\begin{bmatrix} \dot{x}^1(t) \\ \dot{x}^2(t) \end{bmatrix} = \begin{bmatrix} A_1 & B_1C_2 \\ \emptyset & A_2 \end{bmatrix} \begin{bmatrix} x^1(t) \\ x^2(t) \end{bmatrix} + \begin{bmatrix} B_1D_2 \\ B_2 \end{bmatrix} u(t)$$

$$y(t) = [C_1 \quad D_1C_2] \begin{bmatrix} x^1(t) \\ x^2(t) \end{bmatrix} + D_1D_2u(t)$$

and we see that the composed system has state model

$$G_1(s)G_2(s) \stackrel{s}{=} \begin{bmatrix} A & B \\ C & D \end{bmatrix} \tag{8.9}$$

where

$$A = \begin{bmatrix} A_1 & B_1C_2 \\ \emptyset & A_2 \end{bmatrix} \qquad B = \begin{bmatrix} B_1D_2 \\ B_2 \end{bmatrix}$$

$$C = [C_1 \quad D_1C_2] \qquad D = D_1D_2$$

Notice that the constraint $u_1(t) = y_2(t)$ implies that the number of inputs to the system labeled 1 must equal the number of outputs from the system labeled 2. Alternatively, we could connect the systems in reverse order if the number of inputs to the system labeled 2 and the number of outputs from the system labeled 1 are equal. In this case the transfer function for the composed system would be $G_2(s)G_1(s)$. Notice that we get the same composed system independent of the order in which the component systems are connected in series only if the matrices $G_1(s)$ and $G_2(s)$ commute so that $G_1(s)G_2(s) = G_2(s)G_1(s)$. Thus in the SISO case the result of composing systems is independent of the order in which the component systems are connected.

In the remainder of this chapter we will take up the problem of determining the state model for several other types of system operation. One of these problems concerns the determination of the state models for the systems needed in a series connection so that the resulting composed system has a specified transfer function. This problem is known as the system factorization problem. The requirement that the factors have certain properties gives rise to a number of different classes of system factorizations, several of which are of great importance to the development of control theory.

8.2 System Inversion

Before we begin considering system factorization we need to consider the related problem of finding a system inverse.

Suppose we are given two SISO systems having transfer functions

$$G_1(s) = \frac{b_0s^m + b_1s^{m-1} + \cdots + b_m}{s^n + a_1s^{n-1} + \cdots + a_n} \tag{8.10}$$

$$G_2(s) = \frac{s^n + a_1s^{n-1} + \cdots + a_n}{b_0s^m + b_1s^{m-1} + \cdots + b_m} \tag{8.11}$$

Then the series connection of these systems gives a transfer function which is unity, i.e., the series connected system has its output equal to its input. Thus it would appear that the system labeled 1 has the system labeled 2 as its inverse.

However, physical processes have the property that their sinusoidal steady state gain cannot be unbounded at infinite. Therefore we require $G_1(\infty)$ and $G_2(\infty)$ to be finite. Now $G_1(\infty)$ ($G_2(\infty)$) is finite if and only if $m \leq n$ ($n \leq m$). Therefore we can only have both $G_1(\infty)$ and $G_2(\infty)$ finite if $m = n$. This means that a given physical process having transfer function $G_1(s)$ has an inverse if and only if $G_1(s)$ is proper (not strictly proper) or equivalently only if the state model for $G_1(s)$ has a nonzero D matrix.

In the MIMO case, a system having transfer function $G_1(s)$ has an inverse if and only if its state models have a D matrix which is invertible. Thus only systems having the same number of inputs as outputs can have an inverse. The state model for the inverse system of a given invertible system can be developed as follows.

8.2.1 Inverse system state model

Suppose we are given the transfer function and state model for an MIMO system

$$G(s) \stackrel{s}{=} \begin{bmatrix} A & B \\ C & D \end{bmatrix} \qquad D^{-1} \text{ exists}$$

so that state equations for the system are

$$\dot{x}(t) = Ax(t) + Bu(t) \tag{8.12}$$

$$y(t) = Cx(t) + Du(t) \tag{8.13}$$

Then solving (8.13) for $u(t)$ gives

$$u(t) = -D^{-1}Cx(t) + D^{-1}y(t) \tag{8.14}$$

which when substituted in (8.12) gives

$$\dot{x}(t) = (A - BD^{-1}C)x(t) + BD^{-1}y(t) \tag{8.15}$$

Thus we see from (8.14, 8.15) that the inverse system has output $u(t)$, input $y(t)$, and state model given as

$$G^{-1}(s) \stackrel{s}{=} \begin{bmatrix} A^{\times} & B^{\times} \\ C^{\times} & D^{\times} \end{bmatrix} \tag{8.16}$$

where

$$A^{\times} = A - BD^{-1}C \qquad B^{\times} = BD^{-1}$$

$$C^{\times} = -D^{-1}C \qquad D^{\times} = D^{-1}$$

As we will see, the state model for the system inverse is useful in developing results in control theory.

Notice that the transfer functions for an invertible SISO system and its inverse (8.10, 8.11) are reciprocals of each other. Therefore the poles of the given system equal the zeros of the inverse system and vice versa. A similar statement can be made for invertible MIMO systems. To see this we need interpret the idea of a system zero in terms of a system's state model. This will be done by first considering system zeros for SISO systems.

8.2.2 SISO system zeros

Recall that an SISO system has a system zero at $s = s_0$ if its transfer function, $G(s)$, is zero for $s = s_0$ or

$$Y(s_0) = 0 \qquad U(s_0) \neq 0 \tag{8.17}$$

where

$$Y(s) = G(s)U(s)$$

Notice that (8.17) implies that $G(s_0) = 0$.

Now in order to arrive at an interpretation of system zeros which we can use in the MIMO case, we consider a system's zeros in terms of a state model for the system.

Suppose we are given a controllable and observable state model for the system,

$$G(s) \stackrel{s}{=} \begin{bmatrix} A & B \\ C & D \end{bmatrix} \tag{8.18}$$

Then taking $x(0) = \emptyset$ since system zeros are defined in terms of the system's zero state response, the Laplace transform of the state equations for the system yields

$$\begin{bmatrix} sI - A & -B \\ C & D \end{bmatrix} \begin{bmatrix} X(s) \\ U(s) \end{bmatrix} = \begin{bmatrix} \emptyset \\ Y(s) \end{bmatrix} \tag{8.19}$$

Therefore if s_0 is a system zero we have

$$\begin{bmatrix} s_0 I - A \\ C \end{bmatrix} X(s_0) + \begin{bmatrix} -B \\ D \end{bmatrix} U(s_0) = \emptyset \qquad U(s_0) \neq 0 \tag{8.20}$$

which implies that the column

$$\begin{bmatrix} -B \\ D \end{bmatrix}$$

is dependent on the n columns in

$$\begin{bmatrix} s_0 I - A \\ C \end{bmatrix}$$

Therefore $s = s_0$ is a system zero if and only if the $(n+1) \times (n+1)$ matrix on the left side of (8.19) is not invertible or

$$\text{rank}\begin{bmatrix} s_0 I - A & -B \\ C & D \end{bmatrix} < n + 1 \tag{8.21}$$

As illustration, suppose we are given a second order system in controller form with D zero. Then (8.19) is

$$\begin{bmatrix} s+a_1 & a_2 \\ -1 & s \end{bmatrix} \begin{bmatrix} X_1(s) \\ X_2(s) \end{bmatrix} - \begin{bmatrix} 1 \\ 0 \end{bmatrix} U(s) = \begin{bmatrix} 0 \\ 0 \end{bmatrix}$$

$$\begin{bmatrix} c_1 & c_2 \end{bmatrix} \begin{bmatrix} X_1(s) \\ X_2(s) \end{bmatrix} = Y(s)$$

which after eliminating $X_1(s)$ becomes

$$(s^2 + a_1 s + a_2)X_2(s) - U(s) = 0 \tag{8.22}$$

$$(c_1 s + c_2)X_2(s) = Y(s) \tag{8.23}$$

Notice from (8.22) that if $U(s_0) \neq 0$ then $X(s_0) \neq 0$. Consequently (8.23) implies that $Y(s_0) = 0$ for $U(s_0) \neq 0$ only if s_0 is a zero of the polynomial

$$c_1 s + c_2$$

However

$$G(s) = \frac{c_1 s + c_2}{s^2 + a_1 s + a_2}$$

which shows that s_0 is indeed a zero of $G(s)$.

By proceeding in the same fashion, we can show, when $D \neq 0$, that the foregoing state model has a system zero at s_0 which is a root of the polynomial

$$D(s^2 + a_1 s + a_2) + c_1 s + c_2$$

which is the numerator of $G(s)$.

The foregoing ideas are used now to extend the definition of a system zero to MIMO systems.

8.2.3 MIMO system zeros

Suppose the system denoted by (8.18) has m inputs and p outputs with $m \leq p$ and with all m columns of B independent. Then a complex number s_0 is a system zero if

$$G(s_0)U(s_0) = \emptyset \qquad \text{for some } U(s_0) \neq \emptyset \tag{8.24}$$

Notice that unlike the SISO case where (8.17) implies that $G(s_0) = 0$, (8.24) does not imply that $G(s_0) = \emptyset$.

Now with appropriate modification, (8.20) implies that $s = s_0$ is a system zero if at least one of the m columns of

$$\begin{bmatrix} -B \\ D \end{bmatrix}$$

is dependent on the columns of

$$\begin{bmatrix} s_0 I - A \\ C \end{bmatrix}$$

Therefore, recalling that the rank of a matrix cannot exceed its smallest dimension, we see from (8.21) that s_0 is a system zero if

$$\operatorname{rank}\begin{bmatrix} s_0 I - A & -B \\ C & D \end{bmatrix} < n + m \tag{8.25}$$

Alternatively, when $m > p$ the matrix in the foregoing inequality cannot have rank greater than $n + p$ so that (8.25) is satisfied for all s_0 and (8.25) is meaningless.

In order to overcome this problem we define s_0 to be a system zero if

$$\operatorname{rank}\begin{bmatrix} s_0 I - A & -B \\ C & D \end{bmatrix} < \rho$$

where

$$\rho = \max_{s \in C}\left(\operatorname{rank}\begin{bmatrix} sI - A & -B \\ C & D \end{bmatrix}\right)$$

We want now to apply the foregoing ideas to invertible systems. Since invertible systems are square, i.e., $m = p$, we can use (8.25) to define the system zeros for this class of system.

8.2.4 Zeros of invertible systems

Suppose $G(s)$ is invertible. Then we see, in this case, that the condition for s_0 to be a system zero, (8.25), is equivalent to

$$\det\begin{bmatrix} s_0 I - A & -B \\ C & D \end{bmatrix} = 0 \tag{8.26}$$

Notice that if the state model is not controllable the foregoing condition is also satisfied when s_0 is an uncontrollable eigenvalue of the system since we have an eigenvalue-left-eigenvector pair (λ, w) of A which satisfies

$$w^T(\lambda I - A) = \emptyset \quad \text{and} \quad w^T B = \emptyset$$

Therefore

$$\begin{bmatrix} w^T & \emptyset \end{bmatrix}\begin{bmatrix} \lambda I - A & -B \\ C & D \end{bmatrix} = \begin{bmatrix} \emptyset & \emptyset \end{bmatrix} \tag{8.27}$$

and (8.26) holds for $\lambda = s_0$. Thus, in this case, s_0 is both a system zero and a system pole. Alternatively, if (A, C) is an unobservable pair a similar argument can be used to show that at least one of the eigenvalues of A is both a system zero and a system pole.

However, if the state model is both controllable and observable with D is nonsingular, then there is no system zero which equals a system pole. In addition, each system zero of $G(s)$ is a system pole of $G^{-1}(s)$ and vice versa. To see this suppose s_0 is a system zero. Then (8.26) is satisfied and is equivalent to

$$\det\left[\begin{bmatrix} s_0 I - A & -B \\ C & D \end{bmatrix} M\right] = 0 \tag{8.28}$$

where M is any nonsingular matrix of appropriate dimension.

Now suppose we choose M as

$$M = \begin{bmatrix} I & \emptyset \\ -D^{-1}C & I \end{bmatrix}$$

Then the left side of (8.28) becomes

$$\det\left[\begin{bmatrix} s_0 I - A & -B \\ C & D \end{bmatrix} M\right] = \det\begin{bmatrix} s_0 I - A + BD^{-1}C & -B \\ \emptyset & D \end{bmatrix}$$

$$= \det\left[s_0 I - A + BD^{-1}C\right] \det[D] \tag{8.29}$$

However since $G(s)$ is invertible, we have $\det[D] \neq 0$ and (8.29, 8.28) imply

$$\det[s_0 I - A^{\times}] = 0$$

where

$$A^{\times} = A - BD^{-1}C$$

and $A^{\times}$ is the system matrix for $G^{-1}(s)$, (8.16). Therefore s_0 is a pole of $G^{-1}(s)$ as well as being a zero of $G(s)$.

Finally, we will see that invertible systems which are stable with stable inverses play an important role in control theory. These systems, which are referred to as "units", form a subspace $\mathcal{U}_\infty$ of $\mathcal{H}_\infty$ and are characterized in terms of their transfer functions $G(s)$ as

$$G(s) \in \mathcal{U}_\infty \qquad \text{if and only if} \qquad G(s), G^{-1}(s) \in \mathcal{H}_\infty$$

In the case of an SISO system, this definition implies that a system is a unit if it has a transfer function which is proper with no poles or zeros in the closed right-half plane.

8.3 Coprime Factorization

Problems in system factorization are concerned with the determination of two systems, called factors, whose series connection has the same input-output behavior as the system being factored. Additional constraints on the factors determine different classes of factorization.

Coprime factorization can be characterized as a series connection of a stable system and the inverse of a stable system with no unstable pole-zero cancellations between factors. As an example consider the system with transfer function $G(s)$ where

$$G(s) = \frac{(s+1)(s-2)}{(s+3)(s-4)(s+5)} \tag{8.30}$$

Then a coprime factorization of this system is indicated as follows:

$$G(s) = N_1(s)M_1^{-1}(s)$$

where

$$N_1(s) = \frac{(s-2)(s+1)}{(s+3)\alpha(s)} \qquad M_1(s) = \frac{(s-4)(s+5)}{\alpha(s)}$$

with $\alpha(s)$ being any polynomial of degree 2 having no zeros in the closed right half plane.

Notice that this choice of degree for $\alpha(s)$ ensures that $M_1(s)$ is invertible whereas the restriction on the location of the zeros of $\alpha(s)$ ensures that $\alpha(s)$ is not involved in any unstable pole-zero cancellations between $N_1(s)$ and $M_1^{-1}(s)$. Thus even if $\alpha(s)$ has zeros at $s = -1$ and/or -5 the resulting pole-zero cancellations in $N_1(s)$ and $M_1^{-1}(s)$ are allowed since they are stable. More important, notice that the closed right-half plane zero of $N_1(s)$ at $s = 2$ and of $M_1(s)$ at $s = 4$ are different so that there are no unstable pole-zero cancellation between $N_1(s)$ and $M_1^{-1}(s)$. Stable systems which do not share any system zeros in the closed right-half plane are said to be coprime.

Notice, from the foregoing example, that the form of coprime factorization, $N(s)M^{-1}(s)$, bears a striking resemblance to the expression for a transfer function in terms of its numerator polynomial $N(s)$ and denominator polynomial $M(s)$. Of course the "numerator" and "denominator" in the present context are not polynomials but are each rational and can therefore be thought of as transfer functions for "numerator" and "denominator" systems.

8.3.1 Why coprime?

In the foregoing example the numerator $N_1(s)$ and denominator $M_1(s)$ are coprime. In order to better appreciate this fact we give a second factorization of $G(s)$ in which the factors are not coprime

Suppose we re-express $G(s)$, (8.30), as

$$G(s) = N_2(s)M_2^{-1}(s) \tag{8.31}$$

where

$$N_2(s) = \frac{(s-6)(s-2)(s+1)}{(s+3)\beta(s)} \qquad M_2(s) = \frac{(s-4)(s+5)(s-6)}{\beta(s)}$$

with $\beta(s)$ restricted to be any degree 3 polynomial with no zeros in the closed right half plane. Although this factorization has the desired property that $N_2(s), M_2(s) \in \mathcal{H}_\infty$ with

$M_2(s)$ being invertible, it does not satisfy the requirement that $N_2(s)$ and $M_2(s)$ be coprime since they share a right-half plane zero at $s = 6$. Thus (8.31) is not a coprime factorization of $G(s)$. We can demonstrate the importance of not having unstable pole-zero cancellations between factors as follows.

Suppose that we have a strictly proper scalar transfer function $G(s)$ having one of its n poles, $n > 2$, at $s = s_0$ on the positive real axis with the remaining $n - 1$ poles being in the open left-half plane. These specifications imply that $G(s)$ can be written as

$$G(s) = \frac{q(s)[s - (s_0 + \epsilon)]}{(s - s_0)\, p(s)} \qquad 0 \leq \epsilon < \infty, \quad s_0 > 0$$

where the degree of $q(s)$ is less than $p(s)$ with $q(s_0) \neq 0$ and $p(s)$ having no zeros in the closed right half plane. Now the system having this transfer function is stable only if $\epsilon = 0$ so that there is an unstable pole-zero cancellation. However, closer inspection of the consequences of such a pole-zero cancellation reveals that we cannot rely, in practice, on unstable pole-zero cancellations to make an unstable system stable. This becomes immediately evident when we consider the system's impulse response, $\mathcal{L}_+^{-1}[G(s)] = g(t)$

$$g(t) = K_0 e^{s_0 t} + r(t)$$

where

$$K_0 = -\epsilon \frac{q(s_0)}{p(s_0)}$$

with $r(t)$ bounded for $0 \leq \epsilon < \infty$.

Notice that if $\epsilon = 0$, $K_0 e^{s_0 t}$ is missing from $g(t)$. Then the system's impulse response is bounded and the system is stable. However, if $\epsilon \neq 0$, then $K_0 e^{s_0 t}$ is present in $g(t)$ and the system's impulse response tends to infinity with time and the system is unstable. Thus the system's stability is catastrophically sensitive to ϵ.

Conversely, if s_0 is in the open left-half plane then $K_0 e^{s_0 t}$ tends to zero with time and the system is stable for all ϵ.

Comparing the effect on the system behavior of the two types of pole-zero cancellation, stable and unstable, we see that the lack of robustness of a system's stability to unstable pole-zero cancellation makes this type of pole-zero cancellation a detriment to reliable control system design. We will see that the coprimeness and the stability of the factors in a coprime factorization can be exploited to provide a means for determining controllers which not only make a feedback control system input–output stable but in addition prevent unstable pole-zero cancellations between the controller and plant so that the closed loop system is internally stable.

Returning to coprime factorization, we saw earlier in this chapter that system zeros of an invertible system are poles of that system's inverse. Therefore, we see now, that unstable pole-zero cancellations between $N(s)$ and $M^{-1}(s)$ are prevented if and only if $M(s)$ and $N(s)$ have no common system zeros in the closed right-half plane, i.e., if and only if $M(s)$ and $N(s)$, are coprime.

In addition, since $N(s)$ is always stable, $M^{-1}(s)$ is unstable if $G(s)$ is unstable. Moreover, since $M(s)$ and $N(s)$ are coprime, the unstable poles of $G(s)$ are the only unstable poles of $M^{-1}(s)$. Finally, since $M(s)$ is stable, $M^{-1}(s)$ has no zeros in the closed right-half plane. Therefore the closed right-half plane zeros of $G(s)$ are the only closed right-half plane zeros of $N(s)$.

8.3.2 Coprime factorization of MIMO systems

Moving on to MIMO systems, we need to distinguish between two types of coprime factorization depending on the ordering of the factors. Thus a $p \times m$ real rational transfer function matrix $G(s)$ can be coprime factored as

$$G(s) = N(s)M^{-1}(s) \tag{8.32}$$

or as

$$G(s) = \tilde{M}^{-1}(s)\tilde{N}(s) \tag{8.33}$$

where $N(s), \tilde{N}(s) \in \mathcal{H}_\infty$ are both $p \times m$ and $M(s), \tilde{M}(s) \in \mathcal{H}_\infty$ are $m \times m$ and $p \times p$ respectively. In addition both pairs $\{N(s), M(s)\}$, and $\{\tilde{N}(s), \tilde{M}(s)\}$ must be coprime. Notice that the two varieties of factorization (8.32, 8.33) are referred to as right and left coprime factorizations since the denominator is on the right in (8.32) and on the left in (8.33). In the SISO case these two varieties coincide since products of scalars commute.

We give now, in the theorem to follow, an alternative characterization of coprimeness. We will see later that this characterization is also useful in the problem of determining controllers which make the closed loop control system internally stable.

Theorem 8.1 The factors $M(s), N(s) \in \mathcal{H}_\infty$ $\{\tilde{M}(s), \tilde{N}(s) \in \mathcal{H}_\infty\}$ are right {left} coprime if and only if there exists $X(s)$, $Y(s) \in \mathcal{H}_\infty$ $\{\tilde{X}(s), \tilde{Y}(s) \in \mathcal{H}_\infty\}$ which satisfy the following Bezout, (pronounced "bzoo") identity

$$X(s)M(s) + Y(s)N(s) = I \tag{8.34}$$

$$\{\tilde{M}(s)\tilde{X}(s) + \tilde{N}(s)\tilde{Y}(s) = I\} \tag{8.35}$$

Proof (if) Consider the SISO case. Suppose $M(s), N(s)$ are not coprime, i.e., $M(s_0) = N(s_0) = 0$ for some s_0 in the closed right-half plane. Then since $X(s)$ and $Y(s)$ are stable, $X(s_0) < \infty$, $Y(s_0) < \infty$. Therefore we have $X(s_0)M(s_0) = 0$, $Y(s_0)N(s_0) = 0$ and (8.34) cannot be satisfied. This shows, in the SISO case, that satisfaction of the Bezout identity is sufficient for $M(s), N(s)$ to be coprime. The MIMO case is treated as follows.

Rewrite the Bezout identity, (8.34), as

$$L(s)R(s) = I \tag{8.36}$$

where

$$L(s) = [X(s) \quad Y(s)] \qquad R(s) = \begin{bmatrix} M(s) \\ N(s) \end{bmatrix}$$

and $L(s), R(s) \in \mathcal{H}_\infty$ are $m \times (m+p)$ and $(m+p) \times m$ respectively.

Next suppose that $M(s), N(s)$ are not coprime with s_0 being a zero of $M(s)$ and of $N(s)$ which lies in the closed right-half plane. Then s_0 is a system zero of $R(s)$ and we see from (8.24) that there is a complex vector $U(s_0) \neq \emptyset$ such that

$$R(s_0)U(s_0) = \emptyset$$

However since $L(s) \in \mathcal{H}_\infty$ all elements of $L(s_0)$ are finite. Therefore we have

$$L(s_0)R(s_0)U(s_0) = \emptyset$$

and the Bezout identity (8.36) is not satisfied. Therefore we have shown that, in the MIMO case, if the Bezout identity, (8.36) or (8.34), is satisfied then $M(s), N(s)$ are coprime.

We can show, in a similar way, that (8.35) is sufficient for $\tilde{M}(s), \tilde{N}(s)$ to be coprime.

Proof (only if) This will be done in the next section by using certain state models we will determine for the transfer functions $\tilde{M}(s), \tilde{N}(s), M(s), N(s), X(s), Y(s), \tilde{X}(s), \tilde{Y}(s)$. ■

Before going on to the development of state models for the various systems involved in coprime factorization, we need to consider the possibility of there being more than one coprime factorization for a given system. This is done now as follows.

8.3.3 Relating coprime factorizations

The coprime factors in a coprime factorization of a given system are not unique. Moreover the factors in two different coprime factorizations of the same transfer function are related by a unit, (Section 8.2.4). This fact is shown in the following theorem.

Theorem 8.2 Suppose $G(s)$ has right and left coprime factorizations

$$G = N_1(s)M_1^{-1}(s)$$

$$= \tilde{M}_1^{-1}(s)\tilde{N}_1(s)$$

Then $G(s)$ has right and left coprime factorizations given by

$$G = N_2(s)M_2^{-1}(s)$$

$$= \tilde{M}_2^{-1}(s)\tilde{N}_2(s)$$

if and only if there exists units $R(s), L(s) \in \mathcal{U}_\infty$ such that

$$\begin{aligned} N_2(s) &= N_1(s)R(s) & \qquad \tilde{N}_2(s) &= L(s)\tilde{N}_1(s) \\ M_2(s) &= M_1(s)R(s) & \qquad \tilde{M}_2(s) &= L(s)\tilde{M}_1(s) \end{aligned} \tag{8.37}$$

Proof (if) If $R(s) \in \mathcal{H}_\infty$ then we see from (8.37) that $N_2(s), M_2(s) \in \mathcal{H}_\infty$ since the composition (series connection) of stable systems is stable. In addition we see from (8.37) that

$$N_2(s)M_2^{-1}(s) = N_1(s)R(s)R^{-1}(s)M_1^{-1}(s) = G(s) \tag{8.38}$$

However since $N_1(s), M_1(s)$ are coprime the Bezout identity (8.34) is satisfied. Therefore pre and post-multiplying the Bezout identity by $R^{-1}(s)$ and $R(s)$, and using (8.37) we obtain

$$X_2(s)M_2(s) + Y_2(s)N_2(s) = I \tag{8.39}$$

where

$$X_2(s) = R^{-1}(s)X_1(s) \qquad\qquad Y_2(s) = R^{-1}(s)Y_2(s)$$

Finally if $R^{-1}(s) \in \mathcal{H}_\infty$ then we have $X_2(s), Y_2(s) \in \mathcal{H}_\infty$ and $M_2(s)$ and $N_2(s)$ are coprime from Theorem 8.1. Therefore we have shown that $N_2(s)M_2^{-1}(s)$ is an additional right coprime factorization of $G(s)$ if $R(s) \in \mathcal{U}_\infty$.

We can show, in a similar fashion, that if $L(s) \in \mathcal{U}_\infty$ then $\tilde{M}_2^{-1}(s)\tilde{N}_2(s)$ is a second left coprime factorization of $G(s)$.

Proof (only if) Consideration of (8.38) shows that the invertability of $R(s)$ is necessary. Therefore suppose $R(s)$ is invertible but $R(s) \notin \mathcal{U}_\infty$. Then $R(s) \notin \mathcal{H}_\infty$ and/or $R^{-1}(s) \notin \mathcal{H}_\infty$.

Suppose $R(s) \notin \mathcal{H}_\infty$. Then since $N_1(s)M_1^{-1}(s)$ is a coprime factorization of $G(s)$, we see from (8.37) that either $N_2(s), M_2(s) \notin \mathcal{H}_\infty$ or $N_2(s), M_2(s) \in \mathcal{H}_\infty$. Suppose $N_2(s)$, $M_2(s) \notin \mathcal{H}_\infty$. Then $N_2(s)M_2^{-1}(s)$ is not a coprime factorization of $G(s)$. Alternatively, suppose $N_2(s), M_2(s) \in \mathcal{H}_\infty$. Then we see from (8.37) that all unstable poles of $R(s)$ are cancelled by zeros of $N_1(s)$ and $M_1(s)$. This implies that $N_1(s)$ and $M_1(s)$ share closed right-half plane zeros and therefore, contrary to assumption, $N_1(s)M_1^{-1}(s)$ is not a coprime factorization of $G(s)$. Therefore $R(s) \in \mathcal{H}_\infty$

Alternatively, if $R^{-1}(s) \notin \mathcal{H}_\infty$ then $R(s)$ has closed right-half plane zeros. Thus we see from (8.37) that $N_2(s)$ and $M_2(s)$ share closed right-half plane zeros and are therefore not coprime. Thus $N_2(s)M_2^{-1}(s)$ is not a coprime factorization of $G(s)$.

The foregoing shows that $N_2(s)M_2^{-1}(s)$ is an additional right coprime factorizations of $G(s)$ only if $R(s)$ in (8.37) is a unit.

In the same way we can show that $\tilde{M}_2^{-1}(s)\tilde{N}_2(s)$ is a second left coprime factorization of $G(s)$ only if $L(s)$ in (8.37) is a unit. ■

In summary, we have seen in this section that $M(s)$ is always proper since it is always invertible. Alternatively, $N(s)$ is proper when $G(s)$ is proper and strictly proper when $G(s)$ is strictly proper. We will see in the next section that $Y(s)$ is always strictly proper so that $Y(\infty) = \emptyset$. Therefore the product $Y(s)N(s)$ is always strictly proper so that $Y(\infty)N(\infty) = \emptyset$ no matter whether $G(s)$ is proper or strictly proper. Thus we see that a necessary condition for satisfying the Bezout identity, (8.34), is $X(\infty) = M^{-1}(\infty)$ which implies that the state models for $X(s)$ and $M(s)$ must have D matrices which are inverses of each other. Similarly, state models for $\tilde{X}(s)$ and $\tilde{M}(s)$ must have D matrices which are inverses of each other.

8.4 State Models for Coprime Factorization

In what follows we develop, from a given minimal state model for the system being factored, minimal state models for its coprime factors, $M(s), N(s), \tilde{M}(s), \tilde{N}(s)$ as well as the factors $X(s), Y(s), \tilde{X}(s), \tilde{Y}(s)$ needed to satisfy the Bezout identities, (8.34, 8.35).

8.4.1 Right and left coprime factors

Suppose we are given a controllable and observable (minimal) state model for the system to be factored

$$G(s) \stackrel{s}{=} \begin{bmatrix} A & B \\ C & D \end{bmatrix} \tag{8.40}$$

Then applying state feedback

$$u(t) = Kx(t) + v(t) \tag{8.41}$$

with K chosen to make $A + BK$ is stable, yields the following state equations for the closed loop system

$$\dot{x}(t) = (A + BK)x(t) + Bv(t) \tag{8.42}$$

$$y(t) = (C + DK)x(t) + Dv(t) \tag{8.43}$$

Now we can interpret (8.41–8.43) as two systems having transfer functions $N(s)$, $M(s)$ with each system having $v(t)$ as input

$$Y(s) = N(s)V(s) \tag{8.44}$$

$$U(s) = M(s)V(s) \tag{8.45}$$

where

$$N(s) \stackrel{s}{=} \begin{bmatrix} A + BK & B \\ C + DK & D \end{bmatrix} \qquad M(s) \stackrel{s}{=} \begin{bmatrix} A + BK & B \\ K & I \end{bmatrix} \tag{8.46}$$

Notice that $N(s), M(s) \in \mathcal{H}_\infty$ since $A + BK$ is stable. In addition notice that $M(s)$ is invertible since its state model has a D matrix which is an identity matrix. This allows us to solve (8.45) for $V(s)$ and substitute the result in (8.44) to obtain

$$Y(s) = N(s)M^{-1}(s)U(s)$$

The following simple example provides an illustration of the relation of the foregoing state models, (8.46) to coprime factorization.

Suppose we want to obtain a coprime factorization for

$$G(s) = \frac{s - 1}{s(s - 2)}$$

We begin by obtaining a minimal state model for $G(s)$, say in controller form, so that

$$G(s) \stackrel{s}{=} \begin{bmatrix} A & B \\ C & D \end{bmatrix}$$

where

$$A = \begin{bmatrix} 2 & 0 \\ 1 & 0 \end{bmatrix} \qquad B = \begin{bmatrix} 1 \\ 0 \end{bmatrix}$$

$$C = \begin{bmatrix} 1 & -1 \end{bmatrix} \qquad D = 0$$

Next we need to choose K so that $A + BK$, is stable. Suppose we do this so that $A + BK$ has a multiple eigenvalue at -1. Then K is

$$K = \begin{bmatrix} -4 & -1 \end{bmatrix}$$

and $A + BK$, and $C + DK$ in the state models for the factors, (8.46) are

$$A + BK = \begin{bmatrix} -2 & -1 \\ 1 & 0 \end{bmatrix} \qquad C + DK = C = \begin{bmatrix} 1 & -1 \end{bmatrix}$$

Finally, we get the transfer functions for the coprime factors from the state models, (8.46), as

$$N(s) = (C + DK)[sI - (A + BK)]^{-1}B = \frac{s - 1}{(s + 1)^2}$$

$$M(s) = K[sI - (A + BK)]^{-1}B + I = \frac{s^2 - 2s}{(s + 1)^2}$$

and we see that

$$N(s)M^{-1}(s) = G(s)$$

The required coprimeness of $N(s), M(s)$ in this example is seen by inspection.

We can show that the factors, $M(s)$, $N(s)$, (8.46), are coprime in general by using the definition of a system zero given in Section 8.2.3 of this chapter. This is done as follows.

We begin by noting that the state models for the factors, (8.46), have the same system matrix, $A + BK$, and the same input matrix, B. Therefore the states of these systems are equal for all time if their inputs are equal and if their initial states are equal.

Recall from Section 8.2.3 that the system zeros are defined in terms of the zero state response. Therefore in order to investigate the coprimeness of the factors specified by (8.46), we take the initial states of these state models null, i.e., equal. Moreover, we note from, (8.44, 8.45) that the inputs to these state models are the same being equal to $v(t)$. Therefore the states of the state models for the factors $M(s)$ and $N(s)$, (8.46), are assumed equal in the following investigation of the system zeros of the factors.

Suppose s_0 is a common system zero of the factors $M(s)$ and $N(s)$ given by (8.46). Then we see from (8.20) that the following conditions must be satisfied

$$[s_0 I - A - BK]X(s_0) - BU(s_0) = \emptyset \tag{8.47}$$

$$[C + DK]X(s_0) + DU(s_0) = \emptyset \tag{8.49}$$

$$KX(s_0) + U(s_0) = \emptyset \tag{8.49}$$

Thus substituting for $U(s_0)$ from (8.49) in (8.47, 8.48) yields

$$[s_0 I - A]X(s_0) = \emptyset$$
$$CX(s_0) = \emptyset$$

which implies that the pair (A, C) is unobservable. This contradicts our assumption that the state model for $G(s)$ is minimal. Thus s_0 cannot be both a zero of $N(s)$ and a zero of $M(s)$ which is sufficient for $N(s), M(s)$ to be coprime.

Notice, from the foregoing demonstration of the coprimeness of $M(s)$ and $N(s)$, (8.46), that the detectability of the state model used for $G(s)$, (8.40), is both necessary and sufficient for the coprimeness of these factors, i.e., for there to be no common closed right-half plane system zeros for $M(s), N(s)$.

In the next section we will provide additional evidence for the coprimeness of the pair $M(s), N(s)$, (8.46), by finding state models for $X(s), Y(s)$ to satisfy the Bezout identity, (8.34). Before doing this we determine state models for some left coprime factors of $G(s)$ by utilizing the fact that the right coprime factorization of $G^T(s)$, i.e.,

$$G^T(s) = \tilde{N}^T(s)\tilde{M}^{-T}(s) \tag{8.50}$$

becomes the desired left coprime factorization of $G(s)$ after transposition, i.e.,

$$G(s) = \tilde{M}^{-1}(s)\tilde{N}(s)$$

We do this as follows.

From (8.40) we have

$$G^T(s) \stackrel{s}{=} \begin{bmatrix} A^T & C^T \\ B^T & D^T \end{bmatrix}$$

Then from (8.46) we see that the right coprime factors of $G^T(s)$ are given as

$$\tilde{N}^T(s) \stackrel{s}{=} \begin{bmatrix} A^T + C^T L^T & C^T \\ B^T + D^T L^T & D^T \end{bmatrix} \qquad \tilde{M}^T(s) \stackrel{s}{=} \begin{bmatrix} A^T + C^T L^T & C^T \\ L^T & I \end{bmatrix}$$

where L^T replaces K in (8.46). Notice that as K was chosen to make $A + BK$ stable in (8.46), so must L^T be chosen now to make $A^T + C^T L^T$ stable. Therefore after doing the required transposition we obtain the left coprime factors for $G(s)$ as

$$\tilde{N}(s) \stackrel{s}{=} \begin{bmatrix} A + LC & B + LD \\ C & D \end{bmatrix} \qquad \tilde{M}(s) \stackrel{s}{=} \begin{bmatrix} A + LC & L \\ C & I \end{bmatrix} \tag{8.51}$$

Notice that we have $\tilde{N}(s), \tilde{M}(s) \subset \mathcal{H}_\infty$ with $\tilde{M}(s)$ being invertible.

8.4.2 Solutions to the Bezout identities

In this section we will see that there is an important connection between the coprime factors for a controller in an observer based feedback control system (Section 5.2), and the pairs $\{X(s),\ Y(s)\}$ and $\{\tilde{X}(s),\ \tilde{Y}(s)\}$ in the Bezout identities, (8.34, 8.35). This connection is used here to establish state models for these pairs. We will see later that this connection can be used to characterize all controllers which impart stability to a closed loop control system.

In order to obtain a coprime factorization for the controller, we need to recall that the relevant equations defining an observer based controller for the plant, (8.40) are

$$\dot{\hat{x}}(t) = A\hat{x}(t) + Bu(t) + L[\hat{y}(t) - y(t)]$$
$$\hat{y}(t) = C\hat{x}(t) + Du(t)$$
$$u(t) = K\hat{x}(t)$$

Then we see from these equations that the state equations for the controller are

$$\dot{\hat{x}}(t) = \hat{A}\hat{x}(t) + \hat{B}y(t)$$
$$u(t) = \hat{C}\hat{x}(t) \tag{8.52}$$

where

$$\hat{A} = A + BK + LC + LDK \qquad \hat{B} = -L \qquad \hat{C} = K$$

so that the controller transfer function $H(s)$ is given by

$$H(s) \stackrel{s}{=} \begin{bmatrix} \hat{A} & \hat{B} \\ \hat{C} & \emptyset \end{bmatrix}$$

Notice that the controller is strictly proper. Notice also that since we are assuming that the plant has m inputs and p outputs, the controller has p inputs and m outputs. Therefore $H(s)$ is an $m \times p$ matrix.

Next using (8.46, 8.51) we see that the right and left coprime factorizations of the controller are

$$H(s) = N_H(s)M_H^{-1}(s) = \tilde{M}_H^{-1}(s)\tilde{N}_H(s)$$

where

$$N_H(s) \stackrel{s}{=} \begin{bmatrix} \hat{A} + \hat{B}\hat{K} & \hat{B} \\ \hat{C} & \emptyset \end{bmatrix} \qquad M_H(s) \stackrel{s}{=} \begin{bmatrix} \hat{A} + \hat{B}\hat{K} & \hat{B} \\ \hat{K} & I \end{bmatrix}$$

$$\tilde{N}_H(s) \stackrel{s}{=} \begin{bmatrix} \hat{A} + \hat{L}\hat{C} & \hat{B} \\ \hat{C} & \emptyset \end{bmatrix} \qquad \tilde{M}_H(s) \stackrel{s}{=} \begin{bmatrix} \hat{A} + \hat{L}\hat{C} & \hat{L} \\ \hat{C} & I \end{bmatrix}$$

with $\hat{K}$ and $\hat{L}$ being chosen so that the factors are stable. One way of doing this is to choose

$$\hat{K} = C + DK \qquad \hat{L} = -(B + LD)$$

where K, L were used in the previous section to obtain state models for the left and right coprime factorization of $G(s)$, (8.46, 8.51). Notice that this choice provides the required stability since the A matrices for the state models of the right and left coprime factors of $H(s)$ are given as

$$\hat{A} + \hat{B}\hat{K} = A + BK \qquad \hat{A} + \hat{L}\hat{C} = A + LC$$

Therefore using this choice for $\hat{K}$ and $\hat{L}$ we see that the coprime factor state models for the controller become

$$N_H(s) \overset{s}{=} \begin{bmatrix} A+BK & -L \\ K & \emptyset \end{bmatrix} \qquad M_H(s) \overset{s}{=} \begin{bmatrix} A+BK & -L \\ C+DK & I \end{bmatrix} \tag{8.53}$$

$$\tilde{N}_H(s) \overset{s}{=} \begin{bmatrix} A+LC & -L \\ K & \emptyset \end{bmatrix} \qquad \tilde{M}_H(s) \overset{s}{=} \begin{bmatrix} A+LC & -(B+LD) \\ K & I \end{bmatrix} \tag{8.54}$$

We show now that the Bezout identities, (8.34, 8.35), can be satisfied by making use of the foregoing factorizations of the observer based controller for $G(s)$.

Theorem 8.3 If a system has transfer function $G(s)$ with minimal state model specified as (8.40) and coprime factorizations specified as (8.46, 8.51), then the Bezout identities, (8.34, 8.35), are satisfied by

$$X(s) = \tilde{M}_H(s) \qquad Y(s) = -\tilde{N}_H(s) \tag{8.55}$$

$$\tilde{X}(s) = M_H(s) \qquad \tilde{Y}(s) = -N_H(s) \tag{8.56}$$

where $M_H(s), N_H(s), \tilde{M}_H(s)$, and $\tilde{N}_H(s)$ are given by (8.53, 8.54)

Proof We need to establish that

$$\tilde{M}_H(s)M(s) - \tilde{N}_H(s)N(s) = I \tag{8.57}$$

We can do this by recalling the state model for systems connected in series, (8.9) and using the state models for $\tilde{M}_H(s), M(s), \tilde{N}_H(s), N(s)$ (8.54, 8.46) to obtain

$$\tilde{M}_H(s)M(s) \overset{s}{=} \begin{bmatrix} \begin{bmatrix} A+LC & -(B+LD)K \\ \emptyset & A+BK \end{bmatrix} & \begin{bmatrix} -(B+LD) \\ B \end{bmatrix} \\ \begin{bmatrix} K & K \end{bmatrix} & I \end{bmatrix} \tag{8.58}$$

$$\tilde{N}_H(s)N(s) \overset{s}{=} \begin{bmatrix} \begin{bmatrix} A+LC & -L(C+DK) \\ \emptyset & A+BK \end{bmatrix} & \begin{bmatrix} -LD \\ B \end{bmatrix} \\ \begin{bmatrix} K & \emptyset \end{bmatrix} & \emptyset \end{bmatrix} \tag{8.59}$$

Then changing coordinates for the state representation of $\tilde{N}_H(s)N(s)$ by using the coordinate transformation matrix T

$$T = \begin{bmatrix} I & I \\ \emptyset & I \end{bmatrix}$$

yields

$$\tilde{N}_H(s)N(s) \stackrel{s}{=} \begin{bmatrix} \begin{bmatrix} A+LC & -(B+LD)K \\ \emptyset & A+BK \end{bmatrix} & \begin{bmatrix} -(B+LD) \\ B \end{bmatrix} \\ \begin{bmatrix} K & K \end{bmatrix} & \emptyset \end{bmatrix} \tag{8.60}$$

Finally comparing the state models given by (8.60) and by (8.58) we see that

$$\tilde{M}_H(s)M(s) = \tilde{N}_H(s)N(s) + I$$

which shows that (8.57) is satisfied.

We can show that (8.56, 8.51, 8.53) satisfy the Bezout identity, (8.35) by proceeding in a similar fashion. ■

8.4.3 Doubly-coprime factorization

The foregoing right and left coprime factorizations of $G(s)$, (8.46, 8.51) constitute what is referred to as a doubly-coprime factorization. This type of factorization is defined from the following observations.

First recall, in general, that any right and left coprime factorizations of $G(s)$ and of $H(s)$ satisfy

$$G(s) = N(s)M^{-1}(s) = \tilde{M}^{-1}(s)\tilde{N}(s)$$

$$H(s) = N_H(s)M_H^{-1}(s) = \tilde{M}_H^{-1}(s)\tilde{N}_H(s)$$

Therefore it follows that

$$\tilde{M}(s)N(s) - \tilde{N}(s)M(s) = \emptyset \tag{8.61}$$

$$\tilde{M}_H(s)N_H(s) - \tilde{N}_H(s)M_H(s) = \emptyset \tag{8.62}$$

Next recall, from Theorem 8.3, that if we use the particular coprime factorizations for $G(s)$ and $H(s)$ given by (8.46, 8.51, 8.53, 8.54) then we have the following Bezout identities

$$\tilde{M}_H(s)M(s) - \tilde{N}_H(s)N(s) = I \tag{8.63}$$

$$\tilde{M}(s)M_H(s) - \tilde{N}(s)N_H(s) = I \tag{8.64}$$

Finally, notice that (8.61–8.64) can be written as the single matrix equation

$$\begin{bmatrix} \tilde{M}_H(s) & -\tilde{N}_H(s) \\ -\tilde{N}(s) & \tilde{M}(s) \end{bmatrix} \begin{bmatrix} M(s) & N_H(s) \\ N(s) & M_H(s) \end{bmatrix} = \begin{bmatrix} I & \emptyset \\ \emptyset & I \end{bmatrix} \tag{8.65}$$

Now we use this relation to define a doubly-coprime factorization as follows.

Suppose a system with transfer function $G(s)$ has left and right coprime factorizations

$$G(s) = \tilde{M}^{-1}(s)\tilde{N}(s)$$
$$= N(s)M^{-1}(s)$$

Then these factorizations taken together constitute a doubly-coprime factorization of $G(s)$ if there exists a set of transfer functions, $\{\tilde{M}_H(s), \tilde{N}_H(s), M_H(s), N_H(s)\} \in \mathcal{H}_\infty$, which satisfies (8.65).

In the next section, we use the foregoing relation between the doubly-coprime factorization of a plant and the stabilizing controller for the plant to provide a means of characterizing all controllers which stabilize a given plant, i.e., all controllers which render the feedback control system internally stable.

8.5 Stabilizing Controllers

Recall, from the discussion in Section 8.3.1, that achieving stability through unstable pole-zero cancellations is not robust. Therefore we cannot rely, in practice, on unstable pole-zero cancellations between the mathematical models for the plant and controller to lead to the implementation of a stable feedback control system. To avoid this pitfall we need a controller with model $H(s)$ which when used with a plant having model $G(s)$ makes the state model for the closed loop system internally stable. Controller models, $H(s)$, having this property are said to stabilize the plant model, $G(s)$, or, alternatively, are said to be stabilizing.

Alternatively, the internal stability of the state model for the closed loop system is equivalent to the input-output stability from $\{u_1(t), u_2(t)\}$ as input to $\{y_1(t), y_2(t)\}$ as output for the interconnection shown in Figure 8.3. Satisfaction of this condition implies that $\Theta(s) \in \mathcal{H}_\infty$ where

$$\begin{bmatrix} Y_1(s) \\ Y_2(s) \end{bmatrix} = \Theta(s) \begin{bmatrix} U_1(s) \\ U_2(s) \end{bmatrix} \tag{8.66}$$

We will show that $\Theta(s) \in \mathcal{H}_\infty$ is equivalent to $W(s) \in \mathcal{H}_\infty$ where

$$\begin{bmatrix} E_1(s) \\ E_2(s) \end{bmatrix} = W(s) \begin{bmatrix} U_1(s) \\ U_2(s) \end{bmatrix} \tag{8.67}$$

In order to establish conditions on $G(s)$ and $H(s)$ which insure that $W(s) \in \mathcal{H}_\infty$ we need to determine relations between the blocks of $W(s)$ and $G(s)$, $H(s)$. Before doing this consider the following fundamental aspect of feedback control systems.

Recall that transfer functions of physical processes must be bounded at infinity. Therefore to be physically meaningful $G(s)$, $H(s)$ and $W(s)$ must be either proper or strictly proper. Assuming that $G(s)$ and $H(s)$ are proper or strictly proper, the feedback control system is said to be well-posed if $W(s)$ is proper or strictly proper. We will see that $W(s)$ is proper or strictly proper if and only if

$$[I - G(\infty)H(\infty)]^{-1} \qquad \text{exists}$$

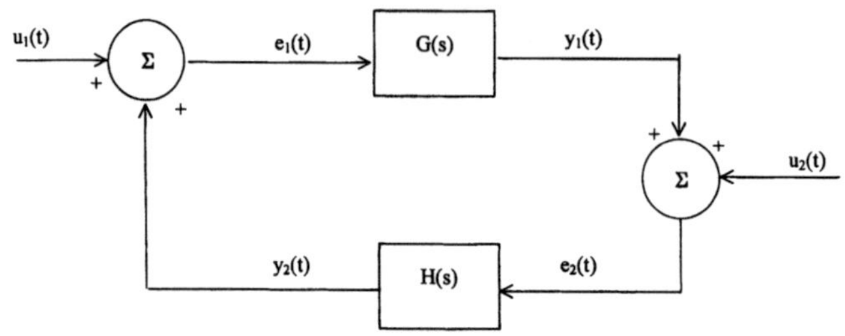

Figure 8.3 Setup for Plant Stabilization Criterion

Notice that if either $G(\infty)$ or $H(\infty)$ is null, i.e., strictly proper, this condition is satisfied and the feedback control system is well-posed. Recall that $H(s)$ is strictly proper, independent of whether $G(s)$ is proper or strictly proper, in the case of an observer based controller, (8.52). Therefore observer based control systems are always well-posed.

We proceed now to establish expressions for the partitions of $W(s)$ in terms of $G(s)$ and $H(s)$.

8.5.1 *Relating* W(s) *to* G(s), H(s)

We begin by noting from Figure 8.3 that

$$E_1(s) = U_1(s) + H(s)E_2(s) \tag{8.68}$$

$$E_2(s) = U_2(s) + G(s)E_1(s) \tag{8.69}$$

or

$$U(s) = W^{-1}(s)E(s)$$

where

$$W^{-1}(s) = \begin{bmatrix} I & -H(s) \\ -G(s) & I \end{bmatrix}$$

Recalling that $G(s), H(s)$ are $p \times m$ and $m \times p$ respectively we see that $W^{-1}(s)$ and $W(s)$ are each $(m+p) \times (m+p)$. Now the determination of $W(s)$, (8.67), in terms of $G(s), H(s)$ proceeds as follows.

First substituting (8.69) in (8.68) and solving for $E_1(s)$ as well as substituting (8.68) in (8.69) and solving for $E_2(s)$ gives

$$E_1(s) = [I - H(s)G(s)]^{-1}U_1(s) + [I - H(s)G(s)]^{-1}H(s)U_2(s) \tag{8.70}$$

$$E_2(s) = [I - G(s)H(s)]^{-1}G(s)U_1(s) + [I - G(s)H(s)]^{-1}U_2(s) \tag{8.71}$$

Alternatively, substituting for $E_1(s)$ from (8.70) in (8.69) gives

$$E_2(s) = G(s)[I - H(s)G(s)]^{-1}U_1(s) + \left[I + G(s)[I - H(s)G(s)]^{-1}H(s)\right]U_2(s) \tag{8.72}$$

Then comparing (8.71, 8.72) yields the following equalities

$$[I - G(s)H(s)]^{-1}G(s) = G(s)[I - H(s)G(s)]^{-1} \tag{8.73}$$

$$[I - G(s)H(s)]^{-1} = I + G(s)[I - H(s)G(s)]^{-1}H(s) \tag{8.74}$$

Alternatively, by substituting for $E_2(s)$ from (8.71) in (8.68) and comparing the resulting expression for $E_1(s)$ with the expression for $E_1(s)$ given by (8.70) leads to equations (8.73, 8.74) with $G(s)$ and $H(s)$ interchanged.

The foregoing results, which can also be obtained using the matrix inversion lemma, (Appendix), enable $W(s)$, (8.67), to be expressed in terms of $G(s), H(s)$ in the following ways

$$W(s) = \begin{bmatrix} W_1(s) & W_2(s) \\ W_3(s) & W_4(s) \end{bmatrix}$$

$$= \begin{bmatrix} [I - H(s)G(s)]^{-1} & [I - H(s)G(s)]^{-1}H(s) \\ G(s)[I - H(s)G(s)]^{-1} & I + G(s)[I - H(s)G(s)]^{-1}H(s) \end{bmatrix} \tag{8.75}$$

$$= \begin{bmatrix} I + H(s)[I - G(s)H(s)]^{-1}G(s) & H(s)[I - G(s)H(s)]^{-1} \\ [I - G(s)H(s)]^{-1}G(s) & [I - G(s)H(s)]^{-1} \end{bmatrix} \tag{8.76}$$

Thus we see from the dependency of the blocks of $W(s)$ on $[I - G(s)H(s)]^{-1}$ that $W(s)$ is proper or strictly proper if and only if $[I - G(s)H(s)]^{-1}$ is proper or strictly proper which requires that $I - D_G D_H$ be invertible where $G(\infty) = D_G$ and $H(\infty) = D_H$.

Finally, notice from (8.67, 8.75, 8.76) that we can write $\Theta(s)$, (8.66), as

$$\Theta(s) = \begin{bmatrix} W_3(s) & W_4(s) - I \\ W_1(s) - I & W_2(s) \end{bmatrix}$$

Therefore we have $\Theta(s) \in \mathcal{H}_\infty$ is equivalent to $W(s) \in \mathcal{H}_\infty$.

8.5.2 A criterion for stabilizing controllers

We showed, in the previous section, that $W(s) \in \mathcal{H}_\infty$ is equivalent to the corresponding closed loop system being internally stable. In this section we give a more useful criterion for deciding if a controller is stabilizing. We will use this criterion in the next section to develop a characterization of all controllers which are capable of stabilizing a given plant. The criterion to be developed here is stated in the following theorem.

Theorem 8.4 A controller having transfer function $H(s)$ with right coprime factorization $H(s) = N_H(s)M_H^{-1}(s)$ stabilizes a plant having transfer function $G(s)$ with left coprime factorization $G(s) = \tilde{M}_G^{-1}(s)\tilde{N}_G(s)$ if and only if

$$\Delta(s) \in \mathcal{U}_\infty \tag{8.77}$$

where

$$\Delta(s) = \tilde{M}_G(s)M_H(s) - \tilde{N}_G(s)N_H(s)$$

Proof Recall that $\Delta(s) \in \mathcal{U}_\infty$ is equivalent to $\Delta(s), \Delta^{-1}(s) \in \mathcal{H}_\infty$. Since $\Delta(s)$ is defined as the sum and product of transfer functions in $\mathcal{H}_\infty$, we have $\Delta(s) \in \mathcal{H}_\infty$ irrespective of the internal stability of the closed loop system. Therefore we need to show that $\Delta^{-1}(s) \in \mathcal{H}_\infty$ is necessary and sufficient for $W(s) \in \mathcal{H}_\infty$.

In order to show that $\Delta^{-1}(s) \in \mathcal{H}_\infty$ is sufficient for $W(s) \in \mathcal{H}_\infty$ we assume $\Delta^{-1}(s) \in \mathcal{H}_\infty$ and proceed as follows.

From the given coprime factorization of the plant and controller we have

$$G(s)H(s) = \tilde{M}_G^{-1}(s)\tilde{N}_G(s)N_H(s)M_H^{-1}(s)$$

or

$$\tilde{M}_G(s)G(s)H(s)M_H(s) = \tilde{N}_G(s)N_H(s)$$

Therefore $\Delta(s)$, (8.77), can be written as

$$\Delta(s) = \tilde{M}_G(s)[I - G(s)H(s)]M_H(s)$$

This allows us to express $[I - G(s)H(s)]^{-1}$as

$$[I - G(s)H(s)]^{-1} = M_H(s)\Delta^{-1}(s)\tilde{M}_G(s)$$

Then we can use this expression to rewrite $W(s)$, (8.75, 8.76) as follows

$$W(s) = \begin{bmatrix} I + N_H(s)\Delta^{-1}(s)\tilde{N}_G(s) & N_H(s)\Delta^{-1}(s)\tilde{M}_G(s) \\ M_H(s)\Delta^{-1}(s)\tilde{N}_G(s) & M_H(s)\Delta^{-1}(s)\tilde{M}_G(s) \end{bmatrix} \tag{8.78}$$

Therefore, assuming that $\Delta^{-1}(s) \in \mathcal{H}_\infty$ we see that $W(s) \in \mathcal{H}_\infty$ since each element of $W(s)$ consists of products and sums of transfer functions in $\mathcal{H}_\infty$. This shows that (8.77) is sufficient for the controller to be stabilizing. In order to show that $\Delta^{-1}(s) \in \mathcal{H}_\infty$ is necessary for $W(s) \in \mathcal{H}_\infty$ we assume $W(s) \in \mathcal{H}_\infty$ and proceed as follows.

Since $H(s) = N_H(s)M_H^{-1}(s)$ is a coprime factorization, the following Bezout identity is satisfied

$$X_H(s)M_H(s) + Y_H(s)N_H(s) = I$$

for some $X_H(s), Y_H(s) \in \mathcal{H}_\infty$. Then post-multiplying this Bezout identity by $\Delta^{-1}(s)$ $\tilde{M}_G(s)$ and by $\Delta^{-1}(s)\tilde{N}_G(s)$ gives the following two equations

$$X_H(s)M_H(s)\Delta^{-1}(s)\tilde{M}_G(s) + Y_H(s)N_H(s)\Delta^{-1}(s)\tilde{M}_G(s) = Q_1(s)$$

$$X_H(s)M_H(s)\Delta^{-1}(s)\tilde{N}_G(s) + Y_H(s)N_H(s)\Delta^{-1}(s)\tilde{N}_G(s) = Q_2(s)$$

where

$$Q_1(s) = \Delta^{-1}(s)\tilde{M}_G(s)$$

$$Q_2(s) = \Delta^{-1}(s)\tilde{N}_G(s)$$

However we can rewrite these equations using $W(s)$, (8.78), as

$$X_H(s)W_4(s) + Y_H(s)W_2(s) = Q_1(s) \tag{8.79}$$

$$X_H(s)W_3(s) + Y_H(s)[W_1(s) - I] = Q_2(s) \tag{8.80}$$

Therefore, since all terms on the left sides of (8.79, 8.80) belong to $\mathcal{H}_\infty$, we have

$$Q_1(s), Q_2(s) \in \mathcal{H}_\infty \tag{8.81}$$

However the factors of an $\mathcal{H}_\infty$ function need not be in $\mathcal{H}_\infty$. Therefore we are not able to use (8.81) to conclude that $\Delta^{-1}(s) \in \mathcal{H}_\infty$.

Since $G(s) = \tilde{M}_G^{-1}(s)\tilde{N}_G(s)$ is a coprime factorization, the following Bezout identity is satisfied

$$\tilde{M}_G(s)\tilde{X}_G(s) + \tilde{N}_G(s)\tilde{Y}_G(s) = I$$

for some $\tilde{X}_G(s), \tilde{Y}_G(s) \in \mathcal{H}_\infty$. Then pre-multiplying this equation by $\Delta^{-1}(s)$ yields

$$Q_1(s)\tilde{X}_G(s) + Q_2(s)\tilde{Y}_G(s) = \Delta^{-1}(s)$$

and we see that $\Delta^{-1}(s) \in \mathcal{H}_\infty$ since all terms on the left side of this equation belong to $\mathcal{H}_\infty$. This shows that (8.77) is necessary for the controller to be stabilizing. ■

By interchanging the role of the plant and controller in the foregoing theorem, we see that a left coprime factored controller, $H(s) = \tilde{M}_H^{-1}(s)\tilde{N}_H(s)$, stabilizes a right coprime factored plant, $G(s) = N_G(s)M_G^{-1}(s)$, if and only if

$$\tilde{\Delta}(s) \in \mathcal{U}_\infty \tag{8.82}$$

where

$$\tilde{\Delta}(s) = \tilde{M}_H(s)M_G(s) - \tilde{N}_H(s)N_G(s)$$

In summary, we see that we can use either (8.77) or (8.82) to determine if a given controller stabilizes a given plant. We can use this fact together with a doubly-coprime factorization of the plant to provide a parametrization of all the plant's stabilizing controllers.

8.5.3 Youla parametrization of stabilizing controllers

We have just shown that if there are coprime factorizations of the transfer functions for the plant and controller which makes $\Delta(s)$, (8.77), or $\tilde{\Delta}(s)$, (8.82), a unit, i.e., $\Delta(s) \in \mathcal{U}_\infty$, or $\tilde{\Delta}(s) \in \mathcal{U}_\infty$, then the corresponding closed loop system is internally stable and the controller is said to be stabilizing. However, if the coprime factorizations of the plant and controller happen to also make both $\Delta(s)$ and $\tilde{\Delta}(s)$ identity matrices, then not only is the controller stabilizing but, in addition, the left and right coprime factorizations of the plant constitute a doubly-coprime factorization, (Section 8.4.3), of the plant.

The foregoing observation is used now to show that we can construct a set of stabilizing controllers by using a doubly-coprime factorization of the plant. The elements of this set are generated by varying an $m \times p$ matrix, $Q(s)$, over $\mathcal{H}_\infty$.

Theorem 8.5 Given a doubly-coprime factorization of a $p \times m$ plant transfer function matrix

$$G(s) = \tilde{M}_G^{-1}(s)\tilde{N}_G(s) = N_G(s)M_G^{-1}(s) \tag{8.83}$$

where

$$\begin{bmatrix} \tilde{M}_H(s) & -\tilde{N}_H(s) \\ -\tilde{N}_G(s) & \tilde{M}_G(s) \end{bmatrix} \begin{bmatrix} M_G(s) & N_H(s) \\ N_G(s) & M_H(s) \end{bmatrix} = \begin{bmatrix} I & \emptyset \\ \emptyset & I \end{bmatrix} \tag{8.84}$$

then either of the following forms for the $m \times p$ controller transfer function matrix stabilize the given plant

$$H(s) = N_c(s)M_c^{-1}(s) \tag{8.85}$$

$$H(s) = \tilde{M}_c^{-1}(s)\tilde{N}_c(s) \tag{8.86}$$

for all $m \times p$ matrices $Q(s)$ satisfying $Q(s) \in \mathcal{H}_\infty$ where

$$\begin{aligned} N_c(s) &= N_H(s) + M_G(s)Q(s) \\ M_c(s) &= M_H(s) + N_G(s)Q(s) \\ \tilde{N}_c(s) &= \tilde{N}_H(s) + Q(s)\tilde{M}_G(s) \\ \tilde{M}_c(s) &= \tilde{M}_H(s) + Q(s)\tilde{N}_G(s) \end{aligned} \tag{8.87}$$

Proof Recall that for (8.85) to be a coprime factorization of $H(s)$, we must have

(i) $N_c(s), M_c(s) \in \mathcal{H}_\infty$
(ii) $N_c(s), M_c(s)$ coprime

Since $N_c(s), M_c(s)$, (8.87), are each dependent on sums and products of matrices in $\mathcal{H}_\infty$, condition (i) is satisfied. In order to show that condition (ii) is satisfied, recall that $N_c(s), M_c(s)$ are coprime if $X(s), Y(s) \in \mathcal{H}_\infty$ satisfy the Bezout identity

$$X(s)M_c(s) + Y(s)N_c(s) = I \tag{8.88}$$

Now substituting for $M_c(s), N_c(s)$ from (8.87) and choosing

$$X(s) = \tilde{M}_G(s) \qquad\qquad Y(s) = -\tilde{N}_G(s) \tag{8.89}$$

yields the following equation

$$X(s)M_c(s) + Y(s)N_c(s) = \Theta_1 + \Theta_2$$

where

$$\Theta_1 = \tilde{M}_G(s)M_H(s) - \tilde{N}_G(s)N_H(s)$$

$$\Theta_2 = \left[\tilde{M}_G(s)N_G(s) - \tilde{N}_G(s)M_G(s)\right]Q(s)$$

Then we see that $\Theta_1 = I$ since the plant transfer function factorization is doubly-coprime, (8.84). Moreover we have $\Theta_2 = \emptyset$ from (8.83). Therefore (8.88) is satisfied and condition (ii) is satisfied. Thus (8.85) is indeed a coprime factorization of $H(s)$.

Finally, in order to show that the controllers, (8.85), are stabilizing, notice that we have just shown that

$$\tilde{M}_G(s)(s)M_c(s) - \tilde{N}_G(s)N_c(s) = I$$

Therefore, referring to Theorem 8.4, we have $\Delta(s) = I \in \mathcal{U}_\infty$ so that the controllers given by (8.85) are stabilizing.

This completes the proof that the controllers given by, (8.85), stabilize the given plant. The proof that the controllers given by (8.86) are stabilizing can be done in a similar fashion. ■

8.6 Lossless Systems and Related Ideas

A linear time-invariant system is said to be lossless if it is stable and if it has zero state response $y(t)$ corresponding to any $u(t) \in \mathcal{L}_2[0,\infty)$ such that

$$\|y(t)\|_2 = \|u(t)\|_2 \tag{8.90}$$

Notice, from Sections 7.4 and 7.5 of the previous chapter, that a lossless system has L_2 gain equal to one. However it is important to note that this condition is necessary but not sufficient for a system to be lossless, i.e., for (8.90) to be satisfied for all $u(t) \in \mathcal{L}_2[0,\infty)$.

In contrast, recall from the previous chapter that $\mathcal{L}_2[0,\infty)$ is isomorphic to $\mathcal{L}_2$ so that the L_2 norms in the time and frequency domains are equivalent. Therefore a system is lossless if its transfer function $G(s) \in \mathcal{H}_\infty$ maps $U(s) \in \mathcal{H}_2$ to $Y(s) \in \mathcal{H}_2$ such that

$$\|Y(j\omega\|_2 = \|U(j\omega)\|_2 \tag{8.91}$$

However since

$$\|Y(j\omega\|_2^2 = \frac{1}{2\pi}\int_{-\infty}^{\infty} U^*(j\omega)G^*(j\omega)G(j\omega)U(j\omega)d\omega$$

we see that (8.91) is satisfied for all $U(s) \in \mathcal{H}_2$ if and only if

$$G^*(j\omega)G(j\omega) = I_m \qquad \text{for all } \omega \in (-\infty,\infty) \tag{8.92}$$

where $G(j\omega)$ is $p \times m$. Notice that $p \geq m$ is necessary for (8.92) to be satisfied.

Mathematically, we can consider lossless systems to be unitary operators. Recall, from Chapter 7, that the Fourier transform is an isometry or unitary operator between the time domain Hilbert space $\mathcal{L}_2(-\infty,\infty)$ and the frequency domain Hilbert space $\mathcal{L}_2$. In the present instance (8.90) implies that an m-input p-output lossless system is an isometry or unitary operator between the m-dimensional input Hilbert space, $\mathcal{L}_2[0,\infty)$, and p-dimensional output Hilbert space, $\mathcal{L}_2[0,\infty)$. Alternatively, (8.91) implies that a lossless system is a unitary operator between the m-dimensional input Hardy space, $\mathcal{H}_2$, and the p-dimensional output Hardy space, $\mathcal{H}_2$.

An important property of lossless systems can be seen as follows. Notice that if we connect two systems with transfer function $G_1(s), G_2(s) \in \mathcal{H}_\infty$ is series with $G_1(s)$ being the transfer function of a lossless system, then we have $\|G_1(s)\|_\infty = 1$ and from Section 7.4.2 we see that

$$\|G_1(s)G_2(s)\|_\infty \leq \|G_1(s)\|_\infty \|G_2(s)\|_\infty = \|G_2(s)\|_\infty \tag{8.93}$$

However (8.92) implies that the L_2 norm of the output from the composite system is independent of $G_1(s)$. Therefore the inequality (8.93) becomes an equality

$$\|G_1(s)G_2(s)\|_\infty = \|G_2(s)\|_\infty$$

In the next chapter we will see that this type of invariance of the H_∞ norm plays an important role in the development of a solution to the H_∞ control problem.

Now since the linear models of the time-invariant physical processes being considered here, e.g., plants, have transfer function matrices whose elements are ratios of polynomials having real coefficients, we have

$$G^*(j\omega) = G^T(-j\omega) \qquad \text{for all } \omega \in (-\infty, \infty)$$

and (8.92) is equivalent to

$$G^T(-s)G(s) = I_m \qquad \text{for all complex } s \tag{8.94}$$

We can gain additional insight into these matters by reviewing the earlier development of these ideas in connection with electric filters.

8.6.1 All pass filters

Recall that a stable time-invariant linear SISO system having transfer function $G(s)$ has a steady-state output given as

$$y(t) = A_o \sin(\omega t + \Theta_0)$$

when its input is given by

$$u(t) = A_i \sin(\omega t)$$

where

$$A_o = A_i\sqrt{G^*(j\omega)G(j\omega)} = |G(j\omega|A_i$$

$$\Theta_0 = \tan^{-1}\left[\frac{\text{Im}[G(j\omega)]}{\text{Re}[G(j\omega)]}\right]$$

This system is referred to as an all pass filter if

$$|G(j\omega| = 1 \qquad \text{for all } \omega \in (-\infty, \infty) \tag{8.95}$$

or if

$$G(-s)G(s) = 1 \qquad \text{for all complex } s \tag{8.96}$$

so that the amplitude A_o of the steady state output equals the amplitude A_i of the input for all frequencies. Thus comparing (8.95, 8.96) with (8.92, 8.94) we see that all pass filters are lossless systems. Moreover if $G(s)$ is the transfer function of an all pass filter, then we see from (8.96) that $G(s)$ is proper, with $G(0) = 1$, and has poles which are mirror images of its zeros across the imaginary axis, i.e., if p_i is a pole of $G(s)$ then $-p_i$ must be a zero of $G(s)$. For example, the first order system having transfer function

$$G(s) = \frac{(s-1)}{(s+1)}$$

is an all pass filter

An all pass transfer function is a generalization of the concept of an all pass filter. This generalization is done by relaxing the requirement of stability. Thus $G(s)$ is said to be an all pass transfer function if (8.96) is satisfied. Notice that the mirror imaging of poles and zeros implied by (8.96) means that when $G(s)$ is an all pass transfer function it has no imaginary axis poles, i.e., $G(s) \in \mathcal{L}_\infty$

8.6.2 Inner transfer functions and adjoint systems

In order to distinguish between stable and unstable all pass transfer functions, stable all pass transfer functions are referred to as being inner. Thus a transfer function is inner when it is the model for a lossless system.

In the case of a transfer function matrix the following generalization is made. If $G(s) \in \mathcal{H}_\infty$ is $p \times m$, then $G(s)$ is said to be inner if (8.97) is satisfied or co-inner if (8.98) is satisfied where

$$G^{\sim}(s)G(s) = I_m \qquad \text{if } p \geq m \tag{8.97}$$

$$G(s)G^{\sim}(s) = I_p \qquad \text{if } p \leq m \tag{8.98}$$

for all complex s with

$$G^{\sim}(s) = G^T(-s)$$

Notice that $p \geq m$ $(p \leq m)$ is necessary for $G(s)$ to be inner (co-inner). Also notice that if the transfer function is square and inner then $G^{-1}(s) = G^{\sim}(s) \in \mathcal{H}_\infty^-$. Moreover, since dimensionally square systems have the property that the poles of $G^{-1}(s)$ are the zeros of $G(s)$, (Section 8.2.4), we see that in this case that all system zeros of $G(s)$ lie "in" the open right half plane, i.e., lie in the region of the complex plane where $G(s)$ is analytic. This is the origin of the terminology "inner". However in the nonsquare case this concentration of zeros in the open right half plane is not necessary for a transfer function to be inner, e.g.,

$$G(s) = \frac{1}{s+\sqrt{2}}\begin{bmatrix} s+1 \\ 1 \end{bmatrix}$$

is inner without having any zeros in the right half plane.

The system whose transfer function is $G^{\sim}(s)$ is referred to as the adjoint system (relative to a system having transfer function $G(s)$). Notice that the state model for $G^{\sim}(s)$ is related to the state model for $G(s)$ as

$$G(s) \stackrel{s}{=} \begin{bmatrix} A & B \\ C & D \end{bmatrix} \qquad G^{\sim}(s) \stackrel{s}{=} \begin{bmatrix} -A^T & -C^T \\ B^T & D^T \end{bmatrix} \tag{8.99}$$

since

$$G^T(-s) = B^T\left\{-\left[sI - (-A^T)\right]^{-1}\right\}C^T + D^T$$

Properties which the parameters of a minimal state model must satisfy in order for a system to have an inner transfer function are given in the following theorem.

Theorem 8.6 $G(s) \in \mathcal{H}_\infty$ is inner if it has a minimal state model

$$G(s) \stackrel{s}{=} \begin{bmatrix} A & B \\ C & D \end{bmatrix}$$

with parameters that satisfy the following conditions:

$$D^T D = I$$
$$B^T W_o + D^T C = \emptyset$$
$$A^T W_o + W_o A + C^T C = \emptyset$$

Proof From the state model for $G^{\sim}(s)$, (8.99), and the rule for composing two systems, (8.9), we obtain

$$G^{\sim}(s)G(s) \stackrel{s}{=} \begin{bmatrix} \begin{bmatrix} -A^T & -C^T C \\ \emptyset & A \end{bmatrix} & \begin{bmatrix} -C^T D \\ B \end{bmatrix} \\ \begin{bmatrix} B^T & D^T C \end{bmatrix} & D^T D \end{bmatrix}$$

Next transforming coordinates using the coordinate transformation matrix

$$T = \begin{bmatrix} I & X \\ \emptyset & I \end{bmatrix}$$

gives

$$G^{\sim}(s)G(s) \stackrel{s}{=} \begin{bmatrix} \begin{bmatrix} -A^T & A_2 \\ \emptyset & A \end{bmatrix} & \begin{bmatrix} B_1 \\ B_2 \end{bmatrix} \\ \begin{bmatrix} C_1 & C_2 \end{bmatrix} & D^T D \end{bmatrix}$$

where $A_2 = -A^T X - XA - C^T C$ and

$$\begin{bmatrix} B_1 \\ B_2 \end{bmatrix} = \begin{bmatrix} -C_2^T \\ C_1^T \end{bmatrix} = \begin{bmatrix} -C^T D - XB \\ B \end{bmatrix}$$

Finally, applying the conditions given in the theorem yields

$$X = W_o \qquad A_2 = \emptyset \qquad B_1 = \emptyset \qquad C_2 = \emptyset \qquad D^T D = I$$

and the state model for $G^{\sim}(s)G(s)$ becomes

$$G^{\sim}(s)G(s) \stackrel{s}{=} \begin{bmatrix} \bar{A} & \bar{B} \\ \bar{C} & \bar{D} \end{bmatrix}$$

where

$$\bar{A} = \begin{bmatrix} -A^T & \emptyset \\ \emptyset & A \end{bmatrix} \qquad \bar{B} = \begin{bmatrix} \emptyset \\ B_2 \end{bmatrix}$$

$$\bar{C} = [\, C_1 \quad \emptyset \,] \qquad \bar{D} = I$$

Thus we have

$$\begin{aligned} G^{\sim}(s)G(s) &= \bar{C}(sI - \bar{A})^{-1}\bar{B} + \bar{D} \\ &= I \end{aligned}$$

and the theorem is shown. ■

In developing a solution to the H_∞ state feedback control problem in the next chapter, we will extend the foregoing characterization of state models so that the resulting system is J-lossless with J-inner transfer function.

8.7 Summary

In this chapter we have established state models for certain interconnections and operations with systems. This included the ideas of system inversion, system zeros, system coprimeness and lossless systems. We will see in the next two chapters that the operation of coprime factorization coupled with the requirements on a system's state model for that system to be lossless, plays an important role in developing a solution to the H_∞ feedback control problem.

8.8 Notes and References

The treatise by Vidyasagar [44], on the consequent ramifications of the coprime factorization approach to feedback control places coprime factorization in the mathematical setting of algebra known as ring theory. The term "unit" refers to elements in a ring that have an inverse in the ring. The development of state models for the coprime factors used here is treated in [14].

9

H_∞ State Feedback and Estimation

9.1 Introduction

In the previous two chapters we developed ideas for characterizing signals and systems. In the next two chapters we use these ideas to develop the design equations for a controller which is required to

(i) stabilize the closed-loop control system and
(ii) constrain the L_∞ norm of the closed-loop transfer function from disturbance input, $u_1(t)$, to desired output, $y_1(t)$, to be no greater than γ, a specified positive scalar.

Figure 9.1 gives the setup for this problem. Before going on to consider this setup in more detail, we make the following more general observations on this control problem.

First, we should recall that to avoid having a system's input–output stability dependent on the physically unrealistic requirement of pole-zero cancellation, we have used the term "stability" to refer to a system's internal stability. Thus when requirement (i) is satisfied so that the closed-loop system is internally stable, the L_∞ norm mentioned in requirement (ii) is an H_∞ norm. Therefore controllers satisfying (i, ii) are said to solve the H_∞ control problem.

Alternatively, recall from Chapter 7, that the H_∞ system norm is induced by the H_2 signal norm in the frequency domain or the L_2 signal norm in the time domain. Therefore when requirement (i) is satisfied, (ii) can be interpreted as the requirement that the closed-loop control system's L_2 gain from $u_1(t)$ to $y_1(t)$ be no greater than γ. Thus the H_∞ control problem is sometimes referred to as the L_2 gain control problem.

In addition, notice that satisfaction of requirement (ii) does not imply that requirement (i) is satisfied since the L_∞ norm is defined for both stable and unstable systems provided only that the system's transfer function have no imaginary axis poles. This differs with both the quadratic and LQG control problems since the performance indices, J_{QC} and J_{GC}, are infinite when the closed-loop system is unstable.

Concerning the setup for the H_∞ control problem, it has become customary to replace the actual plant in the feedback control configuration, Figure 9.1, by a composite system referred to as a generalized plant. The generalized plant consists of the plant to be controlled plus any other systems which the control system designer may want to connect

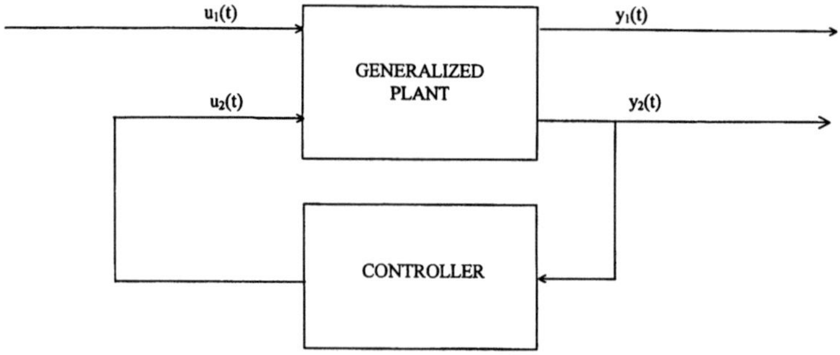

Figure 9.1 Setup for H_∞ Control Problem

$u_1(t)$ = disturbance input $y_1(t)$ = desired output
$u_2(t)$ = controlled input $y_2(t)$ = measured output

where $u_1(t), u_2(t), y_1(t), y_2(t)$ have dimensions m_1, m_2, p_1, p_2 respectively.

in cascade with the plant (referred to as weights or filters). For instance, the filters can be used to take into account any known spectral characteristics of the disturbance signals, e.g., narrow band noise. Then when the feedback control system is implemented, the controller determined by applying the control algorithm to the generalized plant is used with the actual plant.

In what follows the signals $\{u_i(t), y_i(t) : i = 1, 2\}$ shown in Figure 9.1 are assumed to be the various inputs and outputs associated with the generalized plant which has real-rational transfer function $G(s)$ with given real-parameter state model

$$G(s) \stackrel{s}{=} \begin{bmatrix} A & \begin{bmatrix} B_1 & B_2 \end{bmatrix} \\ \begin{bmatrix} C_1 \\ C_2 \end{bmatrix} & \begin{bmatrix} D_{11} & D_{12} \\ D_{21} & D_{22} \end{bmatrix} \end{bmatrix} \tag{9.1}$$

As in the LQG and quadratic control problems, a solution to the H_∞ output feedback control problem takes the form of an observer-based controller and involves the stabilizing solutions to two algebraic Riccati equations. Thus in order to develop a solution to the H_∞ output feedback control problem we need first to consider two simpler problems namely, the H_∞ state feedback control problem, and the H_∞ state estimation problem. These basic H_∞ problems are taken up in this chapter with the H_∞ output feedback control problem being addressed in the next chapter.

As we progress through the development of solutions to these problems, we will see that the generalized plant's D matrix plays a pivotal role. However, only certain parts of the D matrix play a role in each problem. For instance, the existence of a solution to the H_∞ state feedback control problem requires D_{12} to have independent columns and only the first p_1 rows of D, denoted by D_1, are involved in the solution to this problem where

$$D_1 = [D_{11} \quad D_{12}]$$

However, the existence of a solution to the H_∞ state estimation problem requires D_{21} to have independent rows and only the first m_1 columns of D, denoted by D_2, are involved in the solution to this problem where

$$D_2 = \begin{bmatrix} D_{11} \\ D_{21} \end{bmatrix}$$

In the next chapter we will see that the last p_2 rows of D, denoted by D_3, are involved in the solution to the H_∞ output feedback control problem where

$$D_3 = [\, D_{21} \quad D_{22} \,]$$

9.2 H_∞ State Feedback Control Problem

Assuming that the state of the generalized plant, $x(t)$, is known we want to choose the state feedback matrix, K_2, so that setting the controlled input as

$$u_2(t) = K_2 x(t) \tag{9.2}$$

makes the closed-loop system internally stable and the transfer function from $u_1(t)$ to $y_1(t)$ have H_∞ norm no greater than a specified positive constant γ.

Since the measured output, $y_2(t)$, is not needed in this problem, we take the plant as the generalized plant, (9.1), with $y_2(t)$ missing. Thus we assume the parameters of $G_1(s)$ are known where

$$Y_1(s) = G_1(s) \begin{bmatrix} U_1(s) \\ U_2(s) \end{bmatrix} \tag{9.3}$$

with

$$G_1(s) \stackrel{s}{=} \begin{bmatrix} A & B \\ C_1 & D_1 \end{bmatrix}$$

where

$$B = [\, B_1 \quad B_2 \,] \qquad\qquad D_1 = [\, D_{11} \quad D_{12} \,]$$

The relation between the signals and systems in the H_∞ state feedback control problem are shown in Figure 9.2. where the additional signals $\bar{u}_1(t)$ and $K_1 x(t)$ are going to be used in the development of a solution, K_2, to the H_∞ state feedback control problem.

Now after implementing state feedback we have

$$Y_1(s) = T_c(s) U_1(s) \tag{9.4}$$

where

$$T_c(s) \stackrel{s}{=} \begin{bmatrix} A + B_2 K_2 & B_1 \\ C_1 + D_{12} K_2 & D_{11} \end{bmatrix}$$

and we want to find K_2 such that

$$\text{(i)}: \; T_c(s) \in \mathcal{H}_\infty \qquad \text{and} \qquad \text{(ii)}: \; \|T_c(s)\|_\infty \leq \gamma \tag{9.5}$$

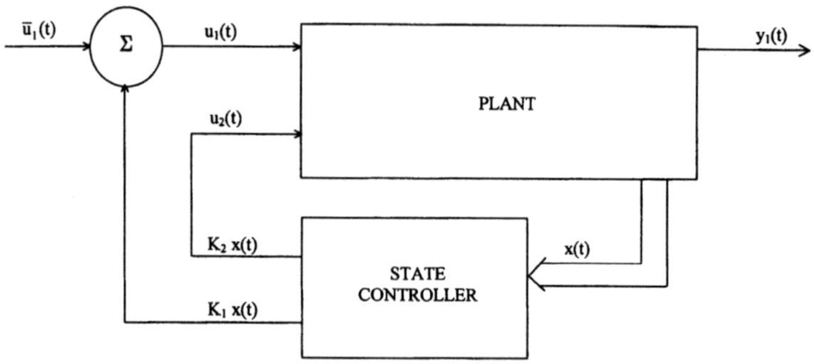

Figure 9.2 Setup for the H_∞ State Feedback Control Problem

Next recall from Chapter 7 that the H_∞ norm is defined by

$$\|T_c(s)\|_\infty = \sup_{\omega\in(-\infty,\infty)} \sigma_{\max}[T_c(j\omega)]$$

where $\sigma_{\max}[T_c(j\omega)]$ is the largest singular value of $T_c(j\omega)$. Now since

$$\lim_{\omega\rightarrow\infty} T_c(j\omega) = D_{11}$$

we see that it is not possible to satisfy condition (ii), (9.5), if

$$\sigma_{\max}[D_{11}] > \gamma$$

where

$$\sigma_{\max}[D_{11}] = \max_i\{\delta_i^{\frac{1}{2}}\} \qquad \delta_i \in \lambda[D_{11}^T D_{11}]$$

In subsequent sections we will see that the more restrictive condition

$$\sigma_{\max}[D_{11}] < \gamma \tag{9.6}$$

must be satisfied in order to solve the H_∞ state feedback control problem. In addition, we will see that this condition is also needed to solve the H_∞ state estimation problem.

We can summarize the strategy we will employ now to develop the design equations for K_2 as follows.

1. Start by assuming stabilizing feedback, i.e., assume K_2 achieves condition (i), (9.5).
2. Define a scalar quadratic performance index, P_γ, in terms of $y_1(t)$ and $u_1(t)$ such that $P_\gamma \leq 0$ implies condition (ii), (9.5), is satisfied.
3. Use $G_1(s)$, (9.3), to introduce a transfer function $\bar{G}_1(s)$ relating $\{u_1(t), y_1(t)\}$ as output to $\{u_1(t), u_2(t)\}$ as input under the constraint of stabilizing state feedback, i.e., $u_2(t) = K_2x(t)$ such that $\bar{G}_1(s) \in \mathcal{H}_\infty$. This enables P_γ to be rewritten as a quadratic in $u_1(t)$ and $u_2(t)$ with $U_1(s) \in \mathcal{H}_2$ and $U_2(s)$ restricted to stabilize $\bar{G}_1(s)$.
4. Convert P_γ to a quadratic in $\{V_1(s), V_2(s)\} \in \mathcal{H}_2$ where $\{V_1(s), V_2(s)\}$ arise as a consequence of using J-inner coprime factorization on $\bar{G}_1(s)$.

We will see that the design equations for K_2 result from carrying out the last step.

9.2.1 Introduction of $\mathbf{P}_\gamma$

Suppose K_2 is chosen so that condition (i), (9.5), is satisfied. Then we see from Chapter 7 that

$$y_1(t) \in \mathcal{L}_2[0,\infty) \qquad \text{if} \qquad u_1(t) \in \mathcal{L}_2[0,\infty)$$

and we can interpret condition (ii), (9.5), in the time domain as a constraint on the L_2 gain

$$\| y_1(t)\|_2 \leq \gamma \|u_1(t)\|_2 \tag{9.7}$$

Next define P_γ as

$$P_\gamma = -\gamma^2 \|u_1(t)\|_2^2 + \|y_1(t)\|_2^2 \tag{9.8}$$

Then condition given in (9.7) is equivalent to

$$P_\gamma \leq 0 \qquad u_1(t) \in \mathcal{L}_2[0,\infty) \tag{9.9}$$

Now since $\mathcal{L}_2[0,\infty)$ is an inner product space we can rewrite P_γ as

$$P_\gamma = \left\langle \begin{bmatrix} u_1(t) \\ y_1(t) \end{bmatrix}, J_{\gamma mp} \begin{bmatrix} u_1(t) \\ y_1(t) \end{bmatrix} \right\rangle \tag{9.10}$$

where

$$J_{\gamma mp} = \begin{bmatrix} -\gamma^2 I_{m_1} & \varnothing \\ \varnothing & I_{p_1} \end{bmatrix}$$

However, recalling from Chapter 7 that $\mathcal{L}_2[0,\infty)$ is isomorphic to $\mathcal{H}_2$, we see that P_γ, (9.10), can be rewritten in the frequency domain as

$$\begin{aligned} P_\gamma &= \left\langle \begin{bmatrix} U_1(s) \\ Y_1(s) \end{bmatrix}, J_{\gamma mp} \begin{bmatrix} U_1(s) \\ Y_1(s) \end{bmatrix} \right\rangle \\ &= -\gamma^2 \|U_1(s)\|_2^2 + \|Y_1(s)\|_2^2 \end{aligned} \tag{9.11}$$

and condition (ii), (9.5), is satisfied if

$$P_\gamma \leq 0 \qquad U_1(s) \in \mathcal{H}_2 \tag{9.12}$$

It is important to remember that the foregoing result assumes stabilizing state feedback.

9.2.2 Introduction of $\bar{\mathbf{G}}_1(\mathbf{s})$

Recalling from (9.3) that $Y_1(s)$ depends on $U_1(s)$ and $U_2(s)$, we see that we can relate

$\{U_1(s), Y_1(s)\}$ to $\{U_1(s), U_2(s)\}$ through the use of a transfer function $\bar{G}_1(s)$ as

$$\begin{bmatrix} U_1(s) \\ Y_1(s) \end{bmatrix} = \bar{G}_1(s) \begin{bmatrix} U_1(s) \\ U_2(s) \end{bmatrix} \tag{9.13}$$

where

$$\bar{G}_1(s) = \left[\begin{array}{c|c} A & \bar{B} \\ \hline \bar{C}_1 & \bar{D}_1 \end{array}\right] \qquad \bar{C}_1 = \begin{bmatrix} \emptyset \\ C_1 \end{bmatrix} \qquad \bar{D}_1 = \begin{bmatrix} I_{m_1} & \emptyset \\ D_{11} & D_{12} \end{bmatrix}$$

Now under the constraint that the state feedback is stabilizing we have

$$X(s), Y_1(s) \in \mathcal{H}_2 \qquad \text{for} \qquad \begin{array}{l} \text{(i): } U_1(s) \in \mathcal{H}_2 \\ \text{(ii): } U_2(s) = K_2 X(s) \end{array}$$

where $\lambda[A + B_2K_2]$ are all in the open left half plane, and $X(s)$ is the Laplace transform of the closed-loop system state.

Therefore we can use (9.13) to rewrite the inner product defining P_γ, (9.11), in terms of $\{U_1(s), U_2(s)\}$ as

$$\begin{aligned} P_\gamma &= \left\langle \bar{G}_1(s) \begin{bmatrix} U_1(s) \\ U_2(s) \end{bmatrix}, J_{\gamma mp} \bar{G}_1(s) \begin{bmatrix} U_1(s) \\ U_2(s) \end{bmatrix} \right\rangle \\ &= \left\langle \begin{bmatrix} U_1(s) \\ U_2(s) \end{bmatrix}, \bar{G}_1^T(-s) J_{\gamma mp} \bar{G}_1(s) \begin{bmatrix} U_1(s) \\ U_2(s) \end{bmatrix} \right\rangle \end{aligned} \tag{9.14}$$

where

$$U_2(s) = K_2 X(s)$$

In what follows, it is important to remember that even though $\bar{G}_1(s)$ may possibly be unstable, the restriction that the controlled input, $U_2(s) = K_2X(s)$, must be stabilizing maintains P_γ, given as either (9.8) or (9.14), finite.

9.2.3 Introduction of J-inner coprime factorization

Recall, from Chapter 8, that $\bar{G}_1(s) \in \mathcal{L}_\infty$ has a right coprime factorization given by

$$\bar{G}_1(s) = \bar{N}(s)\bar{M}^{-1}(s) \tag{9.15}$$

where $\bar{N}(s), \bar{M}(s) \in \mathcal{H}_\infty$ with $\bar{N}(s), \bar{M}(s)$ being coprime. Now we are going to make use of a special type of right coprime factorization in which $\bar{N}(s)$ in (9.15) satisfies

$$\bar{N}^T(-s) J_{\gamma mp} \bar{N}(s) = J_{\gamma mm} \tag{9.16}$$

$$J_{\gamma mp} = \begin{bmatrix} -\gamma^2 I_{m_1} & \emptyset \\ \emptyset & I_{p_1} \end{bmatrix} \qquad J_{\gamma mm} = \begin{bmatrix} -\gamma^2 I_{m_1} & \emptyset \\ \emptyset & I_{m_2} \end{bmatrix}$$

A system having transfer function $\bar{N}(s) \in \mathcal{H}_\infty$ which satisfies (9.16) is said to be J-inner and a right coprime factorization (9.15) which has $\bar{N}(s)$ J-inner is referred to here as a J-inner coprime factorization. The utility of using this type of coprime factorization here can be seen immediately as follows.

Let $\bar{G}_1(s)$ have a J-inner coprime factorization as denoted by (9.15, 9.16). Then using the fact that $\bar{N}(s)$ is J-inner we see that P_γ, (9.14), can be rewritten as

$$P_\gamma = \left\langle \begin{bmatrix} U_1(s) \\ U_2(s) \end{bmatrix}, \left([\bar{M}^{-1}(-s)]^T J_{\gamma mm} \bar{M}^{-1}(s) \right) \begin{bmatrix} U_1(s) \\ U_2(s) \end{bmatrix} \right\rangle \tag{9.17}$$

where

$$U_2(s) = K_2 X(s)$$

Next recall, from Chapter 8, that since $\bar{M}(s), \bar{N}(s) \in \mathcal{H}_\infty$ we can obtain $\{U_1(s), U_2(s)\}$ and $\{U_1(s), Y_1(s)\}$ from $V_1(s), V_2(s) \in \mathcal{H}_2$ as

$$\begin{bmatrix} U_1(s) \\ U_2(s) \end{bmatrix} = \bar{M}(s) \begin{bmatrix} V_1(s) \\ V_2(s) \end{bmatrix} \tag{9.18}$$

$$\begin{bmatrix} U_1(s) \\ Y_1(s) \end{bmatrix} = \bar{N}(s) \begin{bmatrix} V_1(s) \\ V_2(s) \end{bmatrix} \tag{9.19}$$

Then since $\bar{N}(s) \in \mathcal{H}_\infty$ we have

$$\begin{bmatrix} U_1(s) \\ Y_1(s) \end{bmatrix} \in \mathcal{H}_2 \qquad \text{for all} \begin{bmatrix} V_1(s) \\ V_2(s) \end{bmatrix} \in \mathcal{H}_2 \tag{9.20}$$

Finally, by using (9.18) in (9.17) we can rewrite P_γ as

$$\begin{aligned} P_\gamma &= \left\langle \begin{bmatrix} V_1(s) \\ V_2(s) \end{bmatrix}, J_{\gamma mm} \begin{bmatrix} V_1(s) \\ V_2(s) \end{bmatrix} \right\rangle \\ &= -\gamma^2 \| V_1(s) \|_2^2 + \| V_2(s) \|_2^2 \end{aligned} \tag{9.21}$$

Notice that unlike $U_2(s)$, which is restricted to a subspace of $\mathcal{H}_2$ since it must provide stabilizing state feedback, $V_1(s), V_2(s)$ are not restricted to any subspace of $\mathcal{H}_2$. We will use this freedom in the choice of the $V_i s$ to ensure that $P_\gamma \le 0$.

9.2.4 Consequences of J-inner coprime factorization

Suppose we have a J-inner coprime factorized for $\bar{G}_1(s)$, (9.15, 9.16). Then to proceed we need to consider state models for these coprime factors. Therefore, recalling (8.46) and Section 8.3.3 we see that state models for the coprime factors, $\bar{M}(s), \bar{N}(s)$, (9.15), can be determined from the state model for $\bar{G}_1(s)$, (9.13), as

$$\bar{M}(s) \stackrel{s}{=} \begin{bmatrix} A + BK & B\Delta^{-1} \\ K & \Delta^{-1} \end{bmatrix} \tag{9.22}$$

$$\bar{N}(s) \stackrel{s}{=} \begin{bmatrix} A + BK & B\Delta^{-1} \\ \bar{C}_1 + \bar{D}_1 K & \bar{D}_1 \Delta^{-1} \end{bmatrix} \tag{9.23}$$

where K, Δ are chosen so that $\bar{M}(s), \bar{N}(s) \in \mathcal{H}_\infty$ with $\bar{N}(s)$ satisfying (9.16); Δ is a constant nonsingular $(m_1 + m_2) \times (m_1 + m_2)$ matrix which along with K is partitioned so as to conform with the partitioning of B and $\bar{D}_1$, viz.,

$$\Delta = \begin{bmatrix} \Delta_1 & \Delta_2 \\ \Delta_3 & \Delta_4 \end{bmatrix} \qquad K = \begin{bmatrix} K_1 \\ K_2 \end{bmatrix}$$

We show now that the H_∞ state feedback control problem is solved when K and Δ are chosen so that the foregoing coprime factorization is J-inner. The equations needed to determine K and Δ are developed in subsequent sections.

We begin by using (9.18) to express $V_1(s), V_2(s)$ in terms of $U_1(s), U_2(s)$ as

$$\begin{bmatrix} V_1(s) \\ V_2(s) \end{bmatrix} = \bar{M}^{-1}(s) \begin{bmatrix} U_1(s) \\ U_2(s) \end{bmatrix} \tag{9.24}$$

Then using the state model for the inverse system developed in Chapter 8, we see that the state model for the inverse of $\bar{M}(s)$, (9.22), is

$$\bar{M}^{-1}(s) \stackrel{s}{=} \begin{bmatrix} A & B \\ -\Delta K & \Delta \end{bmatrix} \tag{9.25}$$

Notice that the system and input matrices for the generalized plant, $G_1(s)$, (9.3), and for $\bar{M}^{-1}(s)$, (9.24), are the same. Therefore the states of these systems are equal for all time if their initial states are equal. The initial state of each system is the same since each is null. This is because we are basing our development of the H_∞ state feedback controller on transfer functions which relate the system output to the system input under the assumption the initial state is null.

The foregoing observation together with the state model for $\bar{M}^{-1}(s)$, (9.25), allows us to write the $V_i s$ as

$$\begin{bmatrix} V_1(s) \\ V_2(s) \end{bmatrix} = -\Delta K X(s) + \Delta \begin{bmatrix} U_1(s) \\ U_2(s) \end{bmatrix} \tag{9.26}$$

where $X(s)$ is the Laplace transform of the state of the state model for $G_1(s)$, (9.3).

Now suppose we choose $U_1(s)$ and $U_2(s)$ as

$$\begin{bmatrix} U_1(s) \\ U_2(s) \end{bmatrix} = \begin{bmatrix} K_1 \\ K_2 \end{bmatrix} X(s) + \begin{bmatrix} \bar{U}_1(s) \\ \emptyset \end{bmatrix} \tag{9.27}$$

where $\bar{U}_1(s) \in \mathcal{H}_2$ is an external disturbance input whose effect will be seen to be equivalent to the actual disturbance input $U_1(s) \in \mathcal{H}_2$. Then substituting (9.27) in (9.26) yields

$$\begin{bmatrix} V_1(s) \\ V_2(s) \end{bmatrix} = \begin{bmatrix} \Delta_1 \\ \Delta_3 \end{bmatrix} \bar{U}_1(s) \qquad \bar{U}_1(s) \in \mathcal{H}_2 \tag{9.28}$$

We show now that $P_\gamma \leq 0$ for $U_1(s) \in \mathcal{H}_2$, (9.12). This is done by first showing that (9.28) implies that $P_\gamma \leq 0$ for $\bar{U}_1(s) \in \mathcal{H}_2$ and then showing that this implies $P_\gamma \leq 0$ for $U_1(s) \in \mathcal{H}_2$.

Recall that

$$\| V_i(s)\|_2^2 = \frac{1}{2\pi}\int_{-\infty}^{\infty} V_i^*(j\omega)V_i(j\omega)\mathrm{d}\omega$$

so that using (9.28) in P_γ, (9.21) can be written as

$$P_\gamma = \frac{1}{2\pi}\int_{-\infty}^{\infty *} \bar{U}_1^*(j\omega)\left[-\gamma^2\Delta_1^*\Delta_1 + \Delta_3^*\Delta_3\right]\bar{U}_1(j\omega)\mathrm{d}\omega \tag{9.29}$$

Therefore we see from (9.29) that

$$\left.\begin{array}{lll} \text{if} & \left[-\gamma^2\Delta_1^*\Delta_1 + \Delta_3^*\Delta_3\right] < 0 & \\ \text{then} & P_\gamma < 0 & \text{for all non-null } \bar{U}_1(s) \in \mathcal{H}_2 \\ & P_\gamma = 0 & \text{for } \bar{U}_1(s) = \emptyset \end{array}\right\} \tag{9.30}$$

However recall, from the beginning of this section, that the H_∞ state feedback control problem has a solution only if D_{11} satisfies (9.6). We will show in Section 9.3.3 that

$$\left.\begin{array}{ll} \text{if} & \sigma_{\max}[D_{11}] < \gamma \\ \text{then} & -\gamma^2\Delta_1^*\Delta_1 + \Delta_3^*\Delta_3 < 0 \end{array}\right\} \tag{9.31}$$

Therefore (9.31, 9.30) imply that

$$\left.\begin{array}{lll} \text{if} & \sigma_{\max}[D_{11}] < \gamma & \\ \text{then} & P_\gamma < 0 & \text{for all non} - \text{null } \bar{U}_1(s) \in \mathcal{H}_2 \\ & P_\gamma = 0 & \text{for } \bar{U}_1(s) = \emptyset \end{array}\right\} \tag{9.32}$$

We need next to show that (9.32) implies (9.12). To see this notice from (9.27) that

$$\bar{U}_1(s) = U_1(s) - K_1X(s) \tag{9.33}$$

Now we will show in Section 9.3.3 that the state feedback, $u_2(t) = K_2x(t)$ stabilizes the generalized plant by itself without requiring $u_1(t) = K_1x(t)$. Thus $X(s) \in \mathcal{H}_2$ for any input $U_1(s) \in \mathcal{H}_2$ provided $u_2(t) = K_2x(t)$. Therefore we see from (9.33) that $\bar{U}_1(s) \in \mathcal{H}_2$ for all $U_1(s) \in \mathcal{H}_2$. This fact together with (9.32) implies that (9.12) is satisfied, i.e.,

$$\left.\begin{array}{lll} P_\gamma < 0 & \text{for} & U_1(s) \in \mathcal{H}_2 \text{ and } U_1(s) \neq K_1X(s) \\ P_\gamma = 0 & \text{for} & U_1(s) = K_1X(s) \end{array}\right\} \tag{9.34}$$

as is required to satisfy requirement (ii), (9.5), for the solution of the H_∞ state feedback control problem.

Notice that the disturbance input $U_1(s) = K_1X(s)$ maximizes P_γ. Therefore $U_1(s) = K_1X(s)$ is referred to as the worst disturbance input for the H_∞ state feedback control system.

In summary, we have shown that a solution to the H_∞ state feedback problem is obtained when (9.6) is satisfied and when we choose K and Δ so that the right coprime factorization of $\bar{G}_1(s)$, (9.13), has numerator $\bar{N}(s)$, (9.23), which is J-inner. The equations for K which make $\bar{N}(s)$ J-inner are the design equations for a solution to the H_∞ state feedback problem. These equations are developed in the next section.

9.3 H_∞ State Feedback Controller

In order to obtain a J-inner coprime factorization of $\bar{G}_1(s)$, (9.13), we need conditions on the state model parameters of a given system which ensure that the system is J-inner. These conditions are given in Theorem 9.1. A similar set of conditions was given in the previous chapter in connection with inner systems, (Theorem 8.6). Once we have Theorem 9.1 we can apply it to the state model parameters for $\bar{N}(s)$, (9.23). In this way we determine an equation relating Δ to D_1 and an expression for K involving the stabilizing solution X_∞ to an algebraic Riccati equation.

9.3.1 Design equations for K

Notice from (9.13, 9.15) that $\bar{N}(s)$ is an $(m_1 + p_1) \times (m_1 + m_2)$ matrix. Now for $\bar{N}(s)$ to be J-inner, the conditions given in the following theorem must be satisfied.

Theorem 9.1 A system having an $(m_1 + p_1)\times(m_1 + m_2)$ transfer function $N(s)$ is J-inner if $N(s) \in \mathcal{H}_\infty$ and the parameters of a minimal state model for the system

$$N(s) \stackrel{s}{=} \begin{bmatrix} A_N & B_N \\ C_N & D_N \end{bmatrix}$$

satisfy

$$\begin{aligned} \text{(i)} \quad J_{\gamma mm} &= D_N^T J_{\gamma mp} D_N \\ \text{(ii)} \quad \emptyset &= B_N^T X + D_N^T J_{\gamma mp} C_N \\ \text{(iii)} \quad \emptyset &= A_N^T X + X A_N + C_N^T J_{\gamma mp} C_N \end{aligned}$$

where $J_{\gamma mp}$, $J_{\gamma mm}$ are defined as

$$J_{\gamma mp} = \begin{bmatrix} -\gamma^2 I_{m_1} & \emptyset \\ \emptyset & I_{p_1} \end{bmatrix} \qquad J_{\gamma mm} = \begin{bmatrix} -\gamma^2 I_{m_1} & \emptyset \\ \emptyset & I_{m_2} \end{bmatrix}$$

Proof The proof is similar to the proof of Theorem 8.7. Here we need to obtain a state model for $N^T(-s)J_{\gamma mp}N(s)$ and change coordinates with coordinate transformation matrix T given by

$$T = \begin{bmatrix} I & X \\ 0 & I \end{bmatrix}$$

The resulting state model then reveals that in order for $N^T(-s)J_{\gamma mp}N(s) = J_{\gamma mm}$ we need to impose the conditions in the theorem. ■

Now we want to choose Δ and K in the state model for $\bar{N}(s)$, (9.23), so that $\bar{N}(s)$ is J-inner. We do this by equating $\bar{N}(s)$ to $N(s)$ and applying Theorem 9.1.

To begin, recall that the D matrix in a state model is unchanged by a change of coordinates. Therefore from $N(s) = \bar{N}(s)$ we have

$$D_N = \bar{D}_1\Delta^{-1} \tag{9.35}$$

Then using condition (i) in Theorem 9.1, we obtain

$$D_{Jc} = \Delta^T J_{\gamma mm}\Delta \tag{9.36}$$

where

$$D_{Jc} = \bar{D}_1^T J_{\gamma mp}\bar{D}_1 = \begin{bmatrix} -\gamma^2 I_{m_1} + D_{11}^T D_{11} & D_{11}^T D_{12} \\ D_{12}^T D_{11} & D_{12}^T D_{12} \end{bmatrix}$$

Notice that D_{Jc} is square. We will show later in this section that D_{Jc} is invertible if D_{12} has independent columns and D_{11} satisfies (9.6). This inverse is needed to determine K.

Continuing with the development, we see from Theorem 9.1 and (9.23) that further consequences of setting $N(s) = \bar{N}(s)$ are

$$A_N = A + BK \tag{9.37}$$

$$B_N = B\Delta^{-1} \tag{9.38}$$

$$C_N = \bar{C}_1 + \bar{D}_1 K \tag{9.39}$$

Then using (9.35, 9.38, 9.39) in condition (ii) of Theorem 9.1 we obtain

$$\Delta^{-T}B^T X + \Delta^{-T}\bar{D}_1^T J_{\gamma mp}(\bar{C}_1 + \bar{D}_1 K) = \emptyset$$

which after multiplying on the left by Δ^T becomes

$$B^T X + \bar{D}_1^T J_{\gamma mp}\bar{C}_1 + \bar{D}_1^T J_{\gamma mp}\bar{D}_1 K = \emptyset \tag{9.40}$$

Now we see from the definitions of $\bar{C}_1$ and $\bar{D}_1$, (9.13), that

$$\bar{C}_1^T J_{\gamma mp}\bar{D}_1 = C_1^T D_1 \tag{9.41}$$

$$\bar{C}_1^T J_{\gamma mp}\bar{C}_1 = C_1^T C_1 \tag{9.42}$$

Thus using (9.41, 9.36) and assuming D_{Jc} is invertible enables us to solve (9.40) for K as

$$K = -D_{Jc}^{-1}[B^T X + D_1^T C_1] \tag{9.43}$$

Notice that K depends on the state model parameters for the given generalized plant, (9.3), and on X. We will show now that X is a solution to an algebraic Riccati equation.

To begin, we substitute from (9.37, 9.39) in condition (iii) of Theorem 9.1 and use (9.41, 9.42) to obtain

$$A^T X + XA + K^T B^T X + XBK + K^T D_1^T C_1 + C_1^T D_1 K + K^T D_{Jc} K + C_1^T C_1 = \emptyset \quad (9.44)$$

Next grouping terms involving K and K^T and using (9.43) to substitute $-D_{Jc}K$ for $B^T X + D_1^T C_1$ enables (9.44) to be rewritten as

$$A^T X + XA - K^T D_{Jc} K + C_1^T C_1 = \emptyset \quad (9.45)$$

Finally substituting for K from (9.43) gives the equation

$$A_{\infty 1}^T X + XA_{\infty 1} - XR_{\infty 1} X + Q_{\infty 1} = \emptyset \quad (9.46)$$

where

$$A_{\infty 1} = A - BD_{Jc}^{-1} D_1^T C_1$$

$$R_{\infty 1} = BD_{Jc}^{-1} B^T$$

$$Q_{\infty 1} = C_1^T (I - D_1 D_{Jc}^{-1} D_1^T) C_1$$

with

$$D_1 = [D_{11} \quad D_{12}] \qquad D_{Jc} = \begin{bmatrix} -\gamma^2 I_{m_1} + D_{11}^T D_{11} & D_{11}^T D_{12} \\ D_{12}^T D_{11} & D_{12}^T D_{12} \end{bmatrix}$$

Notice that (9.46) is an algebraic Riccati equation like the GCARE (Section 6.2). We refer here to (9.46) as the HCARE. Since $A_{\infty 1} - R_{\infty 1} X = A + BK$ we have K as a stabilizing state feedback matrix if the solution X to (9.46) makes $A_{\infty 1} - R_{\infty 1} X$ stable. As in the quadratic and LQG state feedback control problems, we reserve a special notation for these solutions which in this case is X_∞. Notice that $X_\infty \geq 0$.

In addition, notice that unlike $D_{12}^T D_{12}$, D_{Jc} is not sign definite, i.e., neither non-negative nor non-positive. Thus unlike R_1 in the GCARE, (section 6.2.2), $R_{\infty 1}$ in the HCARE is not sign definite. This makes the determination of conditions on the state model for the generalized plant which ensure the existence of a stabilizing solution to the HCARE more involved than in the case of the QCARE.

9.3.2 On the stability of $A + B_2 K_2$

We have just developed equations for the determination of K to effect a J-inner coprime factorization of $\bar{G}_1(s)$, a system which is made up from the given generalized plant. Thus $A + BK$ is a stability matrix. However the H_∞ state feedback controller, (9.2), affects only $u_2(t)$ through the state feedback equation $u_2(t) = K_2 x(t)$. In addition, recall from Section 9.2.4, that condition (ii), (9.5), is satisfied provided K_2 is the last m_2 rows of the matrix K

which is needed in the J-inner coprime factorization of $\bar{G}_1(s)$. Therefore it remains to establish that condition (i), (9.5), is satisfied when K is determined in the foregoing manner.

More specifically, assuming that the HCARE, (10.96), has a stabilizing solution X_∞, and that K, (9.43), is determined using $X = X_\infty$, we need to show that $A + B_2K_2$ has all its eigenvalues in the open left half plane when K_2 is the last m_2 rows of K. This is done in the proof of the following theorem.

Theorem 9.2 If the pair (A, B_2) is stabilizable and $X_\infty \geq 0$ is a stabilizing solution to the HCARE, (10.96), then $\lambda[A + B_2K_2]$ are all in the open left half plane where

$$K = \begin{bmatrix} K_1 \\ K_2 \end{bmatrix}$$

with K given by (9.43).

Proof Suppose X_∞ is a stabilizing solution to (9.46). Then $A_{\infty 1} - R_{\infty 1}X_\infty$ is stable where

$$A_{\infty 1} - R_{\infty 1}X_\infty = \bar{A}_2 + B_1K_1 \tag{9.47}$$

with $\bar{A}_2 = A + B_2K_2$.

Notice, from Chapter 2, that the stabilizabilty of (A, B_2) is necessary for $\bar{A}_2$ to be stable. In addition, recall from Chapter 3, that a pair (A, C) is detectable if we can find L such that $A + LC$ is stable. Alternatively, when (A, C) is detectable all unstable eigenvalues of A are observable, i.e., if $\mathrm{Re}[\lambda] \geq 0$ when $Av = \lambda v$, then $Cv \neq \emptyset$. Therefore since $\bar{A}_2 + B_1K_1$, (9.47), is stable, we have $(\bar{A}_2, K_1)$ detectable. Thus if $K_1v = \emptyset$ where $\bar{A}_2v = \lambda v$ then $\mathrm{Re}[\lambda] < 0$. We will use this fact later in the proof.

Next we add and subtract both $X_\infty B_2K_2$ and its transpose to and from (9.45) to obtain

$$\bar{A}_2^T X_\infty + X_\infty \bar{A}_2 = \Theta \tag{9.48}$$

where

$$\Theta = X_\infty B_2K_2 + K_2^T B_2^T X_\infty + K^T D_{Jc} K - C_1^T C_1$$

Now we are going to show that $\Theta \leq 0$. Then we will see that this fact plus the fact that $(\bar{A}_2, K_1)$ is detectable implies that $\bar{A}_2$ is stable.

We begin by expanding $K^T D_{Jc} K$

$$\begin{aligned} K^T D_{Jc} K &= \begin{bmatrix} K_1^T & K_2^T \end{bmatrix} \begin{bmatrix} M_1 & M_2 \\ M_2^T & M_4 \end{bmatrix} \begin{bmatrix} K_1 \\ K_2 \end{bmatrix} \\ &= K_1^T M_1 K_1 + K_1^T M_2 K_2 + K_2^T M_2^T K_1 + K_2^T M_4 K_2 \end{aligned}$$

where we see from (9.46) that the $M_i s$ are given as

$$M_1 = -\gamma^2 I_{m_1} + D_{11}^T D_{11} \qquad M_2 = D_{11}^T D_{12} \qquad M_4 - D_{12}^T D_{12}$$

Then Θ, (9.48), becomes

$$\Theta = (X_\infty B_2 + K_1^T M_2)K_2 + K_2^T(B_2^T X_\infty + M_2^T K_1) + K_1^T M_1 K_1 + K_2^T M_4 K_2 - C_1^T C_1 \tag{9.49}$$

Next after rearranging (9.43) as

$$D_{Jc}K = \begin{bmatrix} M_1K_1 + M_2K_2 \\ M_2^TK_1 + M_4K_2 \end{bmatrix} = -\begin{bmatrix} B_1^TX_\infty + D_{11}^TC_1 \\ B_2^TX_\infty + D_{12}^TC_1 \end{bmatrix}$$

we see that

$$B_2^TX_\infty + M_2^TK_1 = -\left(M_4K_2 + D_{12}^TC_1\right) \tag{9.50}$$

Therefore using (9.50) in Θ, (9.49), gives

$$\begin{aligned}\Theta &= -\left(M_4K_2 + D_{12}^TC_1\right)^TK_2 - K_2^T\left(M_4K_2 + D_{12}^TC_1\right) + K_1^TM_1K_1 + K_2^TM_4K_2 - C_1^TC_1 \\ &= -K_2^TM_4K_2 - C_1^TD_{12}K_2 - K_2^TD_{12}^TC_1 + K_1^TM_1K_1 - C_1^TC_1 \\ &= -(C_1 + D_{12}K_2)^T(C_1 + D_{12}K_2) + K_1^TM_1K_1 \end{aligned} \tag{9.51}$$

Finally, recall from (9.6) that we require all the eigenvalues of $D_{11}^TD_{11}$ to be less that γ^2 since this is a necessary condition for the solution of the H_∞ state feedback control problem. We can readily show that under this condition we have $M_1 < 0$. Thus we see from (9.51) that $\Theta \leq 0$. We can now proceed to show that $\bar{A}_2$ is stable.

Pre and post multiplying (9.48) by v^* and v where $\bar{A}_2v = \lambda v$ gives

$$2\text{Re}[\lambda]v^*X_\infty v = v^*\Theta v \tag{9.52}$$

Now since we have just shown that $\Theta \leq 0$ we have either $v^*\Theta v < 0$ or $v^*\Theta v = 0$.

If $v^*\Theta v < 0$ it follows from the condition $X_\infty \geq 0$ that $v^*X_\infty v > 0$ and therefore

$$\text{Re}[\lambda] < 0$$

If $v^*\Theta v = 0$ we see from (9.51) that

$$-v^*(C_1 + D_{12}K_2)^T(C_1 + D_{12}K_2)v + v^*\left(K_1^TM_1K_1\right)v = 0 \tag{9.53}$$

and since $M_1 < 0$ we see that (9.53) requires

$$K_1v = \emptyset \qquad \text{and} \qquad (C_1 + D_{12}K_2)v = \emptyset \tag{9.54}$$

Finally, recalling from the beginning of the proof that the pair $(\bar{A}_2, K_1)$ is detectable, we see that if $K_1v = \emptyset$ then $\text{Re}[\lambda] < 0$. ■

To recap, we have shown that the state feedback matrix K_2 which solves the H_∞ state feedback control problem can be determined as the last m_2 rows of the matrix K, (9.43), when $X = X_\infty$ the stabilizing solution to the HCARE, (9.46). A remarkable feature of this solution, shown in Theorem 9.2, is that while X_∞ makes $A + BK$ stable we also have $A + BK_2$ stable. Before discussing conditions on the generalized plant which ensure the existence of a stabilizing solution to the HCARE, we pose and solve the H_∞ state estimation problem.

9.3.3 Determination of Δ

Recall that in Section 9.2.4 we made use of the condition

$$-\gamma^2\Delta_1^T\Delta_1 + \Delta_3^T\Delta_3^T < 0$$

We will show now that this condition is satisfied when

$$\sigma_{\max}[D_{11}] < \gamma \qquad \text{and} \qquad D_{12}^T D_{12} \text{ is invertible}$$

We do this by first developing an expression for Δ in terms of D_{11}, D_{12} so that (9.36) is satisfied. This is done in the following theorem.

Theorem 9.3 If

$$\sigma_{\max}[D_{11}] < \gamma \qquad \text{and} \qquad D_{12}^T D_{12} \text{ is invertible}$$

then Δ can be taken as

$$\Delta = \begin{bmatrix} \gamma^{-1}\Gamma^{1/2} & \emptyset \\ \Theta^{-T/2}D_{12}^T D_{11} & \Theta^{1/2} \end{bmatrix} = \begin{bmatrix} \Delta_1 & \emptyset \\ \Delta_3 & \Delta_4 \end{bmatrix}$$

where

$$\Theta = \Theta^{T/2}\Theta^{1/2} \qquad \Gamma = \Gamma^{T/2}\Gamma^{1/2}$$

$$\Theta = D_{12}^T D_{12} \qquad \Gamma = \gamma^2 I_{m_1} - D_{11}^T(I_{p_1} - D_{12}\Theta^{-1}D_{12}^T)D_{11}$$

with

$$D_{Jc} = \begin{bmatrix} M_1 & M_2 \\ M_2^T & M_4 \end{bmatrix} = \Delta^T J_{\gamma mm}\Delta \tag{9.55}$$

where $J_{\gamma mm}$ is given in (9.16) and

$$M_1 = -\gamma^2 I_{m_1} + D_{11}^T D_{11} \qquad M_2 = D_{11}^T D_{12} \qquad M_4 = D_{12}^T D_{12}$$

Proof Consider the constant symmetric matrix factorization given by

$$\begin{bmatrix} M_1 & M_2 \\ M_2^T & M_4 \end{bmatrix} = \begin{bmatrix} I & M_2M_4^{-1} \\ \emptyset & I \end{bmatrix}\begin{bmatrix} M_1 - M_2M_4^{-1}M_2^T & \emptyset \\ \emptyset & M_4 \end{bmatrix}\begin{bmatrix} I & \emptyset \\ M_4^{-1}M_2^T & I \end{bmatrix} \tag{9.56}$$

Then identifying the blocks of D_{Jc} with the M_is, (9.55), we see from (9.56) that

$$D_{Jc} = \begin{bmatrix} I & D_{11}^T D_{12}\Theta^{-1} \\ \emptyset & I \end{bmatrix}\begin{bmatrix} -\Gamma & \emptyset \\ \emptyset & \Theta \end{bmatrix}\begin{bmatrix} I & \emptyset \\ \Theta^{-1}D_{12}^T D_{11} & I \end{bmatrix} \tag{9.57}$$

where Γ, Θ are defined in the theorem.

Next assume for the time being that Γ, Θ can be factored in the manner specified in the theorem. Then we can write

$$\begin{bmatrix} -\Gamma & \emptyset \\ \emptyset & \Theta \end{bmatrix} = \Psi^T J_{\gamma mm} \Psi$$

where $J_{\gamma mm}$ was given in (9.16) and

$$\Psi = \begin{bmatrix} \gamma^{-1}\Gamma^{\frac{1}{2}} & \emptyset \\ \emptyset & \Theta^{\frac{1}{2}} \end{bmatrix}$$

Finally refering to (9.57) we have

$$\Psi \begin{bmatrix} I & \emptyset \\ \Theta^{-1} D_{12}^T D_{11} & I \end{bmatrix} = \Delta$$

so that D_{Jc} can be factored as (9.55) provided Θ, Γ have square root factors.

Recall from the proof of Theorem 6.1 that a matrix has a square root factor provided it is symmetric and non-negative. Now Θ and Γ are seen to be symmetric by inspection. However unlike Θ it is not evident that Γ is non-negative.

To show that Γ is non-negative notice that it can be expressed as

$$\Gamma = \gamma^2 I_{m_1} - D_{11}^T D_c D_{11} \tag{9.58}$$

where D_c was used in Section 6.5.2

$$D_c = I_{p_1} - D_{12}(D_{12}^T D_{12})^{-1} D_{12}^T$$

Next recall from Chapter 7 that the singular value σ_i and corresponding singular vectors v^i, u^i for D_{11} are related as

$$D_{11}^T D_{11} v^i = \sigma_i^2 v^i \qquad D_{11} D_{11}^T u^i = \sigma_i^2 u^i$$

where

$$v^{iT} v^i = u^{iT} u^i = 1$$

Therefore noting that

$$D_{11} v^i = \sigma_i u^i \qquad D_{11}^T u^i = \sigma_i v^i$$

we see from (9.58) that

$$v^{iT} \Gamma v^i = \gamma^2 - \sigma_i^2 u^{iT} D_c u^i \tag{9.59}$$

However we saw, following the proof of Theorem 6.1, that D_c is a contraction so that

$$u^{iT} D_c u^i \leq u^{iT} u^i = 1 \tag{9.60}$$

Therefore from (9.59, 9.60) we have

$$v^{iT}\Gamma v^i \geq \gamma^2 - \sigma_i^2$$

and

$$v^{iT}\Gamma v^i \geq 0$$

for all eigenvectors, $v^i : i = 1, 2 \cdots p_1$ of $D_{11}^T D_{11}$ provided the singular values of D_{11} satisfy

$$\sigma_i \leq \gamma \qquad\qquad i = 1, 2 \cdots p_1$$

Finally, since $D_{11}^T D_{11}$ is symmetric, its eigenvectors, v^i, are complete and it follows that if

$$\sigma_{\max}[D_{11}] \leq \gamma$$

then

$$v^T \Gamma v \geq 0$$

for any p_1 dimensional vector v. Therefore $\Gamma \geq 0$ when the necessary condition (9.6) is satisfied, and we can factor Γ in the manner stated in the theorem. ■

The foregoing expression for Δ can now be used to prove (9.31). Thus we see from Theorem 9.3 that

$$-\gamma^2 \Delta_1^T \Delta_1 + \Delta_3^T \Delta_3 = -\gamma^2 I_{m_1} + D_{11}^T D_{11}$$

and (9.31) follows.

Finally, recall that K, (9.43), requires D_{Jc} to be invertible. We can show that D_{Jc} is indeed invertible, by using the factorization of D_{Jc} given in Theorem 9.3 to obtain an expression D_{Jc}^{-1} as

$$D_{Jc}^{-1} = \Delta^{-1} J_{\gamma mm}^{-1} \Delta^{-T} \tag{9.61}$$

Then we see, from the lower triangular structure of Δ, (9.55), that

$$\Delta^{-1} = \begin{bmatrix} \Delta_1^{-1} & \emptyset \\ -\Delta_4^{-1}\Delta_3\Delta_1^{-1} & \Delta_4^{-1} \end{bmatrix} \tag{9.62}$$

$$= \begin{bmatrix} -\Gamma^{-1} & \Gamma^{-1} D_{11}^T D_{12} \Theta^{-1} \\ \Theta^{-1} D_{12}^T D_{11} \Gamma^{-1} & \Theta^{-1} - \Theta^{-1} D_{12}^T D_{11} \Gamma^{-1} D_{11}^T D_{12} \Theta^{-1} \end{bmatrix}$$

where Γ, Θ are given in (9.55).

In the next chapter we will use the form for Δ given in Theorem 9.3 to develop the solution to the H_∞ output feedback control problem. Before we can do this we need to develop a solution to the H_∞ state estimation problem.

9.4 H_∞ State Estimation Problem

Recall from the introduction to this chapter that our ultimate goal is the determination of an H_∞ controller which uses the measured output to determine the controlled input to the generalized plant so that the resulting closed loop system is internally stable and has transfer function from the disturbance input to the desired output, $T_o(s)$, with H_∞ norm no greater than γ.

In this section we treat the problem of determining an estimator or observer for the generalized plant state, $\hat{x}(t)$, from knowledge of the measured output, $y_2(t)$, and controlled input, $u_2(t)$. We need to do this so that we can use this state estimator in conjunction with the H_∞ state feedback controller determined in the previous two sections to obtain the desired H_∞ controller. Notice that since the state of the generalized plant depends on the disturbance input $u_1(t)$, which is unknown, it is not possible to design an observer to give an asymptotic estimate of the generalized plant state. Instead we design the observer so that the L_2 gain from the disturbance input to desired output estimation error, $\tilde{y}_1(t)$, is no greater than γ, i.e., so that

$$\|\tilde{y}_1(t)\|_2 \leq \gamma \|u_1(t)\|_2 \tag{9.63}$$

where

$$\tilde{y}_1(t) = \hat{y}_1(t) - y_1(t) \tag{9.64}$$

and $\hat{y}_1(t)$ is the estimate of the desired output,

$$\hat{y}_1(t) = C_1\hat{x}(t) + D_{12}u_2(t) \tag{9.65}$$

Alternatively, we must determine the observer so that the transfer function $T_e(s)$ from $u_1(t)$ to $\tilde{y}_1(t)$ satisfies

$$(i)\ T_e(s) \in \mathcal{H}_\infty \qquad \text{and} \qquad (ii):\ \|T_e(s)\|_\infty \leq \gamma \tag{9.66}$$

As in the quadratic and LQG state estimation problems where the solution was obtained by dualizing the solution to the corresponding state feedback control problem, we will see now that the solution to the H_∞ state estimation problem can be obtained by dualizing the solution we obtained in previous sections for the H_∞ state feedback control problem.

9.4.1 Determination of $T_e(s)$

Recalling the development of observers given in Chapter 3, we see that an observer which estimates the state of the generalized plant (9.1) using $u_2(t)$ and $y_2(t)$ is constructed as

$$\dot{\hat{x}}(t) = A\hat{x}(t) + L_2\tilde{y}_2(t) + B_2u_2(t) \tag{9.67}$$

where

$$\tilde{y}_2(t) = \hat{y}_2(t) - y_2(t)$$

However, from the state model for the generalized plant, (9.1), we see that

$$y_2(t) = C_2x(t) + D_{21}u_1(t) + D_{22}u_2(t) \qquad \hat{y}_2(t) = C_2\hat{x}(t) + D_{22}u_2(t) \tag{9.68}$$

Thus using (9.67, 9.68) we see that the generalized plant state estimation error, $\tilde{x}(t) = x(t) - \hat{x}(t)$, is the solution to the differential equation

$$\dot{\tilde{x}}(t) = (A + L_2C_2)\tilde{x}(t) + (B_1 + L_2D_{21})u_1(t) \tag{9.69}$$

In addition, we see from (9.1, 9.64, 9.65) that the desired output estimation error, $\tilde{y}_1(t) = \hat{y}_1(t) - y_1(t)$, depends on the state estimation error and disturbance input as

$$\tilde{y}_1(t) = -C_1\tilde{x}(t) - D_{11}u_1(t) \tag{9.70}$$

Therefore we see from (9.69, 9.70) that the transfer function $T_e(s)$ from $u_1(t)$ to $\tilde{y}_1(t)$ has state model given as

$$T_e(s) \stackrel{s}{=} \begin{bmatrix} A + L_2C_2 & (B_1 + L_2D_{21}) \\ -C_1 & -D_{11} \end{bmatrix} \tag{9.71}$$

Recalling the discussion following (9.5), we see from this state model for $T_e(s)$ that condition (ii), (9.66), cannot be satisfied if

$$\sigma_{\max}[D_{11}] > \gamma$$

Thus the same necessary condition, (9.6), which we required to solve the H_∞ state feedback control problem, is needed now to solve the H_∞ state estimation problem namely

$$\sigma_{\max}[D_{11}] < \gamma \tag{9.72}$$

We will see now that the design equations for the determination of L_2 so that conditions (i, ii), (9.66), are satisfied can be obtained by using the fact that the present H_∞ state estimation problem is the dual of the H_∞ state feedback problem solved previously in this chapter.

9.4.2 Duality

Recall, from Chapter 7, that the H_∞ norm of $T_e(s)$ is defined as

$$\|T_e(s)\|_\infty = \sup_{\omega \in (-\infty, \infty)} \sigma_{\max}(j\omega)$$

where $\sigma_{\max}(j\omega)$ is the largest singular value of $T_e(j\omega)$.

More specifically, recall that $\sigma_{\max}(j\omega)$ is the positive square root of the largest eigenvalue of either of the Hermitian matrices $T_e^*(j\omega)T_e(j\omega)$ or $T_e(j\omega)T_e^*(j\omega)$. Therefore the largest singular value of $T_e(j\omega)$ equals the largest singular value of $T_e^*(j\omega)$. As discussed in Chapter 8, the adjoint system corresponding to a system having real-rational transfer function $T_e(s)$ has transfer function $T_e^T(-s)$ and satisfies

$$T_e^*(s)\mid_{s=j\omega}=T_e^T(-s)\mid_{s=j\omega}$$

Therefore condition (*ii*), (9.66), is equivalent to the condition

$$\|T_e^T(-s)\|_\infty \le \gamma \tag{9.73}$$

where we see from (9.71) that

$$T_e^T(-s) \stackrel{s}{=} \begin{bmatrix} -(A^T+C_2^TL_2^T) & -C_1^T \\ -(B_1^T+D_{21}^TL_2^T) & -D_{11}^T \end{bmatrix} \tag{9.74}$$

Notice that condition (*i*), (9.66), is satisfied if and only if $T_e^T(-s)$ is antistable since $T_e(s) \in \mathcal{H}_\infty(s)$ is satisfied if and only if $T_e^T(-s) \in \mathcal{H}_\infty^-(s)$.

Thus we see from the foregoing that the H_∞ state estimation problem and the H_∞ state feedback control problem are dual problems in the sense that a solution to the H_∞ state estimation problem requires that

$$(i):\ T_e^T(-s) \in \mathcal{H}_\infty^- \quad \text{and} \quad (ii):\ \|T_e^T(-s)\|_\infty \le \gamma \tag{9.75}$$

whereas a solution to the H_∞ state feedback control problem requires that

$$(i):\ T_c(s) \in \mathcal{H}_\infty \quad \text{and} \quad (ii):\ \|T_c(s)\|_\infty \le \gamma \tag{9.76}$$

9.4.3 Design Equations for L_2

In order to exploit the foregoing dual nature of the H_∞ state feedback control problem and H_∞ state estimation problem, (9.75, 9.76), we need to establish a symbol correspondence between these problems. This can be done by comparing the state models for $T_c(s)$ and $T^T(-s)$, (9.4, 9.74). This gives the following correspondences between their parameters.

Recall that we determine K_2 in the H_∞ state feedback problem by first solving the HCARE, (10.96), for its stabilizing solution, X_∞. Then K_2 is the last m_2 rows of the matrix K which we determine by setting $X = X_\infty$ in (9.43). Therefore, proceeding in an analogous manner we determine L_2 as the last p_2 columns of the matrix L obtained by using the stabilizing solution Y_∞ to an algebraic Riccati equation. More specifically, the design equations for determining L are obtained by replacing parameters in (9.46) and (9.43) using the parameter equivalences in Table 9.1. Thus doing this and also replacing $-X$ with Y gives

$$A_{\infty2}Y + YA_{\infty2}^T - YR_{\infty2}Y + Q_{\infty2} = \emptyset \tag{9.77}$$

and

$$L = -(YC^T + B_1D_2^T)D_{Jo}^{-1} \tag{9.78}$$

Table 9.1 State model Parameter Correspondences

Control	Estimation
A	$-A^T$
B_2	$-C_2^T$
K_2	L_2^T
B_1	$-C_1^T$
C_1	$-B_1^T$
D_{12}	$-D_{21}^T$
D_{11}	$-D_{11}^T$

where

$$\begin{aligned} A_{\infty 2} &= A - B_1 D_2^T D_{Jo}^{-1} C \\ R_{\infty 2} &= C^T D_{Jo}^{-1} C \\ Q_{\infty 2} &= B_1[I - D_2^T D_{Jo}^{-1} D_2] B_1^T \end{aligned}$$

with

$$D_2 = \begin{bmatrix} D_{11} \\ D_{21} \end{bmatrix} \qquad D_{Jo} = \begin{bmatrix} -\gamma^2 I_{p_1} + D_{11} D_{11}^T & D_{11} D_{21}^T \\ D_{21} D_{11}^T & D_{21} D_{21}^T \end{bmatrix}$$

Notice that (9.77) is an algebraic Riccati equation like the GFARE, (Section 6.3.2) which is referred to here as the HFARE. The solution Y which makes $A_2 - YR_2 = A + LC$ stable is referred to as the stabilizing solution to the HFARE and is denoted Y_∞ and satisfies $Y_\infty \geq 0$. Just as $A + B_2K_2$ is stable when K is determined from (9.43) using X_∞ so too is $A + L_2C_2$ stable when L is determined from (9.78) using Y_∞ where

$$L = [\,L_1 \quad L_2\,]$$

Just as the independence of the columns of D_{12} and condition (9.72) ensure that D_{Jc} is invertible so too does the independence of the rows of D_{21} and condition (9.72) ensure that $D_{Je.}$ is invertible.

Finally, recall that in the H_∞ state feedback problem we required (A, B_2) to be stabilizable and the HCARE to have a solution $X_\infty \geq 0$ in order for $A + B_2K_2$ to be stable. We see by analogy that in the H_∞ state estimation problem we need the HFARE to have a solution $Y_\infty \geq 0$ and (A, C_2) to be detectable in order for $A + L_2C_2$ to be stable.

9.5 Sufficient Conditions

Recall that a given state model for the generalized plant must satisfy certain conditions in order for the GCARE and GFARE to have stabilizing solutions (Section 5 of Chapter 6). A similar set of conditions which ensure the existence of stabilizing solutions to the

HCARE and HFARE are

(A1) (A, B_2) and (A, C_2) are respectively stabilizable and detectable pairs.

(A2)

$$\begin{bmatrix} (A - j\omega I) & B_2 \\ C_1 & D_{12} \end{bmatrix} \quad \text{and} \quad \begin{bmatrix} (A - j\omega I) & B_1 \\ C_2 & D_{21} \end{bmatrix}$$

have respectively independent columns and independent rows for all real ω.

(A3) D_{12} and D_{21} have respectively independent columns and independent rows.

Taken together, **(A1–A3)** constitute a sufficient condition for the solution of both the H_∞ state feedback control problem and the H_∞ state estimation problem.

Notice that **(A1,A2)** ensure that the Hamiltonian matrices corresponding to the HCARE and HFARE have no imaginary axis eigenvalues. This can be seen by noting the similarity of the HCARE with the GCARE in Chapter 6 and referring to Theorem 6.1 in Section 6.5.2.

9.6 Summary

In this chapter we have given a derivation of the design equations for solving the H_∞ state feedback control and state estimation problems for a given linear time-invariant, continuous-time plant. Unlike the solutions to the corresponding quadratic or LQG problems obtained in Chapters 5 and 6, solutions obtained here for the H_∞ problems do not minimize any performance criterion or cost. Rather the solutions to the H_∞ problems achieve a specified upper bound on the L_2 gain from disturbance input to desired output with the resulting closed loop systems being stable. As in the quadratic and LQG control problems, algebraic Riccati equations still play a central role in the solution. In the next chapter we will see how the solutions obtained in this chapter can be used to provide controllers which use the measured plant output instead of the plant state to achieve the just mentioned H_∞ norm bounding and closed loop stability.

9.7 Notes and References

A great deal of work has been done on the H_∞ control problem since its inception over a decade ago. One of the first approaches solved this problem by beginning with a Youla parametrization of all stabilizing controllers for the given plant and finished by having to solve a model matching problem. This work is the subject of [14]. This and other early approaches which are reviewed in [8] and [20] are quite complicated and have controllers with unnecessarily high dimension. Indications for a simpler approach came from research on stability robustness [35], which lead to the two-Riccati-equation solution given in this chapter [12, 16]. Following this similar results were obtained using J type factorizations, e.g., J spectral, J lossless coprime, [19], [41], [20], [40]. Discussion of this approach in connection with the chain-scattering formulation of the generalized plant is given in [24].

10

H_∞ Output Feedback and Control

10.1 Introduction

Having seen how to solve the basic H_∞ problems of state feedback control and state estimation for the given generalized plant, we come now to the problem of determining a controller which generates the controlled input from the measured output such that:

(i) the closed loop system is internally stable and
(ii) the H_∞ norm from the disturbance input to desired output is no greater than a given positive scalar γ.

This problem is referred to as the H_∞ output feedback control problem. The setup for the output feedback control problem was given previously at the start of Chapter 9. Again, as in that chapter, we are assuming that the generalized plant has known transfer function $G(s)$ and state model, i.e.,

$$\begin{bmatrix} Y_1(s) \\ Y_2(s) \end{bmatrix} = G(s)\begin{bmatrix} U_1(s) \\ U_2(s) \end{bmatrix} \qquad G(s) \stackrel{s}{=} \begin{bmatrix} A & B \\ D & C \end{bmatrix} \tag{10.1}$$

where

$$\begin{aligned} U_1(s) &= \text{disturbance input} & Y_1(s) &= \text{desired output} \\ U_2(s) &= \text{controlled input} & Y_2(s) &= \text{measured output} \end{aligned}$$

In the development to follow, we will see that there are many controllers which solve the H_∞ output feedback control problem. Moreover, we will show that these controllers constitute a constrained subset of Youla parameterized controllers for the system having $U_2(s)$ as input and $Y_2(s)$ as output. Recall, from Theorem 8.5, that Youla parameterized controllers are stabilizing. Therefore condition (i) is satisfied. We will show how to constrain this set so as to achieve condition (ii).

10.2 Development

Recall, from Section 9.3.2, that the partition K_2 of K which effects a J-inner coprime factorization of $\bar{G}_1(s)$ stabilizes the generalized plant. In addition, recall from Section 9.2.4 that $P_\gamma \leq 0$ for all $\bar{U}_1(s) \in \mathcal{H}_2$ and that this implies that $P_\gamma \leq 0$ for all $U_1(s) \in \mathcal{H}_2$ where

$$U_1(s) = K_1 X(s) + \bar{U}_1(s) \tag{10.2}$$

This fact is used now to develop a solution to the H_∞ output feedback control problem. The plan for doing this is given as follows.

1. We begin by assuming that the disturbance input $U_1(s)$ is given by (10.2) with K_1 determined in Section 9.3.1.
2. Then we show that the signals $V_1(s)$, $V_2(s)$ can be replaced in the expression for the quadratic performance index P_γ, introduced in Section 9.2.1, by the "external" disturbance input $\bar{U}_1(s)$ and an output-like signal, $Y_o(s)$. This gives $P_\gamma = P_{\gamma o}$ where $P_{\gamma o}$ is used to denote the new expression of P_γ. The output-like signal, $Y_o(s)$, depends on $\bar{U}_1(s)$, $U_2(s)$ and the state of the given state model for the generalized plant.
3. Next we determine an estimator for the generalized plant state which uses $y_o(t)$ and achieves $P_{\gamma o} \leq 0$ for all $\bar{U}_1(s) \in \mathcal{H}_2$. This is accomplished by adapting the H_∞ state estimator obtained in the previous chapter.
4. Following this we replace the actual generalized plant state in the H_∞ state feedback controller determined in Chapter 9, by the estimate of this state which is generated by the state estimator developed in the previous step. This leads to a controller which achieves $P_{\gamma o} \leq 0$ for all $\bar{U}_1(s) \in \mathcal{H}_2$ but requires $y_o(t)$ as an input in addition to the measured output $y_2(t)$.
5. Since $y_o(t)$ is not available for use as an input to the controller, the remainder of the development concentrates on eliminating the need for $y_o(t)$ as an input to the controller.

10.2.1 Reformulation of $\mathbf{P}_\gamma$

Recall from the development of the solution to the H_∞ state feedback control problem (Section 9.2.1) that $P_\gamma \leq 0$ for all $u_1(t) \in \mathcal{L}_2[0, \infty)$ implies that condition (ii) is satisfied. We are going to use this fact to solve the H_∞ output feedback control problem. Also recall, (Section 9.2.3), that P_γ was expressed in terms of the signals $V_1(s)$, $V_2(s)$, which resulted from a J-inner coprime factorization, i.e.,

$$P_\gamma = -\gamma^2 \| V_1(s) \|_2^2 + \| V_2(s) \|_2^2 \tag{10.3}$$

and recalling (9.25) we have

$$\begin{bmatrix} V_1(s) \\ V_2(s) \end{bmatrix} = \bar{M}^{-1}(s) \begin{bmatrix} U_1(s) \\ U_2(s) \end{bmatrix}$$

$$\bar{M}^{-1}(s) \stackrel{s}{=} \begin{bmatrix} A & B \\ -\Delta K & \Delta \end{bmatrix}$$

Now in order to incorporate the solution to the H_∞ state estimation problem we need to re-express P_γ in terms of the generalized plant state. We do this by assuming the state feedback matrix K is determined in the manner specified in Section 9.3.1 and that Δ can be taken, from Theorem 9.3, as

$$\Delta = \begin{bmatrix} \Delta_1 & \emptyset \\ \Delta_3 & \Delta_4 \end{bmatrix} = \begin{bmatrix} \gamma^{-1}\Gamma^{1/2} & \emptyset \\ \Theta^{-T/2}D_{12}^T D_{11} & \Theta^{1/2} \end{bmatrix} \tag{10.4}$$

where

$$\Theta = \Theta^{T/2}\Theta^{1/2} = D_{12}^T D_{12}$$

$$\Gamma = \Gamma^{T/2}\Gamma^{1/2} = \gamma^2 I_{m_1} - D_{11}^T(I_{p_1} - D_{12}\Theta^{-1}D_{12}^T)D_{11}$$

Then substituting for $U_1(s)$ from (10.2) we see that the state models for the generalized plant $G(s)$, (10.1), and $\bar{M}^{-1}(s)$, (10.3), become

$$\tilde{G}(s) \overset{s}{=} \begin{bmatrix} A + B_1K_1 & B \\ C + D_2K_1 & D \end{bmatrix} \tag{10.5}$$

$$\tilde{M}^{-1}(s) \overset{s}{=} \begin{bmatrix} A + B_1K_1 & B \\ \begin{bmatrix} \emptyset \\ -\Delta_4K_2 \end{bmatrix} & \begin{bmatrix} \Delta_1 & \emptyset \\ \Delta_3 & \Delta_4 \end{bmatrix} \end{bmatrix} \tag{10.6}$$

where

$$\begin{bmatrix} Y_1(s) \\ Y_2(s) \end{bmatrix} = \tilde{G}(s)\begin{bmatrix} \bar{U}_1(s) \\ U_2(s) \end{bmatrix}$$

$$\begin{bmatrix} V_1(s) \\ V_2(s) \end{bmatrix} = \tilde{M}^{-1}(s)\begin{bmatrix} \bar{U}_1(s) \\ U_2(s) \end{bmatrix}$$

Notice that the state differential equations and inputs to the foregoing state models for $\tilde{G}(s)$ and $\tilde{M}^{-1}(s)$ are identical. In addition notice that the initial states of each of these models are identical because we are using transfer functions to relate inputs to outputs which requires that the initial state of each model be null. Therefore the states of these models are identical for all time and equal to the state of the generalized plant when the disturbance input is given by (10.2). This observation together with the output equation of the state model for $\tilde{M}^{-1}(s)$, (10.6), reveals that $V_1(s), V_2(s)$ are dependent on the generalized plant state $X(s)$ and are given as

$$V_1(s) = \Delta_1\bar{U}_1(s) \tag{10.7}$$

$$V_2(s) = -\Delta_4K_2X(s) + \Delta_3\bar{U}_1(s) + \Delta_4U_2(s) \tag{10.8}$$

We can use these expressions for the $V_i(s)$s together with the relations for the Δ_is, (10.4) to re-express P_γ, (10.3), in the manner given in the following theorem.

Theorem 10.1 If the state feedback matrix K is determined as in the previous chapter so that its partition K_2 solves the H_∞ state feedback control problem, then P_γ, (10.3), can be re-expressed as $P_\gamma = P_{\gamma o}$ where

$$P_{\gamma o} = -\gamma^2 \| \bar{U}_1(s) \|_2^2 + \| Y_o(s) \|_2^2 \tag{10.9}$$

with

$$Y_o(s) = -D_{12} K_2 X(s) + D_{11} \bar{U}_1(s) + D_{12} U_2(s)$$

and $X(s)$ being the state of the generalized plant.

Proof From (10.7) we see that

$$-\gamma^2 \| V_1(s) \|_2^2 = -\gamma^2 \| \Delta_1 \bar{U}_1(s) \|_2^2 = -\gamma^2 \langle \Delta_1 \bar{U}_1(s), \Delta_1 \bar{U}_1(s) \rangle \tag{10.10}$$

where

$$\langle \Delta_1 \bar{U}_1(s), \Delta_1 \bar{U}_1(s) \rangle = \frac{1}{2\pi} \int_{-\infty}^{\infty} \bar{U}_1^*(j\omega) \Delta_1^T \Delta_1 \bar{U}_1(j\omega) \mathrm{d}\omega$$

Similarly from (10.8) we see that

$$\begin{aligned} \| V_2(s) \|_2^2 = {} & \| \Delta_4 K_2 X(s) \|_2^2 + \| \Delta_4 U_2(s) \|_2^2 + \| \Delta_3 \bar{U}_1(s) \|_2^2 - 2 \langle \Delta_4 K_2 X(s), \Delta_4 U_2(s) \rangle \\ & - 2 \langle \Delta_4 K_2 X(s), \Delta_3 \bar{U}_1(s) \rangle + 2 \langle \Delta_4 U_2(s), \Delta_3 \bar{U}_1(s) \rangle \end{aligned} \tag{10.11}$$

However from the dependency of the $\Delta_i s$ on D_{11} and D_{12}, (10.4), we have

$$\begin{aligned} \Delta_1^T \Delta_1 &= I_{m_1} - \gamma^{-2} D_{11}^T D_{11} + \gamma^{-2} D_{11}^T D_{12} \left(D_{12}^T D_{12} \right)^{-1} D_{12}^T D_{11} \\ \Delta_3^T \Delta_3 &= D_{11}^T D_{12} \left(D_{12}^T D_{12} \right)^{-1} D_{12}^T D_{11} \\ \Delta_4^T \Delta_3 &= D_{12}^T D_{11} \qquad\qquad \Delta_4^T \Delta_4 = D_{12}^T D_{12} \end{aligned}$$

Therefore using these relations in (10.10, 10.11) gives

$$-\gamma^2 \| V_1(s) \|_2^2 = -\gamma^2 \| \bar{U}_1(s) \|_2^2 + \| D_{11} \bar{U}_1(s) \|_2^2 - \| \Delta_3 \bar{U}_1(s) \|_2^2 \tag{10.12}$$

and

$$\begin{aligned} \| V_2(s) \|_2^2 = {} & \| D_{12} K_2 X(s) \|_2^2 + \| D_{12} U_2(s) \|_2^2 + \| \Delta_3 \bar{U}_1(s) \|_2^2 \\ & - 2 \langle D_{12} K_2 X(s), D_{12} U_2(s) \rangle - 2 \langle D_{12} K_2 X(s), D_{11} \bar{U}_1(s) \rangle \\ & + 2 \langle D_{12} U_2(s), D_{11} \bar{U}_1(s) \rangle \end{aligned} \tag{10.13}$$

Finally substituting (10.12) and (10.13) in P_γ ,(10.3), gives $P_\gamma = P_{\gamma o}$, (10.9). ■

Notice that, in the ideal case when the generalized plant state is available for state feedback, we can set $U_2(s) = K_2X(s)$ so that $Y_o(s) = D_{11}\bar{U}_1(s)$ and $P_{\gamma o}$, (10.9), becomes

$$P_{\gamma o} = \langle \bar{U}_1, (-\gamma^2 I_{m_1} + D_{11}^T D_{11})\bar{U}_1 \rangle$$

which satisfies $P_{\gamma o} \leq 0$ for all $\bar{U}_1(s) \in \mathcal{H}_2$ since we are assuming that $\sigma_{\max}[D_{11}] < \gamma$.

Alternatively, since $X(s)$ is unknown in the present case, we use its estimate, $\hat{X}(s)$, to generate the controlled input. Therefore replacing $U_2(s)$ by $K_2\hat{X}(s)$ in $Y_o(s)$ and substituting the result in $P_{\gamma o}$, (10.9), yield

$$P_{\gamma o} = -\gamma^2 \|\bar{U}_1(s)\|_2^2 + \|\tilde{Y}_o(s)\|_2^2 \tag{10.14}$$

where

$$\begin{aligned} \tilde{Y}_o(s) &= \hat{Y}_o(s) - Y_o(s) \\ &= D_{12}K_2\tilde{X}(s) - D_{11}\bar{U}_1(s) \\ \tilde{X}(s) &= X(s) - \hat{X}(s) \end{aligned}$$

Thus we see from (10.14) that in order for $P_{\gamma o} \leq 0$ for all $\bar{U}_1(s) \in \mathcal{H}_2$ we need to develop an estimator for $x(t)$ so that

$$\text{(i)} : T_{oc}(s) \in \mathcal{H}_\infty \qquad \text{(ii)} : \|T_{oc}(s)\|_\infty \leq \gamma \tag{10.15}$$

where

$$\tilde{Y}_0(s) = T_{oc}(s)\bar{U}_1(s) \tag{10.16}$$

which is an H_∞ state estimation problem. We obtain a solution to this problem by adapting the solution to the H_∞ state estimation problem obtain in Section 4 of Chapter 9.

10.2.2 An H_∞ state estimator

A comparison of the H_∞ state estimation problem just posed with the H_∞ state estimation problem solved in Chapter 9, reveals that the role played by the desired output estimation error, $\tilde{y}_1(t)$, is now being played by $\tilde{y}_o(t)$, the error in estimating $y_o(t)$. In addition, in the present situation, we are using $\bar{u}_1(t)$ in place of $u_1(t)$.

Therefore in order to use the solution to the H_∞ state estimation problem given in Chapter 9 we need to develop a state model for a system which relates $\bar{u}_1(t), u_2(t)$ as input to $y_o(t), y_2(t)$ as output and has the same state as the generalized plant. We can obtain such a system by replacing the desired output, $Y_1(s)$ from $\tilde{G}(s)$, (10.5), by $Y_o(s)$, (10.9). Thus doing this and denoting the resulting transfer function by $\tilde{G}_0(s)$, we obtain a system which is specified as

$$\tilde{G}_o(s) \stackrel{s}{=} \begin{bmatrix} A_o & B \\ C_o & D \end{bmatrix} \qquad \begin{bmatrix} Y_o(s) \\ Y_2(s) \end{bmatrix} = \begin{bmatrix} \tilde{G}_{o1}(s) & \tilde{G}_{o2}(s) \\ \tilde{G}_{o3}(s) & \tilde{G}_{o4}(s) \end{bmatrix} \begin{bmatrix} \bar{U}_1(s) \\ U_2(s) \end{bmatrix} \tag{10.17}$$

where

$$A_o = A + B_1K_1 \qquad C_o = \begin{bmatrix} -D_{12}K_2 \\ C_2 + D_{21}K_1 \end{bmatrix}$$

Now we could develop a solution to the present H_∞ state estimation problem by obtaining $T_{oc}(s)$, (10.15), from the state model for $\tilde{G}_o(s)$, (10.17) and proceeding as we did with T_o in the previous chapter in the development of an H_∞ state estimator for the generalized plant. However, since our goal here is to determine a stable estimator for the state of $\tilde{G}_o(s)$ subject to the requirement that

$$\|\tilde{Y}_o(s)\|_2 < \gamma\|\bar{U}_1(s)\|_2 \tag{10.18}$$

we can obtain the design equations for the desired estimator by replacing the parameters in the design equations for the H_∞ state estimator for the generalized plant state model, $G(s)$, (10.1), by the corresponding parameters in the state model for $\tilde{G}_o$, (10.17). Inspection of the state models for $G(s)$ and $\tilde{G}_o(s)$ shows correspondences between the state model parameters as given in Table 10.1.

Therefore the H_∞ state estimator for $\tilde{G}_o(s)$, (10.17), has differential equation

$$\dot{\hat{x}}(t) = A_o\hat{x}(t) + B_2u_2(t) + L_o\begin{bmatrix} \hat{y}_o(t) - y_o(t) \\ \hat{y}_2(t) - y_2(t) \end{bmatrix} \tag{10.19}$$

Notice that we are assuming temporarily that $y_o(t)$ is known and can be used in the state estimator, (10.19). This assumption will be removed later since the final controller we are seeking has only $y_2(t)$ as input and $u_2(t) = K_2\hat{x}(t)$ as output. Notice also that L_o in (10.19) corresponds to L in the H_∞ state estimator for the generalized plant given in Section 9.4.1.

Now the design equations for L_o are obtained by replacing the parameters in the HFARE and the equation for L which we obtained in Section 9.4.3 by the corresponding parameters as indicated in Table 10.1. Thus L_o is obtained (when it exists) by finding the solution $Y_{\infty 3}$ to the following ARE such that $A_o + L_oC_o$ is stable

$$\begin{gathered} A_{\infty 3}Y_{\infty 3} + Y_{\infty 3}A_{\infty 3}^T - Y_{\infty 3}R_{\infty 3}Y_{\infty 3} + Q_{\infty 3} = \emptyset \\ L_o = -[Y_{\infty 3}C_o^T + B_1D_2^T]D_{Jo}^{-1} \end{gathered} \tag{10.20}$$

where

$$\begin{aligned} A_{\infty 3} &= A_o - B_1D_2^TD_{Jo}^{-1}C_o \\ R_{\infty 3} &= C_o^TD_{Jo}^{-1}C_o \\ Q_{\infty 3} &= Q_{\infty 2} = B_1\left(I - D_2^TD_{Jo}^{-1}D_2\right)B_1^T \end{aligned}$$

with

$$D_2 = \begin{bmatrix} D_{11} \\ D_{21} \end{bmatrix} \qquad A_o = A + B_1K_1$$

$$D_{Jo} = \begin{bmatrix} -\gamma^2 I_{p_1} + D_{11}D_{11}^T & D_{11}D_{21}^T \\ D_{21}D_{11}^T & D_{21}D_{21}^T \end{bmatrix} \qquad C_o = \begin{bmatrix} -D_{12}K_2 \\ C_2 + D_{21}K_1 \end{bmatrix}$$

Table 10.1 Parameter correspondences for the generalized plant, $G(s)$, and $\tilde{G}_o(s)$.

$G(s)$	$\tilde{G}_o(s)$
A	A_o
B	B
C	C_o
C_1	$-D_{12}K_2$
C_2	$C_2 + D_{21}K_1$
D	D

As in the H_∞ state estimation problem treated in Chapter 9, the solution, $Y_{\infty 3}$, to the ARE, (10.20), which makes $A_{\infty 3} - Y_{\infty 3}R_{\infty 3} = A_o + L_o C_o$ stable, satisfies $Y_{\infty 3} \geq 0$ and is referred to as the stabilizing solution to equation (10.20). Notice that the parameters of the ARE, (10.20), involve the solution to the H_∞ state feedback control problem. Therefore the existence of a stabilizing solution to equation (10.20) depends on the solution to the H_∞ state feedback control problem. This observation suggests that the existence of solutions to both the H_∞ state feedback control problem and the H_∞ state estimation problem for a given generalized plant, does not guarantee the existence of a solution to the H_∞ output feedback control problem for that plant. This contrasts with the LQG output feedback control problem, which has a solution whenever both the LQG state feedback control and LQG state estimation problems have solutions.

We will see in Section 10.4 that the existence of a solution to the H_∞ output feedback control problem not only requires that both the HCARE and HFARE have stabilizing solutions, X_∞ and Y_∞, but in addition the product of these solutions satisfies the following inequality

$$\lambda_{\max}[Y_\infty X_\infty] < \gamma^2$$

10.2.3 Introducing estimated state feedback

So far we have developed an H_∞ state estimator for the generalized plant under the assumption that the state of the generalized plant is being used to generate the state feedback signal, $u_2(t) = K_2 x(t)$. In this case the estimate of $y_o(t)$, $\hat{y}_o(t)|_{u_2(t)}$, and of $y_2(t)$, $\hat{y}_2(t)|_{u_2(t)}$ based on the estimate of the generalized plant state, $\hat{x}(t)$, are seen from (10.17) to be

$$\hat{y}_o(t)|_{u_2(t)} = -D_{12}K_2\hat{x}(t) + D_{12}u_2(t) \tag{10.21}$$

$$\hat{y}_2(t)|_{u_2(t)} = (C_2 + D_{21}K_1)\hat{x}(t) + D_{22}u_2(t) \tag{10.22}$$

However, if the controlled input is generated as

$$u_2(t) = K_2\widehat{x}(t) \tag{10.23}$$

we see from (10.21, 10.22) that

$$\tilde{y}_o(t) = \hat{y}_o(t) - y_o(t) = -y_o(t) \tag{10.24}$$

$$\tilde{y}_2(t) = \hat{y}_2(t) - y_2(t) = (C_2 + D_{21}K_1 + D_{22}K_2)\hat{x}(t) - y_2(t) \tag{10.25}$$

where we have $\hat{y}_o(t)$, (10.21), null when the controlled input satisfies (10.23). This fact plays an important role in the development of the controller.

Continuing, we see that use of (10.24, 10.25) in (10.19) yields

$$\dot{\hat{x}}(t) = (A + BK + L_{o2}C_2 + L_{o2}D_3K)\hat{x}(t) - L_o\begin{bmatrix} y_o(t) \\ y_2(t) \end{bmatrix} \tag{10.26}$$

where

$$D_3 = \begin{bmatrix} D_{21} & D_{22} \end{bmatrix} \qquad L_o = \begin{bmatrix} L_{o1} & L_{o2} \end{bmatrix}$$

and the controlled input is generated by a controller with transfer function $\Gamma(s)$ as

$$U_2(s) = \Gamma(s)\begin{bmatrix} Y_o(s) \\ Y_2(s) \end{bmatrix} \qquad \Gamma(s) \stackrel{s}{=} \left[\begin{array}{c|c} A + BK + L_{o2}(C_2 + D_3K) & -L_o \\ \hline K_2 & \emptyset \end{array}\right] \tag{10.27}$$

However, since $y_o(t)$ is unknown we need to eliminate it as part of the input to this controller in a manner which preserves $P_{\gamma o} \leq 0$ for all $U_1(s) \in \mathcal{H}_2$. We will do this in the next section.

Before we go on to develop these $y_o(t)$-free controllers, we summarize what has been achieved so far in the development of a solution to the H$_\infty$ output feedback control problem.

1. We began by incorporating the solution to the H$_\infty$ state feedback control problem in the performance index, P_γ used in the previous chapter. The expression we obtained was denoted $P_{\gamma o}$. It was shown that $P_{\gamma o}$ depends on $y_o(t)$, (10.9), an output from a system, (10.17), whose state equals the state of the generalized plant.
2. Next, in Section 10.2.2, we used the solution to the H$_\infty$ state estimation problem obtained in Chapter 9 to obtain an H$_\infty$ state estimator based on the output $y_o(t)$ so that the error in estimating $y_o(t)$, i.e., $\tilde{y}_o(t)$, satisfies $\|\tilde{y}_o(t)\|_2 < \gamma\|\bar{u}_1(t)\|_2$.
3. Finally, in Section 10.2.3, we showed that by using the estimated state to generate the state feedback signal as $u_2(t) = K_2\hat{x}(t)$ and the H$_\infty$ state estimator based on $y_o(t)$ gave a controller which satisfied $P_{\gamma o} \leq 0$ for all $\bar{U}_1(s) \in \mathcal{H}_2$ or equivalently all $U_1(s) \in \mathcal{H}_2$ as required to solve the H$_\infty$ output feedback control problem.

10.3 H$_\infty$ Output Feedback Controllers

As mentioned before, the controller given by equation (10.27) cannot be used as it has the unknown output-like signal, $y_o(t)$, as an input. In this section we will show how this unknown signal can be removed from the input to the controller, (10.27).

One way of doing this which leads to a single controller referred to as the central controller, is to assume that the disturbance input is the worst disturbance, (Section 9.2.4), so that $\bar{U}_1(s)$ is null.

Another way of doing this is to relate $y_o(t)$ to $\bar{u}_1(t)$ through a system. This approach yields a family of controllers any one of which solves the H_∞ output feedback control problem. We will see that this family is a subset of the Youla parametrized stabilizing controllers for the given generalized plant which we encountered in Chapter 8.

10.3.1 Central Controller

Recall, from the previous section, that the error in estimating $y_o(t)$ satisfies $\tilde{y}_o(t) = -y_o(t)$, (10.24). Therefore we can write the performance index $P_{\gamma o}$, (10.14) as

$$P_{\gamma o} = -\gamma^2 \| \bar{U}_1(s) \|_2^2 + \| Y_o(s) \|_2^2 \tag{10.28}$$

and the relation (10.16) as

$$Y_o(s) = -T_{oc}(s)\bar{U}_1(s) \tag{10.29}$$

However recall that we determined the estimator, (10.19), so that

$$\| T_{oc}(s) \|_\infty \leq \gamma \qquad \bar{U}_1(s) \in \mathcal{H}_2 \tag{10.30}$$

Thus we see from (10.28, 10.29, 10.30) that

$$P_{\gamma o} \leq 0 \qquad \bar{U}_1(s) \in \mathcal{H}_2$$

as is required to satisfy condition (ii) in the statement of the H_∞ output feedback control problem.

Now $\bar{U}_1(s)$ is unknown. However suppose we generate the controlled input, $u_2(t)$, from the controller, (10.27), under the assumption that the disturbance input is the worst disturbance, (Section 9.2.4), i.e., under the assumption that $\bar{u}_1(t)$ is null. Then in this situation we see that $Y_o(s)$, (10.29), is null and therefore the controlled output, (10.27), is given as

$$U_2(s) = \Gamma_2(s) Y_2(s) \qquad \Gamma_2(s) \stackrel{s}{=} \begin{bmatrix} A + BK + L_{o2}(C_2 + D_3 K) & -L_{o2} \\ K_2 & \emptyset \end{bmatrix} \tag{10.31}$$

where

$$D_3 = \begin{bmatrix} D_{21} & D_{22} \end{bmatrix}$$

This controller is referred to as the central controller.

In essence the central controller is an H_∞ state estimator of the generalized plant under the restriction that the disturbance input is the worst disturbance and the controlled input depends on the state estimate as $u_2(t) = K_2 \hat{x}(t)$, where K_2 solves the H_∞ state feedback control problem. This is one way of obtaining a controller which is free from $y_o(t)$.

10.3.2 Controller parametrization

Suppose the unknown output $Y_o(s)$ which is present in the controller, (10.27), is replaced by $Y_\Phi(s)$ where

$$Y_\Phi(s) = \Phi(s)\bar{U}_1(s) \tag{10.32}$$

with

$$\text{(i)} : \Phi(s) \in \mathcal{H}_\infty \qquad\qquad \text{(ii)} : \|\Phi(s)\|_\infty \leq \gamma$$

Notice that when $\bar{U}_1(s) \in \mathcal{H}_2$ we have $Y_\Phi(s) \in \mathcal{H}_2$ and $\| Y_\Phi(s)\|_2 \leq \gamma\|\bar{U}_1(s)\|_2$. Thus when $Y_o(s)$ is replaced by $Y_\Phi(s)$ in (10.28) we have

$$P_{\gamma o} = \left\langle \bar{U}_1(s), \left(-\gamma^2 I + \Phi^*(s)\Phi(s)\right)\bar{U}_1(s)\right\rangle$$

which implies that $P_{\gamma o} \leq 0$ for $\bar{U}_1(s) \in \mathcal{H}_2$ as required to satisfy condition (ii) in the statement of the H_∞ output feedback control problem.

Next suppose we have $W(s) \in \mathcal{H}_\infty$ such that

$$\begin{bmatrix} U_2(s) \\ Y_2(s) \end{bmatrix} = \begin{bmatrix} W_1(s) & W_2(s) \\ W_3(s) & W_4(s) \end{bmatrix} \begin{bmatrix} Y_o(s) \\ \bar{U}_1(s) \end{bmatrix} \tag{10.33}$$

Then replacing $Y_o(s)$ in (10.33) by

$$Y_o(s) = \Phi(s)\bar{U}_1(s)$$

and solving (10.33) for $U_2(s)$ gives the desired controller as

$$U_2(s) = H(s)Y_2(s) \tag{10.34}$$

where

$$H(s) = [W_1(s)\Phi(s) + W_2(s)][W_3(s)\Phi(s) + W_4(s)]^{-1}$$

with $\Phi(s)$ constrained as specified in (10.32). The specification of $W(s)$ is given in the following theorem

Theorem 10.2 The state model for the transfer function $W(s)$, (10.33), satisfies $W(s) \in \mathcal{H}_\infty$ and has state model

$$\begin{bmatrix} W_1(s) & W_2(s) \\ W_3(s) & W_4(s) \end{bmatrix} \stackrel{s}{=} \begin{bmatrix} A + BK & \begin{bmatrix} B_2 D_{12}^\dagger & -L_{o2} \end{bmatrix} \\ \begin{bmatrix} K_2 \\ C_2 + D_3 K \end{bmatrix} & \begin{bmatrix} D_{12}^\dagger & \emptyset \\ D_{22} D_{12}^\dagger & I \end{bmatrix} \end{bmatrix} \tag{10.35}$$

where

$$D_{12}^\dagger = \left(D_{12}^T D_{12}\right)^{-1} D_{12}^T$$

Proof Since K was determined in the solution of the H_∞ state feedback control problem so that $A + BK$ is stable, (Section 9.3.1), we have $W(s) \in \mathcal{H}_\infty$.

In order to determine $W_2(s)$ and $W_4(s)$ we choose $\Phi(s)$ null so that (10.34) becomes

$$U_2(s) = W_2(s) W_4^{-1}(s) Y_2(s) \tag{10.36}$$

and comparing this equation with (10.31) yields

$$\Gamma_2(s) = W_2(s) W_4^{-1}(s) \tag{10.37}$$

Now since $W_2(s), W_4(s) \in \mathcal{H}_\infty$, we can determine $W_2(s)$ and $W_4(s)$ by coprime factorizing $\Gamma_2(s)$. More specifically, we choose $W_2(s) = N_\Gamma(s)$ and $W_4(s) = M_\Gamma(s)$ where $\Gamma_2(s) = N_\Gamma(s) M_\Gamma^{-1}(s)$ with $N_\Gamma(s), M_\Gamma(s) \in \mathcal{H}_\infty$ and $N_\Gamma(s), M_\Gamma(s)$ coprime.

Recall from Section 8.4.1 that if the state model for $\Gamma_2(s)$ is denoted as

$$\Gamma_2(s) \stackrel{s}{=} \begin{bmatrix} A_\Gamma & B_\Gamma \\ C_\Gamma & D_\Gamma \end{bmatrix}$$

then state models for the right coprime factors of $\Gamma_2(s)$ are given as

$$N_\Gamma(s) \stackrel{s}{=} \begin{bmatrix} A_\Gamma + B_\Gamma K_\Gamma & B_\Gamma \\ C_\Gamma + D_\Gamma K_\Gamma & D_\Gamma \end{bmatrix} \qquad M_\Gamma(s) \stackrel{s}{=} \begin{bmatrix} A_\Gamma + B_\Gamma K_\Gamma & B_\Gamma \\ K_\Gamma & I \end{bmatrix}$$

where K_Γ is chosen so that $N_\Gamma(s), M_\Gamma(s) \in \mathcal{H}_\infty$.

Now from (10.31) we have

$$A_\Gamma = A + BK + L_{o2}(C_2 + D_3 K)$$

$$B_\Gamma = -L_{o2} \qquad C_\Gamma = K_2 \qquad D_\Gamma = \emptyset$$

so that choosing $K_\Gamma = (C_2 + D_3 K)$ gives the coprime factors as

$$\begin{aligned} N_\Gamma(s) = W_2(s) &\stackrel{s}{=} \begin{bmatrix} A + BK & -L_{o2} \\ K_2 & \emptyset \end{bmatrix} \\ M_\Gamma(s) = W_4(s) &\stackrel{s}{=} \begin{bmatrix} A + BK & -L_{o2} \\ C_2 + D_3 K & I \end{bmatrix} \end{aligned} \tag{10.38}$$

where $N_\Gamma(s), M_\Gamma(s) \in \mathcal{H}_\infty$ since K was determined in the solution of the H_∞ state feedback control problem, (Section 9.3.1) so that $A + BK$ is stable. This completes the determination of state models for $W_2(s)$ and $W_4(s)$.

Next in order to determine $W_1(s)$ we recall from (10.17) and (10.33) that

$$Y_o(s) = \tilde{G}_{o1}(s) \bar{U}_1(s) + \tilde{G}_{o2}(s) U_2(s) \tag{10.39}$$

$$U_2(s) = W_1(s) Y_o(s) + W_2(s) \bar{U}_1(s) \tag{10.40}$$

Now if the system inverse, $\tilde{G}_{o2}^{-1}(s)$, were to exist we could solve (10.39) for $U_2(s)$ as

$$U_2(s) = \tilde{G}_{o2}^{-1}(s)Y_o(s) - \tilde{G}_{o2}^{-1}(s)\tilde{G}_{o1}(s)\bar{U}_1(s) \tag{10.41}$$

and we would have $W_1(s) = \tilde{G}_{o2}^{-1}(s)$ from a comparison of (10.41) with (10.40). However we see from (10.17) that $\tilde{G}_{o2}(s)$ is not invertible in general since its state model,

$$\tilde{G}_{02}(s) \overset{s}{=} \begin{bmatrix} A + B_1K_1 & B_2 \\ -D_{12}K_2 & D_{12} \end{bmatrix}$$

has a D matrix, D_{12}, which is not invertible in general, (Section 8.2). This difficulty can be overcome by recalling that we are assuming $D_{12}^TD_{12}$ is invertible since this is a necessary condition for the existence of a solution to the H_∞ state feedback control problem. Thus $\bar{G}_{02}(s) = D_{12}^T\tilde{G}_{02}(s)$ is invertible where

$$\bar{G}_{02}(s) \overset{s}{=} \begin{bmatrix} A + B_1K_1 & B_2 \\ -D_{12}^TD_{12}K_2 & D_{12}^TD_{12} \end{bmatrix} \tag{10.42}$$

Therefore we rewrite (10.39, 10.40) as

$$\bar{Y}_o(s) = \bar{G}_{o1}(s)\bar{U}_1(s) + \bar{G}_{o2}(s)U_2(s) \tag{10.43}$$

$$U_2(s) = \bar{W}_1(s)\bar{Y}_o(s) + W_2(s)\bar{U}_1(s) \tag{10.44}$$

where

$$\bar{Y}_o(s) = D_{12}^TY_o(s)$$

$$\bar{G}_{oi}(s) = D_{12}^T\tilde{G}_{oi}(s) : \quad i = 1,2$$

$$W_1(s) = \bar{W}_1(s)D_{12}^T$$

Then solving (10.43) for $U_2(s)$

$$U_2(s) = \bar{G}_{o2}^{-1}(s)\bar{Y}_o(s) - \bar{G}_{o2}^{-1}(s)\bar{G}_{o1}(s)\bar{U}_1(s) \tag{10.45}$$

and comparing (10.44) with (10.45) yields

$$\bar{W}_1(s) = \bar{G}_{o2}^{-1}(s) \tag{10.46}$$

Finally, using the form for the state model of a system inverse, (Section 8.2), we see from (10.42) that

$$\bar{G}_{o2}^{-1}(s) \overset{s}{=} \begin{bmatrix} A + BK & B_2(D_{12}^TD_{12})^{-1} \\ K_2 & (D_{12}^TD_{12})^{-1} \end{bmatrix} \tag{10.47}$$

and (10.47) yields

$$W_1(s) = \bar{W}_1(s)D_{12}^T \overset{s}{=} \begin{bmatrix} A + BK & B_2D_{12}^\dagger \\ K_2 & D_{12}^\dagger \end{bmatrix} \tag{10.48}$$

where $D_{12}^\dagger$ is a left inverse of D_{12}, i.e., $D_{12}^\dagger D_{12} = I$, given as

$$D_{12}^\dagger = \left(D_{12}^T D_{12}\right)^{-1} D_{12}^T \tag{10.49}$$

This completes the determination of $W_1(s)$.

In order to determine $W_3(s)$, rewrite the second block equations in (10.33) and in (10.17) as

$$Y_2(s) = W_3(s)Y_o(s) + W_4(s)\bar{U}_1(s) \tag{10.50}$$

$$Y_2(s) = \tilde{G}_{o3}(s)\bar{U}_1(s) + \tilde{G}_{o4}(s)U_2(s) \tag{10.51}$$

Then substituting for $U_2(s)$ from (10.40) in (10.51)

$$Y_2(s) = \tilde{G}_{o4}(s)W_1(s)Y_o(s) + \left(\tilde{G}_{o3}(s) + \tilde{G}_{o4}(s)W_2(s)\right)\bar{U}_1(s) \tag{10.52}$$

and comparing (10.50) with (10.52) reveals that

$$W_3(s) = \tilde{G}_{o4}(s)W_1(s)$$

Then using the state models for $\tilde{G}_{o4}(s)$, (10.17), and $W_1(s)$, (10.48) together with the rule for combining systems in series, (Section 8.1), we obtain

$$W_3(s) \stackrel{s}{=} \begin{bmatrix} A_W & B_W \\ C_W & D_W \end{bmatrix} = \begin{bmatrix} \begin{bmatrix} A + B_1K_1 & B_2K_2 \\ \emptyset & A + BK \end{bmatrix} & \begin{bmatrix} B_2D_{12}^\dagger \\ B_2D_{12}^\dagger \end{bmatrix} \\ \begin{bmatrix} C_2 + D_{21}K_1 & D_{22}K_2 \end{bmatrix} & D_{22}D_{12}^\dagger \end{bmatrix}$$

Next this state model can reduced without affecting $W_3(s)$. We can do this through the use of an appropriate change of coordinates. Therefore if we let the coordinate transformation matrix be

$$T = \begin{bmatrix} I & \emptyset \\ I & I \end{bmatrix}$$

the state model for $W_3(s)$ in the new coordinates is

$$\begin{aligned} W_3(s) &\stackrel{s}{=} \begin{bmatrix} T^{-1}A_WT & T^{-1}B_W \\ C_WT & D_W \end{bmatrix} = \begin{bmatrix} \bar{A}_W & \bar{B}_W \\ \bar{C}_W & D_W \end{bmatrix} \\ &\stackrel{s}{=} \begin{bmatrix} \begin{bmatrix} A + BK & B_2K_2 \\ \emptyset & A + B_1K_1 \end{bmatrix} & \begin{bmatrix} B_2D_{12}^\dagger \\ \emptyset \end{bmatrix} \\ \begin{bmatrix} C_2 + D_3K & D_{22}K_2 \end{bmatrix} & D_{22}D_{12}^\dagger \end{bmatrix} \end{aligned}$$

where

$$D_3 = \begin{bmatrix} D_{21} & D_{22} \end{bmatrix}$$

Finally, notice that $W_3(s)$ is given by

$$W_3(s) = \bar{C}_W(sI - \bar{A}_W)^{-1}\bar{B}_W + D_W$$
$$= (C_2 + D_3K)[sI - A - BK]^{-1}B_2D_{12}^{\dagger} + D_{22}D_{12}^{\dagger}$$

which implies that $W_3(s)$ has a state model given as

$$W_3(s) \stackrel{s}{=} \begin{bmatrix} A+BK & B_2D_{12}^{\dagger} \\ C_2 + D_3K & D_{22}D_{12}^{\dagger} \end{bmatrix} \qquad D_{12}^{\dagger} = \left(D_{12}^TD_{12}\right)^{-1}D_{12}^T$$

■

In summary, in this section we have determined a family of controllers, $H(s)$, (10.34, 10.35), any one of which solves the H$_\infty$ output feedback control problem. In the next section we relate this family of controllers to a subset of the Youla parametrized controllers.

10.3.3 Relation to Youla parametrization

Recall, from Theorem 8.5, that the Youla parametrization of stabilizing controllers, $H(s)$, for a plant, $G(s)$, involves the doubly coprime factorization of $G(s)$, (Section 8.4.3). These controllers are given by

$$H(s) = N_c(s)M_c^{-1}(s) \tag{10.53}$$

where

$$\left.\begin{aligned} N_c(s) &= M_G(s)Q_G(s) + N_H(s) \\ M_c(s) &= N_G(s)Q_G(s) + M_H(s) \end{aligned}\right\} \quad Q_G(s) \in \mathcal{H}_\infty$$

with

$$G(s) = \tilde{M}_G^{-1}(s)\tilde{N}_G(s) = N_G(s)M_G^{-1}(s)$$

$$\begin{bmatrix} \tilde{M}_H(s) & -\tilde{N}_H(s) \\ -\tilde{N}_G(s) & \tilde{M}_G(s) \end{bmatrix} \begin{bmatrix} M_G(s) & N_H(s) \\ N_G(s) & M_H(s) \end{bmatrix} = \begin{bmatrix} I & \emptyset \\ \emptyset & I \end{bmatrix}$$

Now in the present situation, the H$_\infty$ output feedback controller is connected between the generalized plant's measured output, $y_2(t)$ and controlled input, $u_2(t)$. Moreover, this controller must stabilize the generalized plant for all disturbance inputs satisfying $U_1(s) \in \mathcal{H}_2$ including the worst disturbance, $u_1(t) = K_1x(t)$ (Section 9.2.4). Therefore, assuming the disturbance input is the worst disturbance, we see that the controller must stabilize a plant $G(s) = \tilde{G}_{22}(s)$, where from (10.5) we have

$$G(s) = \tilde{G}_{22}(s) \stackrel{s}{=} \begin{bmatrix} A_G & B_G \\ C_G & D_G \end{bmatrix} = \begin{bmatrix} A + B_1K_1 & B_2 \\ C_2 + D_{21}K_1 & D_{22} \end{bmatrix} \tag{10.54}$$

Thus recalling the state models for the systems involved in a doubly coprime factorization, (8.46, 8.53), we see that $\tilde{G}_{22}(s)$ has a doubly coprime factorization given by

$$N_H(s) \overset{s}{=} \begin{bmatrix} A_G + B_G K_G & L_G \\ K_G & \emptyset \end{bmatrix} \qquad M_H(s) \overset{s}{=} \begin{bmatrix} A_G + B_G K_G & L_G \\ C_G + D_G K_G & I \end{bmatrix}$$
$$N_G(s) \overset{s}{=} \begin{bmatrix} A_G + B_G K_G & B_G \\ C_G + D_G K_G & D_G \end{bmatrix} \qquad M_G(s) \overset{s}{=} \begin{bmatrix} A_G + B_G K_G & B_G \\ K_G & I \end{bmatrix} \tag{10.55}$$

Then substituting the state model parameters as specified in (10.54) and setting $K_G = K_2$ and $L_G = -L_{o2}$ in (10.55) gives

$$\begin{bmatrix} M_G(s) & N_H(s) \\ N_G(s) & M_H(s) \end{bmatrix} \overset{s}{=} \begin{bmatrix} A + B_1K_1 + B_2K_2 & \begin{bmatrix} B_2 & -L_{o2} \end{bmatrix} \\ \begin{bmatrix} K_2 \\ C_2 + D_{21}K_1 + D_{22}K_2 \end{bmatrix} & \begin{bmatrix} I & \emptyset \\ D_{22} & I \end{bmatrix} \end{bmatrix} \tag{10.56}$$

However, since $D_{12}^\dagger Q_G(s) \in \mathcal{H}_\infty$ if $Q_G(s) \in \mathcal{H}_\infty$, we can replace $Q_G(s)$ in the general expression for the parametrized stabilizing controller, (10.53), by $D_{12}^\dagger \Phi(s)$ so that

$$H(s) = [\bar{M}_G(s)\Phi(s) + N_H(s)][\bar{N}_G(s)\Phi(s) + M_H(s)]^{-1} \qquad \Phi(s) \in \mathcal{H}_\infty \tag{10.57}$$

where

$$\begin{bmatrix} \bar{M}_G(s) & N_H(s) \\ \bar{N}_G(s) & M_H(s) \end{bmatrix} \overset{s}{=} \begin{bmatrix} A + B_1K_1 + B_2K_2 & \begin{bmatrix} B_2D_{12}^\dagger & -L_{o2} \end{bmatrix} \\ \begin{bmatrix} K_2 \\ C_2 + D_{21}K_1 + D_{22}K_2 \end{bmatrix} & \begin{bmatrix} D_{12}^\dagger & \emptyset \\ D_{22}D_{12}^\dagger & I \end{bmatrix} \end{bmatrix}$$

Thus controllers satisfying (10.57) stabilize the generalized plant and have the same form as the H_∞ output feedback controllers determined in the previous section, (10.34, 10.35). Notice that in addition to the restriction $\Phi(s) \in \mathcal{H}_\infty$ which ensures that the control system is internally stable, in the present problem we also require $\|\Phi(s)\|_\infty \leq \gamma$ in order that the control system satisfies the condition $\|y_1(t)\|_2 \leq \gamma\|u_1(t)\|_2$.

10.4 H_∞ Separation Principle

Recall, from Chapter 6, that we obtained the controller which solves LQG output feedback control problem directly from the corresponding solutions to the basic LQG plant state estimation and plant state feedback control problems. We referred to this fact as the separation principle. However in the H_∞ case, being able to solve the corresponding basic H_∞ state feedback control and state estimation problems is only a necessary condition for being able to solve the H_∞ output feedback control problem.

In this section we will show that the additional condition needed to ensure the existence of the solution to the H_∞ output feedback control problem is

$$\lambda_{\max}[Y_\infty X_\infty] < \gamma^2 \tag{10.58}$$

where X_∞, Y_∞ are stabilizing solutions to the HCARE and HFARE, (9.46, 9.77). Before doing this we show now that $Y_{\infty 3}$ and L_o, (10.20), needed to construct the controllers, (10.34, 10.35) can be calculated directly from the stabilizing solutions to the HCARE and HFARE as follows:

$$Y_{\infty 3} = ZY_\infty \tag{10.59}$$

$$L_o = \begin{bmatrix} Z(L_1 - \gamma^2 Y_\infty C_{1K}^T) & ZL_2 \end{bmatrix} \tag{10.60}$$

where

$$Z = \left(I - \gamma^{-2} Y_\infty X_\infty\right)^{-1}$$
$$C_{1K} = D_{11}K_1 + D_{12}K_2 + C_1$$

with L being the solution to the H$_\infty$ state estimation problem for the given generalized plant (9.78).

The approach used to show these results relies on a similarity relation between the Hamiltonian matrices associated with the HFARE, (9.77), and the ARE, (10.20).

10.4.1 A relation between Hamiltonians

We are going to make use of the following ideas from Chapter 5.

The Hamiltonian matrix H_P associated with the ARE

$$AP + PA^T - PRP + Q = \emptyset \tag{10.61}$$

is given as

$$H_P = \begin{bmatrix} A^T & R \\ Q & -A \end{bmatrix} \tag{10.62}$$

The eigenvalues of H_P exhibit mirror image symmetry across the imaginary axis.

$$\lambda[H_P] = \lambda[(A - PR)^T] \cup \lambda[-(A - PR)]$$

This fact is seen from the following similarity transformation of H_P

$$\bar{H}_P = T^{-1} H_P T = \begin{bmatrix} (A - PR)^T & R \\ AP + PA^* - PRP + Q & -(A - PR) \end{bmatrix} \tag{10.63}$$

where

$$T = \begin{bmatrix} I & \emptyset \\ -P & I \end{bmatrix}$$

with $\bar{H}_P$ being block-upper triangular when P satisfies (10.61).

Notice that equating the first n columns on either side of (10.63) yields

$$H_P\begin{bmatrix} I \\ -P \end{bmatrix} = \begin{bmatrix} I \\ -P \end{bmatrix}(A - PR)^T \tag{10.64}$$

so that if P is a stabilizing solution to (10.61) then

$$\text{range}\begin{bmatrix} I \\ -P \end{bmatrix}$$

is the stable subspace of H_P.

Now the relation between the Hamiltonian matrices associated with the HFARE, (9.77), and the ARE, (10.20) which is given in the following theorem will be used to show (10.59, 10.60).

Theorem 10.3 The Hamiltonian matrix H_2 corresponding to the HFARE (9.77) and the Hamiltonian matrix H_3 corresponding to the ARE (10.20) are similar and related as

$$H_2 = \Phi^{-1}H_3\Phi \tag{10.65}$$

where

$$H_i = \begin{bmatrix} A^T_{\infty i} & R_{\infty i} \\ Q_{\infty i} & -A_{\infty i} \end{bmatrix} \qquad i = 2,3 \qquad \Phi = \begin{bmatrix} I & \gamma^{-2}X_\infty \\ \varnothing & I \end{bmatrix}$$

Proof Suppose (10.65) is valid. Then carrying out the matrix multiplication which is indicated on the right side of (10.65) enables (10.65) to be rewritten as

$$H_2 = \begin{bmatrix} (A_{\infty 3} - \gamma^{-2}Q_{\infty 3}X_\infty)^T & \gamma^{-2}(A^T_{\infty 3}X_\infty + X_\infty A_{\infty 3}) - \gamma^{-4}X_\infty Q_{\infty 3}X_\infty + R_{\infty 3} \\ Q_{\infty 3} & -(A_{\infty 3} - \gamma^{-2}Q_{\infty 3}X_\infty) \end{bmatrix} \tag{10.66}$$

Then equating like positioned blocks on either side of (10.66) yields

$$Q_{\infty 2} = Q_{\infty 3} \tag{10.67}$$

$$A_{\infty 2} = A_{\infty 3} - \gamma^{-2}Q_{\infty 3}X_\infty \tag{10.68}$$

$$R_{\infty 2} = \gamma^{-2}(A^T_{\infty 3}X_\infty + X_\infty A_{\infty 3}) - \gamma^{-4}X_\infty Q_{\infty 3}X_\infty + R_{\infty 3} \tag{10.69}$$

Now we show that (10.65) is satisfied by showing that (10.67–10.69) are satisfied.

Recall from (10.20) and the HFARE, (9.77), that

$$\begin{aligned} A_{\infty 3} &= A_o - B_1D_2^TD_{Jo}^{-1}C_o & A_{\infty 2} &= A - B_1D_2^TD_{Jo}^{-1}C \\ R_{\infty 3} &= C_o^TD_{Jo}^{-1}C_o & R_{\infty 2} &= C^TD_{Jo}^{-1}C \\ Q_{\infty 3} &= B_1(I - D_2^TD_{Jo}^{-1}D_2)B_1^T & Q_{\infty 2} &= B_1(I - D_2^TD_{Jo}^{-1}D_2)B_1^T \end{aligned} \tag{10.70}$$

where

$$D_2 = \begin{bmatrix} D_{11} \\ D_{21} \end{bmatrix} \qquad A_o = A + B_1 K_1$$

$$D_{Jo} = \begin{bmatrix} -\gamma^2 I_{p_1} + D_{11}D_{11}^T & D_{11}D_{21}^T \\ D_{21}D_{11}^T & D_{21}D_{21}^T \end{bmatrix} \qquad C_o = \begin{bmatrix} -D_{12}K_2 \\ C_2 + D_{21}K_1 \end{bmatrix}$$

and (10.67) is satisfied.

Next in order to show (10.68) we use (10.70) so that

$$A_{\infty 3} - A_{\infty 2} = B_1 K_1 + B_1 D_2^T D_{Jo}^{-1} \begin{bmatrix} C_1 + D_{12}K_2 \\ -D_{21}K_1 \end{bmatrix} \tag{10.71}$$

Now we are going to re-express the right side of this equation using results from the H_∞ state feedback control problem.

Recall from the solution to the H_∞ state feedback control problem given in Chapter 9, (9.43), that

$$D_{Jc}K = -(B^T X_\infty + D_1^T C_1) \tag{10.72}$$

Now writing D_{Jc} as

$$D_{Jc} = D_1^T D_1 - \begin{bmatrix} \gamma^2 I_{m_1} & \emptyset \\ \emptyset & \emptyset \end{bmatrix} \tag{10.73}$$

where

$$D_1 = [D_{11} \quad D_{12}]$$

enables (10.72) to be rewritten as

$$D_{11}^T C_{1K} = \gamma^2 K_1 - B_1^T X_\infty \tag{10.74}$$

$$D_{12}^T C_{1K} = -B_2^T X_\infty \tag{10.75}$$

where

$$C_{1K} = D_1 K + C_1 \tag{10.76}$$

Then K_1 is obtained from (10.74) as

$$K_1 = \gamma^{-2}(B_1^T X_\infty + D_{11}^T C_{1K}) \tag{10.77}$$

Next, we can readily show that

$$\begin{bmatrix} C_1 + D_{12}K_2 \\ -D_{21}K_1 \end{bmatrix} = \begin{bmatrix} C_{1K} \\ \emptyset \end{bmatrix} - D_2 K_1 = C - C_0$$

where

$$D_2 = \begin{bmatrix} D_{11} \\ D_{21} \end{bmatrix}$$

which after substituting for K_1 from (10.77) becomes

$$\begin{bmatrix} C_1 + D_{12}K_2 \\ -D_{21}K_1 \end{bmatrix} = \begin{bmatrix} C_{1K} \\ \emptyset \end{bmatrix} - \gamma^{-2} D_2 (B_1^T X_\infty + D_{11}^T C_{1K})$$

$$= \begin{bmatrix} I_{p_1} - \gamma^{-2} D_{11} D_{11}^T \\ -\gamma^{-2} D_{21} D_{11}^T \end{bmatrix} C_{1K} - \gamma^{-2} D_2 B_1^T X_\infty \tag{10.78}$$

However writing D_{Jo}, (10.70), as

$$D_{Jo} = -\gamma^2 \begin{bmatrix} I_{p_1} - \gamma^{-2} D_{11} D_{11}^T & -\gamma^{-2} D_{11} D_{21}^T \\ -\gamma^{-2} D_{21} D_{11}^T & -\gamma^{-2} D_{21} D_{21}^T \end{bmatrix}$$

gives

$$\begin{bmatrix} I_{p_1} - \gamma^{-2} D_{11} D_{11}^T \\ -\gamma^{-2} D_{21} D_{11}^T \end{bmatrix} = -\gamma^{-2} D_{Jo} \begin{bmatrix} I_{p_1} \\ \emptyset \end{bmatrix}$$

and (10.78) can be rewritten as

$$\begin{bmatrix} C_1 + D_{12}K_2 \\ -D_{21}K_1 \end{bmatrix} = -\gamma^{-2} D_{Jo} \begin{bmatrix} I_{p_1} \\ \emptyset \end{bmatrix} C_{1K} - \gamma^{-2} D_2 B_1^T X_\infty \tag{10.79}$$

Returning to (10.71) we see that use of (10.79, 10.77) on the right side of (10.71) gives

$$A_{\infty 3} - A_{\infty 2} = \gamma^{-2} B_1 (B_1^T X_\infty + D_{11}^T C_{1K}) - \gamma^{-2} B_1 D_2^T \left(\begin{bmatrix} I_{p_1} \\ \emptyset \end{bmatrix} C_{1K} + D_{Jo}^{-1} D_2 B_1^T X_\infty \right)$$

$$= \gamma^{-2} (B_1 B_1^T X_\infty - B_1 D_2^T D_{Jo}^{-1} D_2 B_1^T X_\infty)$$

and we see from the definition of $Q_{\infty 3}$, (10.70), that (10.68) is satisfied.

Finally it remains to show that (10.69) is satisfied, i.e., that

$$R_{\infty 2} = R_{\infty 3} + \Omega \tag{10.80}$$

where

$$\Omega = \gamma^{-2} (A_{\infty 3}^T X_\infty + X_\infty A_{\infty 3}) - \gamma^{-4} X_\infty Q_{\infty 3} X_\infty$$

We begin by using (10.67, 10.68) to express Ω as

$$\Omega = \gamma^{-2} (A_{\infty 2}^T X_\infty + X_\infty A_{\infty 2}) + \gamma^{-4} X_\infty Q_{\infty 2} X_\infty$$

Then using the definitions for $A_{\infty 2}$ and $Q_{\infty 2}$, (10.70), enables Ω to be written as

$$\Omega = \gamma^{-2}(A^T X_\infty + X_\infty A - C^T M^T X_\infty - X_\infty MC) + \gamma^{-4}(X_\infty B_1 B_1^T X_\infty - X_\infty MD_2 B_1^T X_\infty) \tag{10.81}$$

where

$$M = B_1 D_2^T D_{jo}^{-1}$$

Next we are going to re-express $A^T X_\infty + X_\infty A$ for use in (10.81).

Recall from the development of the HCARE given in the previous chapter that

$$A^T X_\infty + X_\infty A - K^T D_{Jc} K + C_1^T C_1 = \emptyset$$

which we can rewrite using the expression for D_{Jc}, (10.73), as

$$A^T X_\infty + X_\infty A = K^T D_1^T D_1 K - \gamma^2 K_1^T K_1 - C_1^T C_1 \tag{10.82}$$

However, from the definition of C_{1K}, (10.76), we see that expanding $C_{1K}^T C_{1K}$ and $C_1^T C_{1K}$ gives

$$K^T D_1^T D_1 K = C_{1K}^T C_{1K} - C_1^T C_{1K} - C_{1K}^T C_1 + C_1^T C_1$$

which when substituted in (10.82) yields

$$A^T X_\infty + X_\infty A = C_{1K}^T C_{1K} - C_1^T C_{1K} - C_{1K}^T C_1 - \gamma^2 K_1^T K_1$$

Finally we can expand this relation by substituting for K_1 from (10.77) to obtain

$$A^T X_\infty + X_\infty A = F_{CK} - \gamma^{-2} F_X - \gamma^{-2} X_\infty B_1 B_1^T X_\infty \tag{10.83}$$

where

$$F_{CK} = C_{1K}^T C_{1K} - C_1^T C_{1K} - C_{1K}^T C_1$$
$$F_X = X_\infty B_1 D_{11}^T C_{1K} + C_{1K}^T D_{11} B_1^T X_\infty + C_{1K}^T D_{11} D_{11}^T C_{1K}$$

Returning to (10.81), we can use (10.83) to rewrite (10.81) as

$$\Omega = \gamma^{-2}(F_{CK} - C^T M^T X_\infty - X_\infty MC) - \gamma^{-4}(F_X + X_\infty MD_2 B_1^T X_\infty) \tag{10.84}$$

This completes the expansion of Ω for use in showing (10.80)

Finally using (10.79) yields

$$C_o = C + \gamma^{-2} D_{Jo} \begin{bmatrix} I \\ \emptyset \end{bmatrix} C_{1K} + \gamma^{-2} D_2 B_1^T X_\infty \tag{10.85}$$

and from $R_{\infty 3}$, (10.70), we have

$$R_{\infty 3} = R_{\infty 2} - \gamma^{-2}[F_{CK} - C^T M^T X_\infty - X_\infty MC] \\ + \gamma^{-4}\left[F_X + X_\infty B_1 D_2 D_{Jo}^{-1} D_2 B_1^T X_\infty\right] \tag{10.86}$$

where M, F_{CK} and F_X are given in (10.81, 10.83)

Finally we see from (10.86) and (10.84) that (10.80) is satisfied so that (10.69) is valid. ■

10.4.2 Relating stabilizing solutions

The relation between Hamiltonian matrices, (10.65), is used now to show that (10.59) is satisfied and that condition (10.58) is required in order to solve the H_∞ output feedback control problem.

Recall from (10.64) that the Hamiltonian matrix, H_2, corresponding to the HFARE satisfies

$$H_2\begin{bmatrix} I \\ -Y_\infty \end{bmatrix} = \begin{bmatrix} I \\ -Y_\infty \end{bmatrix}(A_{\infty 2} - Y_\infty R_{\infty 2})^T \tag{10.87}$$

so that

$$\text{range}\begin{bmatrix} I \\ -Y_\infty \end{bmatrix}$$

is the stable subspace of H_2.

Therefore premultiplying (10.87) by Φ, inserting $\Phi^{-1}\Phi$ following H_2 and using (10.65) yields

$$H_3\begin{bmatrix} I - \gamma^{-2}X_\infty Y_\infty \\ -Y_\infty \end{bmatrix} = \begin{bmatrix} I - \gamma^{-2}X_\infty Y_\infty \\ -Y_\infty \end{bmatrix}(A_{\infty 2} - Y_\infty R_{\infty 2})^T$$

which indicates that

$$\text{range}\begin{bmatrix} I - \gamma^{-2}X_\infty Y_\infty \\ -Y_\infty \end{bmatrix}$$

is the stable subspace of H_3, since $A_{\infty 2} - Y_\infty R_{\infty 2}$ is stable.

Now since the range of a matrix is unchanged by post-multiplication by a nonsingular matrix, we have

$$\text{range}\begin{bmatrix} I - \gamma^{-2}X_\infty Y_\infty \\ -Y_\infty \end{bmatrix} = \text{range}\begin{bmatrix} I \\ -Y_\infty(I - \gamma^{-2}X_\infty Y_\infty)^{-1} \end{bmatrix}$$

provided $I - \gamma^{-2}X_\infty Y_\infty$ is invertible. Therefore we see that

$$\text{range}\begin{bmatrix} I \\ -Y_\infty(I - \gamma^{-2}X_\infty Y_\infty)^{-1} \end{bmatrix}$$

is the stable subspace of H_3. However we saw in the previous subsection that

$$\text{range}\begin{bmatrix} I \\ -Y_{\infty 3} \end{bmatrix}$$

is the stable subspace of H_3. Thus we see that

$$Y_{\infty 3} = Y_\infty\left(I - \gamma^{-2}X_\infty Y_\infty\right)^{-1} = (I - \gamma^{-2}Y_\infty X_\infty)^{-1}Y_\infty = ZY_\infty \tag{10.88}$$

and (10.59) is established. Notice in (10.88) that the rightmost equality is a consequence of the matrix inversion lemma, (Appendix). We show next that (10.58) needed to ensure the existence of the solution to the H_∞ output feedback control problem follows from (10.88).

Recall that $Y_{\infty 3}$ and Y_∞ are each stabilizing solutions to AREs so that we have $Y_{\infty 3}, Y_\infty \geq 0$. Thus we see from (10.88) that $Z \geq 0$. However in the derivation of (10.88) we required Z to be invertible. Therefore we can conclude that $Z > 0$ is required for the existence of a solution to the H_∞ output feedback control problem. Now the condition on Y_∞, X_∞ which ensure that $Z > 0$ is given in the following theorem.

Theorem 10.4 $Z > 0$ if and only if

$$\lambda_{\max}[Y_\infty X_\infty] < \gamma^2 \tag{10.89}$$

where

$$Z = \left(I - \gamma^{-2}Y_\infty X_\infty\right)^{-1}$$

Proof Recall from Chapter 4, that $Z > 0$ if and only if $Z^{-1} > 0$. In addition recall that $Z^{-1} > 0$ if and only if $\lambda_{Z^{-1}} > 0$ for all $\lambda_{Z^{-1}} \in \lambda[Z^{-1}]$.

Let $\lambda_{Z^{-1}}$ be any eigenvalue of Z^{-1}. Then

$$\det\left(\lambda_{Z^{-1}}I - Z^{-1}\right) = 0 \tag{10.90}$$

and from the dependency of Z^{-1} on $Y_\infty X_\infty$ we have

$$\det(\lambda_{YX}I - Y_\infty X_\infty) = 0 \tag{10.91}$$

where $\lambda_{YX} \in \lambda[Y_\infty X_\infty]$ and

$$\lambda_{YX} = \gamma^2(1 - \lambda_{Z^{-1}})$$

Then solving for $\lambda_{Z^{-1}}$ gives

$$\lambda_{Z^{-1}} = 1 - \gamma^{-2}\lambda_{YX} \tag{10.92}$$

and we see that $\lambda_{Z^{-1}} > 0$ if and only if $\lambda_{YX} < \gamma^2$.

Finally we see from (10.92) that $\lambda_Z > 0$ for all $\lambda_Z \in \lambda[Z]$ if and only if the largest eigenvalue of $Y_\infty X_\infty$, $\lambda_{\max}[Y_\infty X_\infty]$, satisfies the condition given in the theorem. ■

10.4.3 Determination of L_o

The relation between L_o and (L, Z), (10.60), can now be established directly form (10.88). As mentioned earlier, this relation enables $Y_{\infty 3}$ to be calculated directly from the stabilizing solutions to the HCARE and the HFARE since L depends on Y_∞ and Z depends on both X_∞ and Y_∞.

In order to establish (10.60), recall that L_o, was determined in (10.20) to be

$$L_o = -[Y_{\infty 3}C_o^T + B_1 D_2^T]D_{Jo}^{-1} \tag{10.93}$$

Therefore substituting for $Y_{\infty 3}$ and C_o from (10.59) and (10.85) yields

$$L_o = -\left[ZY_\infty C^T + \left(I + \gamma^{-2}ZY_\infty X_\infty\right)B_1 D_2^T\right]D_{Jo}^{-1} - \gamma^{-2}ZY_\infty C_{1K}^T\left[I_{p_1} \quad \emptyset\right] \tag{10.94}$$

where

$$C_{1K} = C_1 + D_{11}K_1 + D_{12}K_2$$

However from the definition of Z, (10.88), we have

$$I + \gamma^{-2}ZY_\infty X_\infty = Z\left(Z^{-1} + \gamma^{-2}Y_\infty X_\infty\right) = Z$$

so that (10.94) becomes

$$\begin{aligned}
L_o &= -Z\left(Y_\infty C^T + B_1 D_2^T\right)D_{Jo}^{-1} - \gamma^{-2}ZY_\infty C_{1K}^T[I \quad \emptyset] \\
&= ZL - \gamma^{-2}ZY_\infty C_{1K}^T[I \quad \emptyset] \\
&= \left[Z\left(L_1 + \gamma^{-2}Y_\infty C_{1K}^T\right) \qquad ZL_2\right]
\end{aligned} \tag{10.95}$$

where L was obtained in Section 9.4.3.

10.5 Summary

In this chapter we have shown that if the HCARE and HFARE developed in the previous chapter have stabilizing solutions, X_∞, Y_∞ and provided these solutions satisfy

$$\lambda_{\max}[Y_\infty X_\infty] < \gamma^2$$

then we can solve the H_∞ output feedback control problem. In addition we have shown that when these conditions are satisfied the controllers, $H(s)$, which solve the H_∞ output feedback control problem are given by

$$U_2(s) - H(s)Y_2(s) \qquad\qquad H(s) = N_c(s)M_c^{-1}(s)$$

where

$$\left.\begin{aligned} N_c(s) &= W_1(s)\Phi(s) + W_2(s) \\ M_c(s) &= W_3(s)\Phi(s) + W_4(s) \end{aligned}\right\} \qquad \Phi(s) \in \mathcal{H}_\infty \qquad\qquad \|\Phi(s)\|_\infty < \gamma$$

with $W(s)$ having state model

$$\begin{bmatrix} W_1(s) & W_2(s) \\ W_3(s) & W_4(s) \end{bmatrix} \stackrel{s}{=} \left[\begin{array}{cc} A+BK & \begin{bmatrix} B_2 D_{12}^{\dagger} & ZL_2 \end{bmatrix} \\ \begin{bmatrix} K_2 \\ C_2 + D_3 K \end{bmatrix} & \begin{bmatrix} D_{12}^{\dagger} & \emptyset \\ D_{22} D_{12}^{\dagger} & I \end{bmatrix} \end{array} \right]$$

$$D_{12}^{\dagger} = \left(D_{12}^T D_{12}\right)^{-1} D_{12}^T \qquad Z = \left(I - \gamma^{-2} Y_\infty X_\infty\right)^{-1}$$

and with K, L being determined from

$$K = -D_{Jc}^{-1}\left(B^T X_\infty + D_1^T C_1\right)$$
$$L = -\left(Y_\infty C^T + B_1 D_2^T\right) D_{Jo}^{-1}$$

where X_∞, Y_∞ are stabilizing solution of the HCARE and HFARE, viz.,

$$A_{\infty 1}^T X + X A_{\infty 1} - X R_{\infty 1} X + Q_{\infty 1} = 0$$
$$A_{\infty 2} Y + Y A_{\infty 2}^T - Y R_{\infty 2} Y + Q_{\infty 2} = 0$$

with

$$\begin{array}{ll} A_{\infty 1} = A - B D_{Jc}^{-1} D_1^T C_1 & A_{\infty 2} = A - B_1 D_2^T D_{Jo}^{-1} C \\ R_{\infty 1} = B D_{Jc}^{-1} B^T & R_{\infty 2} = C^T D_{Jo}^{-1} C \\ Q_{\infty 1} = C_1^T \left(I - D_1 D_{Jc}^{-1} D_1^T\right) C_1 & Q_{\infty 2} = B_1 [I - D_2^T D_{Jo}^{-1} D_2] B_1^T \\ D_1 = \begin{bmatrix} D_{11} & D_{12} \end{bmatrix} & D_2 = \begin{bmatrix} D_{11} \\ D_{21} \end{bmatrix} \end{array}$$

and

$$D_{Jc} = \begin{bmatrix} -\gamma^2 I_{m_1} + D_{11}^T D_{11} & D_{11}^T D_{12} \\ D_{12}^T D_{11} & D_{12}^T D_{12} \end{bmatrix}$$

$$D_{Jo} = \begin{bmatrix} -\gamma^2 I_{p_1} + D_{11} D_{11}^T & D_{11} D_{21}^T \\ D_{21} D_{11}^T & D_{21} D_{21}^T \end{bmatrix}$$

Conditions on the generalized plant state model parameters which ensure that the HCARE and the HFARE have stabilizing solutions were given at the end of Chapter 9.

10.6 Notes and References

H_∞ control can be used in conjunction with the small gain theorem (Section 7.4.3) to provide a means for designing controllers to stabilize plants having inexactly specified models [47, pp. 221–229]. These ideas were extended recently to plants having a class of nonlinear state models [31]. Additional material on the ideas presented in this chapter can be found in [18].

Appendix A: Linear Algebra

Most of the current "first" books on control systems with terms like "automatic control" or "feedback control" in their titles have an appendix containing a summary of matrix theory. These can be consulted for this purpose. The topics treated in this appendix are selected to give the reader additional insight into issues alluded to in the text.

A.1 Multiple Eigenvalues and Controllability

When some of the roots of the characteristic polynomial, $\det[\lambda I - A]$ for an $n \times n$ matrix A are repeated, A may not have a complete set of n independent eigenvectors. Notice that if

$$\det[\lambda I - A] = (\lambda - \lambda_1)^{n_1}(\lambda - \lambda_2)^{n_2} \cdots (\lambda - \lambda_m)^{n_m} \tag{A.1}$$

where the λ_ks are unique $\lambda_i \neq \lambda_j \quad i \neq j \quad i,j \in [1,m]$, then we say that λ_k has multiplicity n_k or is repeated n_k times. Since the degree of the characteristic polynomial, (A.1), is n, the dimension of the A matrix, and since an nth degree polynomial always has n roots, counting multiplicities, we have

$$\sum_{k=1}^{m} n_k = n$$

Now there is at least one eigenvector, v^{k1}, corresponding to each eigenvalue $\lambda_k : k \in [1,m]$. However since λ_k is repeated n_k times, we obtain a complete set of eigenvectors only if there are n_k independent eigenvectors, $v^{ki} : i = 1, 2, \cdots n_k$, corresponding to each $\lambda_k, k \in [1,m]$, i.e., only if

$$Av = \lambda_k v$$

has n_k independent solutions $v = v^{ki}$, $i = 1, 2, \cdots n_k$ for each $k \in [1,m]$.

Any square matrix of size n having a complete set of n eigenvectors gives rise to a diagonal matrix under the transformation

$$V^{-1}AV = \Lambda$$

where the n columns of V are eigenvectors of A, i.e.,

$$V = \begin{bmatrix} v^{11} & v^{12} & \cdots & v^{1n_1} & v^{21} & v^{22} & \cdots & v^{2n_2} & \cdots & v^{m1} & v^{m2} & \cdots & v^{mn_m} \end{bmatrix}$$

and Λ is a block diagonal matrix having $n_i \times n_i$ diagonal blocks, $\Lambda_i : i = 1, 2, \cdots m$, with λ_i at each diagonal position, i.e.,

$$\Lambda = \begin{bmatrix} \Lambda_1 & 0 & \cdots & 0 \\ 0 & \Lambda_2 & \cdots & 0 \\ \vdots & \vdots & \ddots & \vdots \\ 0 & 0 & \cdots & \Lambda_m \end{bmatrix} \qquad \Lambda_i = \begin{bmatrix} \lambda_i & 0 & \cdots & 0 \\ 0 & \lambda_i & \cdots & 0 \\ \vdots & \vdots & \ddots & \vdots \\ 0 & 0 & \cdots & \lambda_i \end{bmatrix}$$

Matrices having a complete set of n eigenvectors are referred to as diagonalizable matrices. Any matrix whose eigenvalues are simple, i.e., do not repeat, is a diagonalizable matrix. Alternatively, a matrix A that does not have a complete set of eigenvectors cannot be diagonalized by any invertible matrix V in the transformation $V^{-1}AV$. In this case A is said to be defective, i.e., defective matrices are not diagonalizable matrices. For more detail the reader is referred to [17, pp. 338–339].

Concerning the controllability of a state model having an A matrix with multiple eigenvalues, it turns out that when A has more than one eigenvector corresponding to any eigenvalue it is not possible to control the state from one input, i.e., the pair (A, B) is uncontrollable for all B having only one column.

The foregoing fact can be shown as follows. Whenever A has two independent right-eigenvectors corresponding an eigenvalue λ, A has two independent left-eigenvectors, w^1, w^2 corresponding to this eigenvalue. Then any linear combination of these left-eigenvectors is also a left-eigenvector of A, i.e.,

$$w^T A = \lambda w^T$$

where

$$w = \alpha_1 w^1 + \alpha_2 w^2$$

with the $\alpha_i s$ being arbitrary, constant, non-zero scalars. Therefore we can always choose the $\alpha_i s$ for any vector B of length n so that

$$w^T B = 0$$

showing that λ is an uncontrollable eigenvalue for all vectors B.

A.2 Block Upper Triangular Matrices

Block upper triangular matrices play an important role in many different situations in control theory. They are directly related to invariant subspaces. A subspace W of V is said to be an A-invariant subspace if for all $x \in W$ we have $Ax \in W$.

Suppose we are given a square matrix of size n which has a block upper triangular structure, e.g.,

$$A = \begin{bmatrix} A_1 & A_2 \\ A_3 & A_4 \end{bmatrix} \qquad A_3 = \emptyset \tag{A.2}$$

where A_1, A_2, A_3, and A_4 are $n_1 \times n_1, n_1 \times n_2, n_2 \times n_1$, and $n_2 \times n_2$ with $n = n_1 + n_2$. Then any x of the form

$$x = \begin{bmatrix} x^1 \\ \emptyset \end{bmatrix}$$

where x^1 is any vector of length n_1, lies in an A-invariant subspace, since $x \in W$

$$Ax = \begin{bmatrix} A_1 x^1 \\ \emptyset \end{bmatrix} \in W$$

In this case W is called an eigenspace corresponding to the eigenvalues of A which are also eigenvalues of A_1. To see this consider the eigenvalue-eigenvector equation for A. We can rewrite this equation in the following way

$$A_1 v_1^i + A_2 v_2^i = \lambda_i v_1^i \tag{A.3}$$

$$A_4 v_2^i = \lambda_i v_2^i \tag{A.4}$$

where

$$v^i = \begin{bmatrix} v_1^i \\ v_2^i \end{bmatrix}$$

is the partitioned form for the eigenvector v^i corresponding to eigenvalue λ_i.

Suppose that v_2^i is null. Then we see from (A.3) that v_1^i must be an eigenvector of A_1 corresponding to eigenvalue λ_i. Thus the eigenvalues of A_1 are eigenvalues of A i.e. $\lambda(A_1) \subset \lambda(A)$ with the corresponding eigenvectors of A being

$$v^i = \begin{bmatrix} v_1^i \\ \emptyset \end{bmatrix}$$

with $A_1 v_1^i = \lambda_i v_1^i$

Real Schur decomposition [17, p. 362], provides a computationally robust method for obtaining an orthogonal matrix Q such that

$$Q^T A Q$$

has the form (A.2). In addition there are ways of doing this so that the eigenvalues of A_1 are a specified subset of the eigenvalues of A.

A.3 Singular Value Decomposition (SVD)

In this section we give a proof of Theorem 7.1, viz.,

Theorem 7.1 Any $p \times m$ matrix of complex constants, M, which has rank r can be decomposed as

$$M = U\Sigma V^* \tag{A.5}$$

where U, and V are $p \times p$ and $m \times m$ unitary matrices with

$$U = [\,U_1 \quad U_2\,] \qquad \Sigma = \begin{bmatrix} \Sigma_0 & \emptyset \\ \emptyset & \emptyset \end{bmatrix} \qquad V = [\,V_1 \quad V_2\,]$$

$$U_1 \text{ is } p \times r \qquad U_2 \text{ is } p \times p - r$$

$$V_1 \text{ is } m \times r \qquad V_2 \text{ is } m \times m - r$$

and Σ_0 is an $r \times r$, real, positive definite, diagonal matrix denoted as

$$\Sigma_0 = \begin{bmatrix} \sigma_1 & 0 & \dots & 0 \\ 0 & \sigma_2 & \dots & 0 \\ \vdots & \vdots & \ddots & \vdots \\ 0 & 0 & \dots & \sigma_r \end{bmatrix} = \text{diag}[\sigma_1, \sigma_2, \cdots, \sigma_r]$$

with diagonal entries referred to as singular values and ordered so that

$$\sigma_i \geq \sigma_{i+1} \qquad i = 1, 2, \cdots, r-1$$

Proof Let (σ_i^2, v_i) denote an eigenvalue–eigenvector pair for M^*M an $m \times m$ Hermitian non-negatve matrix. Thus we have

$$M^*Mv_i = \sigma_i^2 v_i \tag{A.6}$$

with σ_i^2 real and positive so that we can always order the eigenvalues as

$$\sigma_1^2 \geq \sigma_2^2 \geq \cdots \geq \sigma_r^2 > 0$$

$$\sigma_{r+1}^2 = \sigma_{r+2}^2 = \cdots = \sigma_m^2 = 0$$

Notice that since rank$[M] = r$, only r of the eigenvalues of M^*M are non-zero.

In addition, since M^*M is Hermitian, we can always choose the eigenvectors so that they are orthonormal

$$v_i^* v_i = 1 \qquad i = 1, 2, \cdots m \tag{A.7}$$

$$v_i^* v_j = 0 \qquad \text{for } i \neq j \tag{A.8}$$

Next we form a matrix V from the foregoing orthonormal eigenvectors of M^*M so that (A.6) can be written as the matrix equations

$$M^*MV_1 = V_1\Sigma_0^2 \tag{A.9}$$

$$M^*MV_2 = \varnothing \tag{A.10}$$

where

$$V = [V_1 \quad V_2] \qquad \Sigma_0^2 = \text{diag}[\sigma_1^2, \sigma_2^2, \cdots, \sigma_r^2]$$
$$V_1 = [v_1 \quad v_2 \quad \cdots \quad v_r] \qquad V_2 = [v_{r+1} \quad v_{r+2} \quad \cdots \quad v_m]$$

Notice that the orthonormal nature of the eigenvectors, (A.7, A.8), makes the $m \times r$ matrix V_1 satisfy $V_1^*V_1 = I$, and the $m \times m - r$ matrix V_2 satisfy $V_2^*V_2 = I$. Thus we have $V^*V = VV^* = I$, i.e., the $m \times m$ matrix V is unitary.

Next, pre-multiplying (A.9) by V_1^* and then pre- and post-multiplying the resulting equation by Σ_0^{-1} gives

$$\Sigma_0^{-1}V_1^*M^*MV_1\Sigma_0^{-1} = I$$

which we can interpret as

$$U_1^*U_1 = I \tag{A.11}$$

where

$$U_1 = MV_1\Sigma_0^{-1}$$

Notice from (A.11) that the $p \times r$ matrix U_1 has orthonormal columns. Therefore we can always choose a $p \times p - r$ matrix U_2 having orthonormal columns so that the $p \times p$ matrix U

$$U = [U_1 \quad U_2]$$

is unitary, i.e., $U^*U = UU^* = I$.

We show next that the foregoing unitary matrices, U, V satisfy (A.5) in the statement of Theorem 7.1.

We begin by expanding U^*MV in terms of its blocks

$$U^*MV = \begin{bmatrix} U_1^*MV_1 & U_1^*MV_2 \\ U_2^*MV_1 & U_2^*MV_2 \end{bmatrix} \tag{A.12}$$

Then we see that the 1,1 and 1,2 blocks can be reduced by substituting for U_1 from (A.11) and using (A.9, A.10). Doing this yields

$$U_1^*MV_1 = \Sigma_0^{-1}V_1^*M^*MV_1 = \Sigma_0^{-1}\Sigma_0^2 = \Sigma_0$$
$$U_1^*MV_2 = \Sigma_0^{-1}V_1^*M^*MV_2 = \Sigma_0V_1^*V_2 = \varnothing$$

Next the reduction of the 2,1 block in (A.12) relies on using the expression for U_1,

(A.11) to enable MV_1 to be replaced by $U_1\Sigma_0$ so that

$$U_2^*MV_1 = U_2^*U_1\Sigma_0 = \emptyset$$

Finally in order to reduce the 2,2 block in (A.12) we use the fact that $MV_2 = \emptyset$. This is seen by noting, from the construction of V_2, (A.10), that

$$M^*Mv_i = \emptyset$$

where v_i is any column of V_2. This implies that

$$v_i^*M^*Mv_i = 0$$

and letting $z_i = Mv_i$ we see that

$$z_i^*z_i = 0$$

which is only possible if z_i is null. Thus Mv_i must be null and since v_i is any column of V_2 we have $MV_2 = \emptyset$. Therefore we see that the 2,2 block of (A.12) becomes

$$U_2^*MV_2 = U_2^*\emptyset = \emptyset$$

Thus from the foregoing we see that (A.12) is reduced to

$$U^*MV = \begin{bmatrix} \Sigma_0 & \emptyset \\ \emptyset & \emptyset \end{bmatrix}$$

and we have the form for the SVD of M which is given in the theorem. ■

A.4 Different Forms for the SVD

There are occasions where constraints imposed on the SVD by the dimensions of the matrix being decomposed can be taken into account so as to provide a more explicit form for the SVD. This possibility arises from the fact that the rank of a rectangular matrix cannot exceed the smaller of its dimensions. Therefore since M is $p \times m$, we have

$$\text{rank}[M] = r \leq \min(p, m)$$

and the SVD of M, (A.5), can be expressed in one of the following forms depending on which of the dimensions of M is larger.

If $p < m$, i.e., M has more columns than rows, then (A.5) can be written as

$$M = U_p\begin{bmatrix} \Sigma_p & \emptyset \end{bmatrix}\begin{bmatrix} V_{p1}^* \\ V_{p2}^* \end{bmatrix}$$

$$= U_p\Sigma_pV_{p1}^*$$

where U, V in (A.5) equal U_p and $\begin{bmatrix} V_{p1} & V_{p2} \end{bmatrix}$ respectively. In addition V_{p1}, V_{p2} are

$m \times p$ and $m \times (m-p)$ with Σ_p being a $p \times p$ diagonal matrix

$$\Sigma_p = \text{diag}[\sigma_1, \sigma_2, \cdots, \sigma_r, 0, \cdots, 0]$$

In this case it is customary to refer to the p diagonal entries in Σ_p as the singular values of M and the p columns of U_p and of V_{p1} as the left and right singular vectors of M respectively.

Similarly, if $p > m$, i.e., M has more rows than columns, then (A.5) can be written as

$$\begin{aligned} M &= \begin{bmatrix} U_{m1} & U_{m2} \end{bmatrix} \begin{bmatrix} \Sigma_m \\ \emptyset \end{bmatrix} V_m^* \\ &= U_{m1} \Sigma_m V_m^* \end{aligned}$$

where U, V in (A.5) equal $[\, U_{m1} \; U_{m2} \,]$ and V_m respectively. In addition U_{m1}, U_{m2} are $p \times m$ and $p \times (p-m)$ with Σ_m being the $m \times m$ diagonal matrix

$$\Sigma_m = \text{diag}[\sigma_1, \sigma_2, \cdots, \sigma_r, 0, \cdots 0]$$

Again it is customary to refer to the m diagonal entries in Σ_m as the singular values of M and the m columns of U_{m1} and of V_m as the left and right singular vectors of M respectively.

A.5 Matrix Inversion Lemma (MIL)

The following relation has been of considerable importance in the development of control theory.

Lemma If Ω and Σ are nonsingular matrices then

$$L^{-1} = R$$

where

$$\begin{aligned} L &= \Omega + \Psi \Sigma \Phi \\ R &= \Omega^{-1} - \Omega^{-1} \Psi \left(\Phi \Omega^{-1} \Psi + \Sigma^{-1} \right)^{-1} \Phi \Omega^{-1} \end{aligned}$$

Proof Expanding the product RL yields

$$RL = I + \Omega^{-1} \Psi \Sigma \Phi - \Gamma_1 - \Gamma_2 \tag{A.13}$$

where

$$\begin{aligned} \Gamma_1 &= \Omega^{-1} \Psi (\Phi \Omega^{-1} \Psi + \Sigma^{-1})^{-1} \Phi \\ \Gamma_2 &= \Omega^{-1} \Psi (\Phi \Omega^{-1} \Psi + \Sigma^{-1})^{-1} \Phi \Omega^{-1} \Psi \Sigma \Phi \end{aligned}$$

Then we see that

$$\begin{aligned}\Gamma_1 + \Gamma_2 &= \Omega^{-1}\Psi(\Phi\Omega^{-1}\Psi + \Sigma^{-1})^{-1}[I + \Phi\Omega^{-1}\Psi\Sigma]\Phi \\ &= \Omega^{-1}\Psi(\Phi\Omega^{-1}\Psi + \Sigma^{-1})^{-1}[\Sigma^{-1} + \Phi\Omega^{-1}\Psi]\Sigma\Phi \\ &= \Omega^{-1}\Psi\Sigma\Phi \qquad \text{(A.14)}\end{aligned}$$

Finally substituting (A.14) in (A.13) gives $RL = I$. ■

Appendix B: Reduced Order Model Stability

As mentioned at the end of Section 4.6, since the full order model is stable, we require the reduced order model to also be stable. Otherwise the input–output behavior of the reduced order model cannot approximate the full order model behavior as the difference between the outputs of the full and reduced order models diverges in response to a common bounded input. Since the reduced order model has a system matrix A_1 which is a partition of the full order system matrix A_b, in balanced coordinates,

$$A_b = \begin{bmatrix} A_1 & A_2 \\ A_3 & A_4 \end{bmatrix} \begin{matrix} \updownarrow n_1 \\ \updownarrow n - n_1 \end{matrix}$$

we need a criterion which can be applied to ensure that A_1 has all its eigenvalues in the open left half plane. A practical criterion which ensures this property was given, without proof, as Theorem 4.3 in Section 4.6. In this appendix we are going to prove this theorem.

Theorem 4.3 If in balanced coordinates, no diagonal entry in Σ_{b1} is the same as any diagonal entry in Σ_{b2} then A_1 will be stable where

$$W_c = W_o = \Sigma_b = \begin{bmatrix} \Sigma_{b1} & \emptyset \\ \emptyset & \Sigma_{b2} \end{bmatrix} \begin{matrix} \updownarrow n_1 \\ \updownarrow n - n_1 \end{matrix}$$

Proof Suppose A_1 is not stable. Then we have $\lambda \in \lambda[A_1]$ such that

$$A_1 v = \lambda v \qquad w^* A_1 = \lambda w^* \qquad \mathrm{Re}[\lambda] \geq 0 \tag{B.1}$$

Now we are going to show that under the conditions of the theorem, A_b is unstable. However this is impossible since a balanced realization is only possible if the given system is stable.

To begin, consider expanding the Lyapunov equations for the controllability and observability Gramians as

$$\begin{bmatrix} A_1 & A_2 \\ A_3 & A_4 \end{bmatrix}\begin{bmatrix} \Sigma_{b1} & 0 \\ 0 & \Sigma_{b2} \end{bmatrix} + \begin{bmatrix} \Sigma_{b1} & 0 \\ 0 & \Sigma_{b2} \end{bmatrix}\begin{bmatrix} A_1^* & A_3^* \\ A_2^* & A_4^* \end{bmatrix} = -\begin{bmatrix} B_1 \\ B_2 \end{bmatrix}[B_1^* \quad B_2^*] \tag{B.2}$$

$$\begin{bmatrix} A_1^* & A_3^* \\ A_2^* & A_4^* \end{bmatrix}\begin{bmatrix} \Sigma_{b1} & 0 \\ 0 & \Sigma_{b2} \end{bmatrix} + \begin{bmatrix} \Sigma_{b1} & 0 \\ 0 & \Sigma_{b2} \end{bmatrix}\begin{bmatrix} A_1 & A_2 \\ A_3 & A_4 \end{bmatrix} = -\begin{bmatrix} C_1^* \\ C_2^* \end{bmatrix}[C_1 \quad C_2] \tag{B.3}$$

Then we see that Σ_{b1} satisfies the following Lyapunov equations

$$A_1\Sigma_{b1} + \Sigma_{b1}A_1^* = -B_1B_1^* \tag{B.4}$$

$$A_1^*\Sigma_{b1} + \Sigma_{b1}A_1 = -C_1^*C_1 \tag{B.5}$$

Now since Σ_{b1} is diagonal with all diagonal entries being positive and real we have $\Sigma_{b1} > 0$. Therefore we see from the proofs of Theorems 4.7 and 4.8 that the assumed unstable eigenvalue, λ, of A_1 must lie on the imaginary axis and be an unobservable and uncontrollable eigenvalue of the pairs (A_1, C_1) and (A_1, B_1) respectively. Therefore the corresponding right and left eigenvectors, (B.1), satisfy

$$w^*B_1 = 0 \tag{B.6}$$

$$C_1v = 0 \tag{B.7}$$

Next we show that there are many eigenvectors corresponding to this eigenvalue. In what follows we show how these eigenvectors are related.

To begin, postmultiply (B.5) by the eigenvector v, and use (B.7) to obtain

$$A_1^*\Sigma_{b1}v = -\Sigma_{b1}\lambda v \qquad \text{or} \qquad (\Sigma_{b1}v)^*A_1 = -\lambda^*(\Sigma_{b1}v)^* \tag{B.8}$$

However, λ is on the imaginary axis. Therefore $-\lambda^* = \lambda$ and we see from (B.8) that $\Sigma_{b1}v$ is an additional left eigenvector of A_1 corresponding to the uncontrollable eigenvalue λ and therefore we have

$$v^*\Sigma_{b1}B_1 = 0 \tag{B.9}$$

Next premultiply (B.4) by $v^*\Sigma_{b1}$, and use (B.8, B.9) to obtain

$$\lambda v^*\Sigma_{b1}^2 = -v^*\Sigma_{b1}^2A_1^* \qquad \text{or} \qquad A_1\Sigma_{b1}^2v = \lambda\Sigma_{b1}^2v$$

Thus we see that both v and Σ_{b1}^2v are (independent) right eigenvectors of A_1 corresponding to the eigenvalue λ.

Continuing in this fashion we can generate sets of left and right eigenvectors corresponding to λ as

$$S_w = \{\Sigma_{b1}^iv : i = 1, 3, 5, \cdots, 2k+1\} \tag{B.10}$$

$$S_v = \{\Sigma_{b1}^iv : i = 0, 2, 4, \cdots, 2k\} \tag{B.11}$$

where k is an integer which is chosen large enough so that we can find a matrix Q to satisfy

$$\Sigma_{b1}^2 M = MQ \tag{B.12}$$

where M is the $n_1 \times (k+1)$ matrix

$$M = \begin{bmatrix} v & \Sigma_{b1}^2 v & \cdots & \Sigma_{b1}^{2k} v \end{bmatrix}$$

Notice from (B.12) that under the transformation Σ_{b1}^2, the range of M is unchanged. This fact plays an important role in what is to follow.

Now we saw in Theorem 5.5, that (B12) implies that every eigenvalue of Q is an eigenvalue of Σ_{b1}^2

$$\lambda[Q] \subset \lambda\left[\Sigma_{b1}^2\right] \tag{B.13}$$

This important fact will be used to show that $A_3 M = \emptyset$. Notice that since the elements in S_w, S_v are left and right eigenvectors, respectively, of A_1 corresponding to the uncontrollable and unobservable eigenvalue λ, we have

$$v^* \Sigma_{b1}^i B_1 = 0 \qquad i = 1, 3, \cdots, 2k+1 \tag{B.14}$$

$$C_1 \Sigma_{b1}^i v = 0 \qquad i = 0, 2, \cdots, 2k \tag{B.15}$$

Returning to (B2, B.3) we have

$$A_2 \Sigma_{b2} + \Sigma_{b1} A_3^* = -B_1 B_2^* \tag{B.16}$$

$$A_2^* \Sigma_{b1} + \Sigma_{b2} A_3 = -C_2^* C_1 \tag{B.17}$$

Then premultiplying (B.16) by $v^* \Sigma_{b1}^i$, and postmultiplying (B.17) by $\Sigma_{b1}^i v$ and using (B.14, B.15) gives

$$\Sigma_{b2} A_2^* \Sigma_{b1}^i v = -A_3 \Sigma_{b1}^{i+1} v \qquad i = 1, 3, \cdots, 2k+1 \tag{B.18}$$

$$A_2^* \Sigma_{b1}^{i+1} v = -\Sigma_{b2} A_3 \Sigma_{b1}^i v \qquad i = 0, 2, \cdots, 2k \tag{B.19}$$

Next we can rewrite (B.18) as

$$\Sigma_{b2} A_2^* \begin{bmatrix} \Sigma_{b1}^1 v & \Sigma_{b1}^3 v & \cdots & \Sigma_{b1}^{2k+1} v \end{bmatrix} = -A_3 \Sigma_{b1}^2 M \tag{B.20}$$

and (B19) as

$$\Sigma_{b2} A_2^* \begin{bmatrix} \Sigma_{b1}^1 v & \Sigma_{b1}^3 v & \cdots & \Sigma_{b1}^{2k+1} v \end{bmatrix} = -\Sigma_{b2}^2 A_3 M \tag{B.21}$$

where Σ_{b2} was used to premultiply (B.19) and M was defined previously in (B.12).

Now since the left sides of (B.20, B.21) are equal we have

$$A_3 \Sigma_{b1}^2 M = \Sigma_{b2}^2 A_3 M \tag{B.22}$$

Moreover we can use (B.12) to rewrite the expression on the left side of this equation. Doing this gives rise to the following Lyapunov equation

$$\Delta Q - \Sigma_{b2}^2 \Delta = \emptyset \tag{B.23}$$

where

$$\Delta = A_3 M$$

Now it turns out if Q and Σ_{b2}^2 do not have any eigenvalues in common then $\Delta = \emptyset$ is the only solution to (B.23). Indeed Q and Σ_{b2}^2 do not share any eigenvalues since the eigenvalues of Q were shown to be a subset of the eigenvalues of Σ_{b1}^2, (B.13), and from the statement of the theorem we have that no eigenvalue of Σ_{b1}^2 is an eigenvalue of Σ_{b2}^2. Thus we have $A_3 M = \emptyset$.

Since the columns of M consist of all the right-eigenvectors of A_1 corresponding to the imaginary axis eigenvalue λ of A_1, we have that (λ, q) is an eigenvalue right-eigenvector pair for the full system matrix A_b where

$$q = \begin{bmatrix} v \\ \emptyset \end{bmatrix}$$

since

$$A_b q = \begin{bmatrix} A_1 & A_2 \\ A_3 & A_4 \end{bmatrix} \begin{bmatrix} v \\ 0 \end{bmatrix} = \lambda \begin{bmatrix} v \\ 0 \end{bmatrix} = \lambda q$$

However, since λ lies on the imaginary axis, A_b is unstable. This contradicts the fact that A_b is stable. Therefore our assumption at the start of the proof that A_1 is unstable is false and the proof is complete. ■

Appendix C: Problems

The following problems illustrate the basic ideas introduced in the earlier chapters of the book. Problems dealing with more advanced principles are under development.

C.1 Problems Relating to Chapter 1

1. Use the fact that any matrices M_2 and M_3 have the same eigenvalues as the matrix M_1 if

$$M_2 = M_1^T \qquad\qquad M_3 = S^{-1} M_1 S$$

for any nonsingular matrix S, to determine relations between the unspecified elements in the matrices $A_k : k = 1, 2, 3, 4$, so that each matrix has the same set of eigenvalues. If this set is $\{1, -2, 3\}$ specify the values for the a_{kj}s.

$$A_1 = \begin{bmatrix} a_{11} & a_{12} & a_{13} \\ 1 & 0 & 0 \\ 0 & 1 & 0 \end{bmatrix} \qquad A_2 = \begin{bmatrix} a_{11} & 1 & 0 \\ a_{21} & 0 & 1 \\ a_{31} & 0 & 0 \end{bmatrix}$$

$$A_3 = \begin{bmatrix} 0 & 0 & a_{13} \\ 1 & 0 & a_{23} \\ 0 & 1 & a_{33} \end{bmatrix} \qquad A_4 = \begin{bmatrix} 0 & 1 & 0 \\ 0 & 0 & 1 \\ a_{31} & a_{32} & a_{33} \end{bmatrix}$$

2. Find the eigenvalues $\{\lambda_{ki} : i = 1, 2\}$ for each of the following system matrices $A_k : k \in [1, 6]$. Then calculate the corresponding right eigenvectors, v_k^i for $A_k : k = 1, 2$ letting the first component of each eigenvector be one.

$$A_1 = \begin{bmatrix} -3 & -1 \\ 2 & 0 \end{bmatrix} \quad A_2 = \begin{bmatrix} -1 & 3 \\ 0 & -2 \end{bmatrix} \quad A_3 = \begin{bmatrix} -2 & 4 \\ 0 & -1 \end{bmatrix}$$

$$A_4 = \begin{bmatrix} -\frac{3}{2} & \frac{1}{2} \\ \frac{1}{2} & -\frac{3}{2} \end{bmatrix} \quad A_5 = \begin{bmatrix} -5 & 2 \\ -6 & 2 \end{bmatrix} \quad A_6 = \begin{bmatrix} 2 & 6 \\ -2 & -5 \end{bmatrix}$$

3. Using Laplace transforms, and without calculating numerical values, determine which of the transition matrices corresponding to the $A_k s$, $k \in [1,6]$, given in the previous question have entries dependent on only one mode, $e^{\lambda_i t}$.
4. A certain system has a zero-input state response of

$$x(t) = \begin{bmatrix} x_1(t) \\ x_2(t) \end{bmatrix}$$

where

$$x_1(t) = 6e^{8t} + e^{-2t} \qquad x_2(t) = 8e^{8t} - 2e^{-2t}$$

Find, without using Laplace transforms, the eigenvalues and eigenvectors of the A matrix for the system's state model.
5. Using the trajectories obtained when the initial state is an eigenvector, sketch the state trajectory for the given initial states $x^i(0) : i = 1, 2$ and system matrix A. Use your result to determine a relation between the elements of C, namely, c_1, and c_2, so that the system output is bounded for all initial states.

$$A = \begin{bmatrix} 4 & 3 \\ 8 & 2 \end{bmatrix} \qquad x^1(0) = \begin{bmatrix} -1 \\ 2 \end{bmatrix} \qquad x^2(0) = \begin{bmatrix} 3 \\ 4 \end{bmatrix}$$

6. Calculate the eigenvalue left-eigenvector pairs, $\{\lambda_i, w^i : i \in [1,2]\}$, for A given in the previous question. From this result determine an output matrix C such that for any initial state the zero-input response $y(t)$ depends only on the mode e^{-2t}. What is the relation between the C matrix you obtain here and the C matrix you obtained in the previous question.
7. Suppose a state model has system matrix, A, which is partially specified as

$$A = \begin{bmatrix} \alpha & -2 \\ 1 & 0 \end{bmatrix}$$

If the last component of each right-eigenvector is 1, find:

(i) the initial state, $x(0)$, which gives a zero input response of

$$y(t) = 4e^{-t} \qquad \text{when } C = [-1 \quad 1]$$

(ii) the output matrix C which gives a zero input response of

$$y(t) = 4e^{-t} \qquad \text{when } x(0) = \begin{bmatrix} 1 \\ -2 \end{bmatrix}$$

8. Recall that if $\phi(t)$ is a transition matrix then

$$\text{(i)} \quad \phi(0) = I \qquad\qquad \text{(ii)} \quad \dot{\phi}(t) = A\phi(t)$$

If $M(t)$ is a transition matrix

$$M(t) = \frac{1}{3}\begin{bmatrix} \alpha_1 e^t + 2e^{-2t} & \alpha_2 e^t - 2e^{-2} \\ \alpha_3 e^t - e^{-2t} & \alpha_4 e^t + e^{-2t} \end{bmatrix}$$

find the $\alpha_i s$ and the corresponding system matrix A.

9. Determine diagonal form state models for systems having transfer functions

$$G_1(s) = \frac{3s - 1}{s^2 + 3s + 2} \qquad G_2(s) = \frac{s^2 - 3s + 2}{s^2 + 3s + 2}$$

10. If the transfer function for a certain system is

$$G(s) = \frac{s + 4}{(s + 1)(s + 1 + j)(s + 1 - j)}$$

find the real parameters $a_1, a_2, a_3, c_1, c_2, c_3$ in the state model for this system where

$$A = \begin{bmatrix} a_1 & 0 & 0 \\ 0 & a_2 & a_3 \\ 0 & 1 & 0 \end{bmatrix} \qquad B = \begin{bmatrix} 1 \\ 1 \\ 0 \end{bmatrix}$$

$$C = \begin{bmatrix} c_1 & c_2 & c_3 \end{bmatrix} \qquad D = 0$$

11. Use the eigenvectors of A to find B (other than the null vector) so that the state goes to the origin with time when the input is a unit impulse, $u(t) = \delta(t)$ and the initial state is null. It is known that one of the eigenvalues of A is at $+1$ and

$$A = \begin{bmatrix} +2 & +1 & -2 \\ 1 & 0 & 0 \\ 0 & 1 & 0 \end{bmatrix}$$

12. If the steady-state output is zero when the input is a unit impulse, $u(t) = \delta(t)$, find (without using Laplace transforms) the values for b_2 and b_3 when

$$A = \begin{bmatrix} 2 & -3/2 & 4/3 \\ 0 & -1 & 8/3 \\ 0 & 0 & 3 \end{bmatrix} \qquad B = \begin{bmatrix} 1 \\ b_2 \\ b_3 \end{bmatrix}$$

$$C = \begin{bmatrix} 1 & 1 & 1 \end{bmatrix} \qquad D = 0$$

C.2 Problems Relating to Chapter 2

1. A certain plant is known to have a controller form state model with a system matrix, A, which has eigenvalues at $\{-1, 1, 2, 3\}$. Further, this state model for the plant is known to have direct feedthrough matrix, D, which is zero. If the plant's steady state output is $y(\infty) = 5$ when the plant input and initial state are $u(t) = 1$ and $x(0) = \emptyset$, find the controller form state model for the plant.
2. Suppose we are given the transfer function $G(s)$ for some plant. If the state model for the plant is in controller form, determine the initial state, $x^i(0)$, in each case, so that

the corresponding output, $y_i(t)$, is as specified. The input is zero.

$$y_1(t) = 12e^{-t} \qquad\qquad y_2(t) = 6e^{-t} + 12e^{-2t}$$

$$G(s) = \frac{(s+4)(s+5)}{(s+1)(s+2)(s+3)}$$

3. Given that $\lambda[A] = \{1, 2, -3\}$ determine left-eigenvectors of A, w^i, assuming the first entry in each left-eigenvector is one. Use your result to find α and β in B so that $x(\infty)$ is null when $u(t) = \delta(t)$ and $x(0) = \emptyset$

$$A = \begin{bmatrix} 0 & 1 & 0 \\ 7 & 0 & 1 \\ -6 & 0 & 0 \end{bmatrix} \qquad B = \begin{bmatrix} 1 \\ \alpha \\ \beta \end{bmatrix}$$

4. Suppose there is a set of three state models each having the same A matrix but different B and C matrices as given. Determine which models if any can be transformed to controller form. Do this without using Laplace transforms.

$$\text{case(i)} \quad B_1 = \begin{bmatrix} -1 \\ 2 \end{bmatrix} \quad C_1 = [1 \quad 1]$$

$$A = \begin{bmatrix} 4 & 3 \\ 8 & 2 \end{bmatrix} \qquad \text{case(ii)} \quad B_2 = \begin{bmatrix} 3 \\ 4 \end{bmatrix} \quad C_2 = [2 \quad 1]$$

$$\text{case(iii)} \quad B_3 = \begin{bmatrix} 1 \\ 1 \end{bmatrix} \quad C_3 = [-4 \quad 3]$$

5. Specify for each state model, $\{(A, B_i, C_i) : i = 1, 2, 3\}$, given in the previous problem: (i) the controllable eigenvalues, (ii) the uncontrollable eigenvalues. In addition, specify which of these three state models is stabilizable.
6. Given the state models

$$A_1 = \begin{bmatrix} 0 & 0 & -5 \\ 1 & 0 & -4 \\ 0 & 1 & -4 \end{bmatrix} \quad B_1 = \begin{bmatrix} 1 \\ 2 \\ 0 \end{bmatrix} \quad A_2 = \begin{bmatrix} 0 & -1 & 0 \\ 1 & 0 & 1 \\ 0 & 0 & 0 \end{bmatrix} \quad B_2 = \begin{bmatrix} 0 \\ 0 \\ 1 \end{bmatrix}$$

$$C_1 = [1 \quad 0 \quad 0] \qquad D_1 = 0 \qquad C_2 = [-1 \quad 1 \quad 1] \qquad D_2 = 0$$

find, without using Laplace transforms, a coordinate transformation matrix T in each case so that the state models in the new coordinates are each in controller form. Specify T and the controller form state model, $(\bar{A}, \bar{B}, \bar{C})$ in each case.
7. Use your knowledge of the controllable decomposed form and the eigenvalues of lower triangular matrices to rapidly determine the polynomials

$$p_1(s) = \det[sI - A] \qquad\qquad p_2(s) = C\,\text{adj}[sI - A]B$$

if the state model parameters are

$$A = \begin{bmatrix} 0 & 7 & -6 & 0 & 0 \\ 1 & 0 & 0 & 0 & 0 \\ 0 & 1 & 0 & 0 & 0 \\ 3 & 5 & -2 & -3 & -2 \\ -4 & 6 & 4 & 1 & 0 \end{bmatrix} \qquad B = \begin{bmatrix} 0 \\ 0 \\ 0 \\ 1 \\ 0 \end{bmatrix}$$

$$C = [3 \quad -1 \quad 2 \quad 1 \quad 1] \qquad D = 0$$

Determine the transfer function which is free from coincident poles and zeros, i.e., whose numerator and denominator are coprime.

8. Determine B so that the steady state output is $y(\infty) = 10$ when $u(t) = 5$ for all positive t and

$$A = \begin{bmatrix} +2 & 1 & 0 \\ +1 & 0 & 1 \\ -2 & 0 & 0 \end{bmatrix} \qquad C = [1 \quad 0 \quad 0]$$

C.3 Problems Relating to Chapter 3

1. Find the numerator polynomial, $Num(s)$, and denominator polynomial, $Den(s)$, of the transfer function $G(s)$ corresponding to the following state model

$$A = \begin{bmatrix} -5 & 1 & 0 & 0 \\ -10 & 0 & 1 & 0 \\ -10 & 0 & 0 & 1 \\ -4 & 0 & 0 & 0 \end{bmatrix} \qquad B = \begin{bmatrix} 0 \\ 0 \\ 1 \\ 4 \end{bmatrix}$$

$$C = [1 \quad 0 \quad 0 \quad 0] \qquad D = 1$$

where

$$G(s) = \frac{Num(s)}{Den(s)}$$

2. Suppose there is a set of three state models each having the same A matrix but different B and C matrices as given. Determine which models if any can be

transformed to observer form. Do this without using Laplace transforms.

$$A = \begin{bmatrix} 4 & 3 \\ 8 & 2 \end{bmatrix} \qquad \begin{array}{lll} \text{case(i)} & B_1 = \begin{bmatrix} -1 \\ 2 \end{bmatrix} & C_1 = [1 \quad 1] \\ \text{case(ii)} & B_2 = \begin{bmatrix} 3 \\ 4 \end{bmatrix} & C_2 = [2 \quad 1] \\ \text{case(iii)} & B_3 = \begin{bmatrix} 1 \\ 1 \end{bmatrix} & C_3 = [-4 \quad 3] \end{array}$$

3. Specify for each state model, $\{(A, B_i, C_i) : i = 1, 2, 3\}$, given in the previous problem: (i) the observable eigenvalues, (ii) the unobservable eigenvalues. Specify which of these state models allow the design of an observer to estimate their state .
4. Suppose a plant state model is in controller form and is 3 dimensional with A having eigenvalues at 1,2,-3. If the C matrix in this model is partially specified as

$$C = [1 \quad -2 \quad q]$$

determine values for q which would make it impossible to design an observer which could be used to obtain an asymptotic estimate of the plant state.

C.4 Problems Relating to Chapter 4

1. If every entry in a square matrix is either $1/2$ or $-1/2$ determine the size of the matrix and the sign pattern of its entries if the matrix is orthogonal.
2. Find real numbers a, b, c_1, c_2, and c_3 such that M is orthogonal where

$$M = \begin{bmatrix} a & b & c_1 \\ a & 0 & c_2 \\ a & -b & c_3 \end{bmatrix}$$

3. M_1 and M_2 are square matrices having eigenvalue - right eigenvector pairs as follows:

$$\text{for } \mathrm{M}_1\text{:} \quad \{0, 1\} \quad \text{and} \quad \{[1, 2]^T, [2 - 1]^T\}$$
$$\text{for } \mathrm{M}_2\text{:} \quad \{0, 1\} \quad \text{and} \quad \{[1, 2]^T, [2 \quad 1]^T\}$$

respectively. Is either of these matrices symmetric? If so which one. Justify your answer.
4. For each matrix, specify which of the following properties holds: positive definite, non-negative, indefinite, i.e., not positive definite, not nonnegative, not negative definite. (Hint: use the distinguishing properties of the eigenvalues of each type of matrix)

$$M_1 = \begin{bmatrix} 1 & -2 \\ -2 & 5 \end{bmatrix} \quad M_2 = \begin{bmatrix} 3 & 2 \\ 2 & 1 \end{bmatrix}$$

$$M_3 = \begin{bmatrix} 1 & 0 & 0 \\ 0 & 2 & 0 \\ 0 & 0 & 3 \end{bmatrix} \quad M_4 = \begin{bmatrix} 1 & 0 & 2 \\ 0 & 0 & 0 \\ 2 & 0 & 1 \end{bmatrix}$$

5. Which of the following matrices could be either a controllability or an observability Gramian. Give reasons for your answer in each case.

$$M_1 = \begin{bmatrix} 1 & 2 \\ 2 & -1 \end{bmatrix} \qquad M_2 = \begin{bmatrix} 1 & -2 \\ 2 & -1 \end{bmatrix} \qquad M_3 = \begin{bmatrix} 2 & 3 \\ 3 & 0 \end{bmatrix}$$

6 The observability Gramian for a system is determined to be

$$W_o = \begin{bmatrix} 2 & -4 \\ -4 & 8 \end{bmatrix}$$

Is there an initial state vector, $x(0)$ other than the null vector, which when the input, $u(t)$, is always zero, produces an output, $y(t)$, which is always zero? If "no" justify. If "yes", give an example, i.e., give an $x(0)$, other than the null vector, which has this property.

7. If, in the previous problem, the initial state vector is restricted such that $x^T(0)x(0) = 1$, (and $u(t)$ is always zero), find the largest value possible for the integral from 0 to ∞ of the square of the output signal $y(t)$ under this restriction. Specify $x(0)$ (subject to the restriction) which would produce this value.

8. Determine, in each case, if the pairs (A, C) are observable. Calculate the observability Gramian manually by solving the appropriate Lyapunov equation in each case. Determine, in each case, if the corresponding Gramian is positive definite. Relate your findings on the positive definiteness to your findings on the observability in each case

$$\text{case(i):} \qquad A = \begin{bmatrix} -1 & 2 \\ 0 & -2 \end{bmatrix} \qquad C = [1 \quad 1]$$

$$\text{case(ii):} \qquad A = \begin{bmatrix} -1 & 2 \\ 0 & -2 \end{bmatrix} \qquad C = [0 \quad 1]$$

9. Determine the observability Gramian for the given system assuming that the r_is are real.

$$A = \begin{bmatrix} 0 & 1 & 0 \\ r_1 & 0 & 1 \\ 0 & r_2 & r_3 \end{bmatrix} \qquad B = \begin{bmatrix} 0 \\ 0 \\ 1 \end{bmatrix}$$

$$C = [0 \quad 0 \quad 1] \qquad D = 0$$

What values of the r_is make the Gramian singular? What values of the r_is make the system unobservable? Are these values related? if so explain.

10. A certain plant has state model matrices (A, B) and observability Gramian W_0 given as

$$A = \begin{bmatrix} -3 & \alpha \\ 1 & 0 \end{bmatrix} \quad B = \begin{bmatrix} 1 \\ 0 \end{bmatrix} \quad W_o = \begin{bmatrix} 3 & 1 \\ 1 & 2 \end{bmatrix}$$

Given that $D = 0$ and recalling that the integral defining W_o does not exist when A is unstable, find the transfer function of the plant using the Lyapunov equation relating A, C, W_0 to find α and C.

11. Suppose that a certain system is known to have maximum output energy equal to 2 for all possible initial states satisfying $x_1^2(0) + x_2^2(0) = 1$, when $u(t) \equiv 0$. Suppose the initial state that achieves this is $x^T(0) = [\frac{1}{\sqrt{2}}, \frac{1}{\sqrt{2}}]$. In addition suppose that this system has minimum output energy equal to 1 for some initial state which also satisfies $x_1^2(0) + x_2^2(0) = 1$ when $u(t) \equiv 0$. Find the system's observability Gramian and the initial state that gives this minimum output energy.

C.5 Problems Relating to Chapter 5

1. An observer based controller (Fig. 5.1) generates a feedback signal, $f(t)$, from

$$f(t) = Hz(t)$$
$$\dot{z}(t) = Fz(t) + G_1 v(t) + G_2 y(t)$$

where

$$F = \begin{bmatrix} -6 & 1 & 0 \\ -11 & 0 & 1 \\ -18 & 0 & 0 \end{bmatrix} \quad G_1 = \begin{bmatrix} 0 \\ 0 \\ 1 \end{bmatrix} \quad G_2 = \begin{bmatrix} 0 \\ 0 \\ 12 \end{bmatrix} \quad H^T = \begin{bmatrix} 12 \\ 0 \\ 0 \end{bmatrix}$$

If the plant output equals the first component of the plant state vector, find: (i) the closed loop transfer function from $v(t)$ to $y(t)$, (ii) the characteristic polynomial for the observer.

2. If in the previous problem the plant state model is in controller form and the plant output is the sum of the first three components of the plant state, determine the closed loop transfer function from v to y for the observer based feedback control system where

$$F = \begin{bmatrix} 0 & 4 & 0 & 0 \\ -1 & -4 & 0 & 0 \\ 5 & 7 & -13 & -8 \\ 0 & 0 & 13 & 0 \end{bmatrix} \quad G_1 = \begin{bmatrix} 0 \\ 0 \\ 1 \\ 0 \end{bmatrix} \quad G_2 = \begin{bmatrix} 0 \\ 0 \\ 10 \\ -12 \end{bmatrix} \quad H^T = \begin{bmatrix} -1 \\ 1 \\ 6 \\ 0 \end{bmatrix}$$

3. If in the observer based feedback control system shown in the Figure 5.1 we are given the transfer functions from $v(t)$ to $y(t)$, $G_{vy}(s)$, and from $y(t)$ to $f(t)$, $G_{yf}(s)$, where

$f(t)$ is the feedback signal, determine the controller state model matrices (Q, R, S, T) when

$$G_{vy}(s) = \frac{1}{s^3 + 3s^2 + 2s + 1} \qquad G_{yf}(s) = \frac{-2}{s^3 + 3s^2 + 2s + 3}$$

with plant model parameters given as

$$A = \begin{bmatrix} -3 & -2 & 0 \\ 1 & 0 & 0 \\ 0 & 1 & 0 \end{bmatrix} \qquad B = \begin{bmatrix} 1 \\ 0 \\ 0 \end{bmatrix} \qquad C^T = \begin{bmatrix} 0 \\ 0 \\ 1 \end{bmatrix}$$

and controller equations

$$\dot{z}(t) = Qz(t) + Rv(t) + Sy(t) \qquad f(t) = Tz(t)$$

Appendix D: MATLAB Experiments

D.1 State Models and State Response

The intention of this lab is to introduce some of the ways MATLAB can be used in connection with state models and to provide further experience in thinking in terms of the state space.

D.1.1 Controller form

Given the transfer function model for some system, the command

$$[a, b, c, d] = tf2ss(num, den)$$

can be invoked to find an equivalent state model, (use the help facility), where *num* and *den* are entered as row vectors containing coefficients of the numerator and denominator of the transfer function. Try the following examples

$$G_1(s) = \frac{1}{s^2 + 3s + 2}$$

$$G_2(s) = \frac{s^2 + 2s + 3}{3s^2 + 2s + 1}$$

Check your result by using the command "ss2tf", (use the help facility), to convert back to the transfer function models.

D.1.2 Second order linear behavior

Recall that the state at any time is represented by a point in an n-dimensional space and the path traced out by this point is referred to as a state trajectory. Unlike cases where the system order is greater that 2 , we can readily display the trajectory for a second order system as a graph with the two components of the state forming the axes of the graph.

Consider the system whose behavior is governed by the following second order linear differential equation

$$y^{(2)}(t) + 2\zeta\omega_n y^{(1)}(t) + \omega_n^2 y(t) = \omega_n^2 u(t)$$

where

$$y^{(i)}(t) = \frac{d^i y(t)}{dt^i}$$

Then the controller state model is

$$A = \begin{bmatrix} -2\zeta\omega_n & -\omega_n^2 \\ 1 & 0 \end{bmatrix} \quad B = \begin{bmatrix} 1 \\ 0 \end{bmatrix}$$

$$C = \begin{bmatrix} 0 & \omega_n^2 \end{bmatrix} \qquad D = 0$$

Now we are going to investigate the zero-input state response by obtaining the state trajectories for one value of natural frequency, say $\omega_n = 1$ and different values of the damping ratio, say

$$\zeta = 0.3, 1, 10, -0.3$$

each for some initial state, say $x^T(0) = [2\ 0]$

One way to do this is to invoke the command, (use the help facility),

$$initial(A, B, C, D, x0)$$

Then from the resulting graph of the output vs. time that is obtained on the screen, you can decide over what interval of time you will need the state.

Suppose you decide you need to know the state in the time interval $[0, 5]$ seconds every 0.5 seconds. We can create a vector, t, of these times by incorporating this data in a vector as start:increment:end. This is done here by typing

$$t = 0 : 0.5 : 5$$

and

$$[y, x, t] = initial(A, B, C, D, x0)$$

This produces a column y, followed by two columns, one for each component of the state. The entries in these columns give the values of the output, and components of the state at 0.5 s intervals from 0 to 5 s. Now you can obtain the graph of the state trajectory by entering

$$plot(x(:, 1), x(:, 2))$$

where $x(:, i)$ is all rows in the ith column of x. Your graph can be labelled using the procedure (on line help) where you enter each line (there are four) on the same line separated by commas.

Explain why these plots are markedly different for different signs on ζ.

D.1.3 Second order nonlinear behavior

Consider the system whose behavior is governed by the second order nonlinear differential equation, referred to as a van der Pol equation,

$$y^{(2)}(t) - (1 - y^2(t))y^{(1)}(t) + y(t) = 0$$

Then if we assign componets of the state to derivates of the output in the same manner as done to get the controller form in the linear case, the behavior of the components of the state are seen to be governed by the following first order nonlinear differential equations

$$\dot{x}_1(t) = (1 - x_2^2(t))x_1(t) - x_2(t)$$

$$\dot{x}_2(t) = x_1(t)$$

Now the state trajectory obtained for a specific initial state can be determined using MATLAB by creating an m-file called vdpol.m. To create this file open an editor and type

$$functionyp = vdpol(t, q);$$
$$yp = [(1 - q(2) * q(2)) * q(1) - q(2); q(1)];$$

Save the file and invoke one of the ordinary (vector) differential equation solvers, say ODE23

$$[t, q] = ode23(\text{'}vdpol\text{'}, [t0, tf], q0)$$

where *t0* is the start time, *tf* is the finish time and *q0* is the initial state, entered as a column vector by typing

$$q0 = [x_1(0); x_2(0)]$$

The sample values of the components of the state are returned as the columns of q.

Use $t0 = 0$, $tf = 25$ and try getting the state for each of the initial states

$$x(0) = \begin{bmatrix} 0.2 \\ 0 \end{bmatrix} \quad \text{and} \quad \begin{bmatrix} 2 \\ 0 \end{bmatrix}$$

Obtain a graph of the state trajectory in each case by typing

$$plot(q(:, 1), q(:, 2))$$

Notice that $q(:, 1)$ is all the elements in the first column of q produced by the simulation as the values of the first component of the state.

In the second case where $x^T(0) = [2\ 0]$ it is instructive to concentrate on the latter part of the trajectory, say from time step 40 to the end of the simulation. To find out what this end value is type

$$size(q)$$

Then if this end value is 200 type

$$plot(q(40:200,1),q(40:200,2))$$

to obtain the trajectory from time 40 to time 200.

Generalize the nature of the trajectories in terms of their dependency on the initial state. Compare the state trajectories you obtain for the present nonlinear system with those you obtained for the linear system in the previous section.

D.1.4 Diagonal form

So far we have only considered the controller form state model. There are several other important canonical forms for the state model of a linear dynamic system. One of these is the diagonal or normal form. This form is characterized by having its system matrix **A** which is diagonal. This simplifies solving the state differential equation since the derivative of each component of the state is dependent on itself only.

Recall that if $[A, B, C, D]$ is a state model for a given system so is $[Q^{-1}AQ, Q^{-1}B, CQ, D]$ for any nonsingular matrix Q of appropriate size, where Q is a coordinate transformation matrix. Also recall that if the eigenvalues of A are distinct, we can change coordinates so that A in the new coordinates is diagonal. This is done by letting Q have columns equal to the eigenvectors of A. We are going to use these facts.

Use MATLAB to determine the parameters, (A, B, C, D), of the controller form state model for the system having transfer function

$$G(s) = \frac{s^2 - s + 20}{s^4 + 5s^3 + 5s^2 - 5s - 6}$$

Use the command "roots" to decide if all the roots of the denominator are poles of $G(s)$. Calculate the eigenvalues and eigenvectors of A by typing

$$[Q, E] = eig(A)$$

and the characteristic polynomial of A by typing

$$poly(A)$$

How do the poles of $G(s)$ compare with the eigenvalues of A, i.e., diagonal entries in **E**? Should each of these sets be the same and if not why not?

Use **Q** to transform the controller form state model to a diagonal form state model.

This can be done by typing

$$aa = \text{inv}(Q) * A * Q$$
$$bb = \text{inv}(Q) * B$$
$$cc = CA * Q$$
$$dd = D$$

How do the diagonal entries of aa compare with the poles of $G(s)$?

Finally, invoke the command "residue" to determine the residues, poles and any constant term **D** (nonzero only if $G(s)$ is proper). This is done by typing

$$[res, poles, d] = residue(num, den)$$

How do the residues of $G(s)$ relate to the entries in bb and cc?

D.2 Feedback and Controllability

In order for you to proceed as rapidly as possible with the tasks described in what follows, you should plan a subscripting scheme for the various matrices you will be entering in the computer.

D.2.1 Controllable state models

Enter the state model

$$A = \begin{bmatrix} 2 & 1 & 5 & 9 & 3 \\ 0 & -3 & -6 & 3 & 2 \\ 0 & 0 & 1 & 0 & 0 \\ 0 & 0 & 4 & a_{44} & 0 \\ 0 & 0 & 5 & 7 & -6 \end{bmatrix} \qquad B = \begin{bmatrix} 1 \\ b_2 \\ b_3 \\ b_4 \\ b_5 \end{bmatrix}$$

$$C = [1 \quad 1 \quad 1 \quad 1 \quad 1] \qquad D = 0$$

with the following assignment to the variable entries

$$b_i = 1 : i = 2, 3, 4, 5 \qquad \text{and} \qquad a_{44} = -2$$

Obtain the poles and the zeros of the transfer function using the command "ss2zp". Are there any pole-zero cancellations indicating that part of the state space is superfluous to the input-output behavior?

Next obtain the controllability matrix using the command "ctrb" and check its rank using the command "rank".

Using the command "eig" find the eigenvalues of A. Then find the state feedback gain matrix K which shifts the unstable eigenvalues to -1. Try doing this with both the command "place" and the command "acker". Check using the command "eig" to see that the eigenvalues of the state feedback system matrix $A + BK$ are those you specified.

Obtain the poles and zeros of the state feedback system using the command "ss2zp". Are there any pole-zero cancellations? What effect does the state feedback have on the transfer function zeros? Do other experiments if need be so that you can give a general reasons for your answers to the foregoing questions.

Obtain the feedback matrix which produces a closed loop transfer function which has no finite zeros (they are all cancelled) and poles which are all in the open left half-plane. Does the rank of the controllability matrix change. Explain your answer by using eigenvectors to identify uncontrollable eigenvalues.

D.2.2 Uncontrollable state models

In this section we are going to examine the use of state feedback when the given state model for the plant is not controllable. One way to obtain an uncontrollable system to enable this study, is to form the input matrix B so that it is linearly dependent on a subset of the eigenvectors of A. We do this as follows.

Invoke "eig" to get the eigenvectors of A. These will be displayed in a square matrix you specify in the argument for "eig". Suppose this matrix is V and D is used as the diagonal matrix displaying the eigenvalues. Create a new input matrix which we call $B1$ by combining a subset of the columns of V, i.e., some of the eigenvectors of A. Suppose we do this by combining all the eigenvectors corresponding to the unstable eigenvalues of A with all the weights used to combine these eigenvectors equal to 1. This can be done using the MATLAB subscripting rules (note the use of :).

Obtain the controllability matrix ("ctrb") for $(A, B1)$ as well as the rank of this controllability matrix. Obtain the transfer function for the state model $(A, B1, C, D)$ and comment, with an explanation, on any pole-zero cancellations. What is the order of the system indicated by the transfer function and does the transfer function indicate that this system is stable?

To gain more insight into the nature of uncontrollable state models, invoke "ctrbf" to transform coordinates so that the state model is in controllable decomposed form, (Section 2.5). In this case we see that the block matrix consisting of the last two rows and columns of A and the last two elements of B in these coordinates has special significance. If (A_{bar}, B_{bar}) are the system and input matrices in the controllable decomposed form then we can extract the subsystem corresponding to these rows and columns by typing

$$AC = A_{bar}(4:5, 4:5)$$

$$BC = B_{bar}(4:5, 1)$$

$$CC = C_{bar}(1, 4:5)$$

Then calculate the transfer function using the command"ss2zp" for the system (AC, BC, CC) and compare it with the transfer function you obtained earlier for the system in the original coordinates. Explain what you see.

Next can you find a state feedback matrix K in these coordinates so that the resulting system is internally stable?

Repeat the foregoing by making up a new B matrix from a linear combination of the eigenvectors corresponding to the stable eigenvalues of A. Notice that the new AC matrix will be a 3×3.

Controllability and Repeated Eigenvalues

Recall that we showed in Appendix A that single-input systems are uncontrollable for any input matrix B when the system matrix A has more than one eigenvector associated with any repeated eigenvalue. To investigate this fact we proceed as follows.

Assume A is as given at the beginning of this lab. We can recall A by typing A. Then use the command "eig" to determine the eigenvalues and eigenvectors of A. Are the eigenvalues distinct? Next change the 4,4 element in A by typing

$$A(4,4) = A(4,4) + 3$$

Find the eigenvectors and eigenvalues of the new A matrix. Are the eigenvalues distinct? Is there a complete set of eigenvectors? Construct an input matrix **B** so that the new (A, B) pair is controllable. Explain.

D.3 Observer Based Control Systems

Recall that an observer is used to provide an estimate of the plant state from measurements of the plant's input and output. In this lab you will examine the design and behavior of observers and the effect of using an observer's estimate of the plant state in place of the actual plant state in a state feedback control system.

In order for you to proceed as rapidly as possible with the tasks described in what follows, read the lab through and plan the subscripting scheme you will use in connection with the various matrices you will need to enter in the computer. This is especially important in connection with the second section where you will need to make a block diagram of the control system with each input and output numbered for use with the commands "blkbuild" and "connect".

Observer Estimates of the Plant State

[To Be Done Prior to Using Computer]

Suppose you are given the unstable second order plant

$$\frac{d^2y(t)}{dt^2} + a_1\frac{dy(t)}{dt} + a_2y(t) = u(t)$$

with $a_1 = 0$ and $a_2 = -1$. Write the observer form state model for the plant and design an observer (Section 3.3)

$$\dot{z}(t) = Fz(t) + G\begin{bmatrix} u(t) \\ y(t) \end{bmatrix}$$

where

$$F = A + LC \qquad G = [\,B + LD \quad -L\,]$$

with F having eigenvalues at -1, -2.

Now you will attempt to verify experimentally using MATLAB that the observer state and plant state approach each other with time. Use the eigenvalues of the observer and plant state model to determine a suitable length of time over which to let the simulation operate.

[To Be Done Using Computer]

Begin by entering the observer form plant state model matrices: A, B, C, D. Then use the command "acker" or "place" to determine the observer matrix **L** so that the observer eigenvalues are at -1,-2. This is done by typing

$$L = acker(A^T, C^T, p)^T$$

where p is a column vector of the desired observer eigenvalue locations. Construct F and G and check the eigenvalues of F using the command "eig". Now we are going to check this observer's ability to track the components of the state of the plant by generating both the plant and observer states when the plant input is a unit step.

Invoke the command "step" to obtain the plant output, y and state x. Try using a time interval of ten seconds in steps of 0.1 s. In order to do this you will need to create a row vector **t** having elements equal to the computation times. This can be done by typing

$$t = 0 : 0.1 : 10$$

You will need **t** in the last section of this lab.

Next use the command "lsim" to obtain the state of the observer when the plant input is a step and y is the plant output in response to a step. This can be done by typing

$$[y1, z1] = lsim(F, G, C_1, DB, UB, t)$$

where

$$UB = [uu, y] \qquad DB = [0; 0]$$

and **uu** is a column vector of 1's of the same length as **y** with **y** being a column vector of samples of the plant output you just computed. The input sequence, uu, could be generated by typing

$$uu = ones(101, 1)$$

Once you have $z1$, the state of the observer, (a matrix having two columns of samples of the components of the observer's state) you can compare the observer state, $z1$, just obtained, with the plant state x by generating the error

$$err = (z1 - x)$$

However recall that since the plant is in observer form, the plant output is the first component of the plant state. Therefore we need only compare the second components of the plant and observer states in order to characterize the performance of the observer as a state estimator. In addition in attempting to comment on this performance, you might find it helpful to examine the relative error, i.e.,

$$z1(:, 2) - x(:, 2)/x(:, 2)$$

For instance if we want to compute this for time 10 we would type

$$[z1(101,2) - x(101,2)]/x(101,2)$$

Now the initial state of the plant in the foregoing generation of the plant output is null. However in practice we do not know the initial state of the plant. Therefore to test the observer's ability to estimate the plant state when the initial state of the plant and observer are not equal, suppose you thought the initial plant state was $x^T(0) = [1, 1]$. Then setting the initial state of the observer equal to your assumption, use the command "lsim" with the observer, i.e., type

$$z0 = [1, 1]$$

$$[y2, z2] = lsim(F, G, C, DB, UB, t, z0)$$

Next obtain the history of the estimation error over the simulation time and obtain a plot of this history. This can be done using element by element division by typing

$$rele = (z2 - x)./x$$

Notice that since the actual initial plant state is null, the elements in the first row of the matrix xs are zero, so that the first listing in "rele" has NaN (not a number) or ∞ as its first entry. Try other guesses for the "unknown" initial plant state, e.g., set the initial observer state to $z0 = [100, 100]$ and obtain a plot of the resulting relative error in the estimate of the plant state.

D.3.1 Observer based controllers

[To Be Done Prior to Using Computer]

Suppose you want to use the observer studied in the previous section to implement a state feedback controller. In this section you will begin to study the interaction which takes place between the plant and the observer when they are connected via a feedback matrix K in a closed loop configuration.

Choose the feedback matrix K so that if exact state feedback were used the closed loop system would have eigenvalues at -3, -4. Do this for the plant state model in observer form which you used in previous sections.

Next referring to Section 5.2, obtain the system equations for the feedback system, i.e., let the plant input u be given as $u = Kz + v$. The state vector q for the closed loop system has four entries, viz.,

$$q = [x^T, z^T]^T$$

where x is the state for the plant (in observer form) and z is the state for the observer you designed in the previous section.

[To Be Done Using Computer]

Use the command "acker" to find the feedback matrix K so that the plant state model (in observer form) has its eigenvalues shifted to -3, -4. Then denote the observer based controller state model as

$$[a_3, b_3, c_3, d_3] = [F, G, K, d_3] \quad d_3 = [0, 0]$$

where F,G, were determined in the previous section and K was just determined.

Now you are going to use the commands "blkbuild" and "connect" to form the observer based feedback control system by interconnecting the various subsystems involved. This is done in two steps. First enter each subsystem, either as transfer functions, $(ni_, di)$ or as state models (ai, bi, ci, di). In each case i should identify a different subsystem and should be the integers starting with one. In the present case there are three subsystems: the input-controller $(i = 1)$, the plant $(i = 2)$ and the observer-controller $(i = 3)$. The input-controller is trivial and is done by entering its transfer function as $n1 = 1$ and $d1 = 1$. Then the plant and observer-controller can be entered in terms of their state models as [a2,b2,c2,d2] and [a3,b3,c3,d3] respectively.

Once you have entered all the subsystems, type

$$nblocks = 3$$
$$blkbuild$$

A message appears "state model $[a, b, c, d]$ of the block diagram has 4 inputs and 3 outputs". You are now ready to proceed to the second step which involves specifying how to interconnect the subsystems you have just identified. This is done through the use of a connection matrix, Q.

The connection matrix, Q, specifies how the subsystems inputs and outputs are to be connected to form the observer based control system. Notice that the outputs $y1$, $y2$, and $y3$ are the outputs from the input-controller, the plant and the observer-controller respectively. The inputs $u1$, $u2$, $u3$ and $u4$ are the input to the input-controller, the plant and the two inputs to the observer. It is important to keep the ordering of the inputs to the observer consistent with the ordering used in the previous section in forming the state differential equation for the observer.

Draw a block diagram showing how all inputs and outputs relate to one another.

The connection matrix, Q, has rows corresponding to subsystem inputs with the integer in the first position of each row identifying the subsystem input. Subsequent entries in a given row are signed integers which identify how the subsystem outputs are to be combined to make up the subsystem input. Zero entries are used when a subsystem output does not connect to a subsystem input. The sign associated with each integer entry in Q indicates the sign on the corresponding output in the linear combination of subsystem outputs making up a given input. Since there are three inputs which depend on the subsystem outputs, Q will have three rows.

Using your block diagram, enter the matrix Q.

Once Q is entered, the inputs and outputs of the feedback control system must be declared by typing

$$inputs = [1]$$

$$outputs = [2]$$

where here the input to the input-controller, $i = 1$, is the observer based feedback control system input and the output from the plant, $i = 2$, is the output from the feedback control system.

At this stage you are ready to use the connection information in the Q matrix to form the observer based feedback control system from the multi-input multi-output system $[a, b, c, d]$ you formed using "blkbuild". The connected system $[ac, bc, cc, dc]$, i.e., the observer based feedback control system, is obtained by typing

$$[ac, bc, cc, dc] = connect(a, b, c, d, Q, inputs, outputs)$$

Check your result with the state model you obtained in the pencil-and-paper work you did before starting this computing session.

Once you have the correct closed-loop state model, calculate the poles and zeros of the transfer function for the closed-loop system by using "ss2zp" (don't use k on the left side if you have already used it for the feedback matrix). Are there any common poles and zeros indicated? If so, why? Compare this transfer function with the transfer function of the closed loop system assuming exact state feedback (without the observer). Discuss your observations by relating them to the theory of observer based control systems.

Finally, check the controllability and observability of the closed loop state model by using the commands"ctrb" and "obsv" and computing ranks by using "rank".

D.3.2 Observer based control system behavior

[To Be Done Prior To Using Computer]

For the plant introduced in the first section, and assuming the states are available for feedback, calculate the output for a step input for the closed loop system with state feedback such that the closed loop system has poles at -3, -4. Do this for the initial condition $x0 = [0, 0]$ and $x0 = [10, 10]$. Make a rough sketch of the output in each case for comparison with the results you will obtain using MATLAB.

[To Be Done Using Computer]

Using "lsim" obtain the state and output of the state model of the observer based control system that you obtained in the previous section as a result of using "blkbuild" followed by "connect", i.e., $[ac, bc, cc, dc]$. Do this over a ten second interval in steps of 0.1 s (use the **t** vector and step input vector, **uu**, which you created for use in the first part of this lab). Obtain a plot of the output when the plant state is $x(0) = [0, 0]$ and $[10, 10]$ when the observer state is $z0 = [0, 0]$ in each case.

D.4 State Model Reduction

In previous labs on state feedback and state estimation we saw that controllability and observability play a key role in being able to (i) assign feedback system poles (ii) assign observer poles. Recall that the transfer function model of a system has order less than the dimension of the state model when the state model is unobservable and/or uncontrollable. Therefore the input–output behavior of a given state model which is "almost" uncontrollable and/or unobservable should be able to be approximated using a state model whose dimension is less than the dimension of the given state model. This possibility was introduced theoretically in Chapter 4. In this lab we are going to do some simulation studies to further investigate the effectiveness of this model order reduction technique.

D.4.1 Decomposition of uncontrollable and/or unobservable systems

Recall that we can always transform the coordinates so that a given state model that is not controllable is transformed to controllable decomposed form

$$A_c = \begin{bmatrix} A_{c1} & \emptyset \\ A_{c3} & A_{c4} \end{bmatrix} \qquad B_c = \begin{bmatrix} \emptyset \\ B_{c2} \end{bmatrix}$$

$$C_c = [\,C_{c1} \quad C_{c2}\,]$$

where (A_{c2}, B_{c2}) is controllable (done using "ctrbf"). Alternatively if the given model is unobservable we can change coordinates so the state model becomes

$$A = \begin{bmatrix} A_{o1} & A_{o2} \\ \emptyset & A_{o4} \end{bmatrix} \qquad B = \begin{bmatrix} B_{o1} \\ B_{o2} \end{bmatrix}$$

$$C = [\,\emptyset \quad C_{o2}\,]$$

where (A_{o4}, C_{o2}) is observable (done using "obsvf")

Therefore when a given state model is both uncontrollable and unobservable we can use the foregoing decompositions to obtain a state model which has a state space which is divided into four subspaces. In the following list of these subspaces C, $\bar{C}$ indicates controllable and uncontrollable, respectively with a similar notation for observable and unobservable. Thus the state space splits up into

1. a subspace CO of dimension n_1
2. a subspace $\bar{\text{C}}$O of dimension n_2
3. a subspace C$\bar{\text{O}}$ of dimension n_3
4. a subspace $\bar{\text{C}}\bar{\text{O}}$ of dimension n_4

Notice that if the dimension of the given state model is n then

$$n = \sum_{i=1}^{4} n_i$$

An example demonstrating the use of MATLAB to do this decomposition can be carried out as follows. Enter the system state model matrices

$$A1 = \begin{bmatrix} -1 & 5 & 7 & 9 \\ 0 & -2 & 0 & 8 \\ 0 & 0 & -3 & 6 \\ 0 & 0 & 0 & -4 \end{bmatrix} \qquad B1 = \begin{bmatrix} 1 \\ 1 \\ 0 \\ 0 \end{bmatrix} l$$

$$C1 = [\,0 \quad 1 \quad 0 \quad 1\,] \qquad D1 = 0$$

and determine the corresponding transfer function by entering

$$[z1, p1, k1] = ss2zp(A1, B1, C1, D1)$$

What is the order of the system? What can be said about the controllability and observability of the state model? We can use the eigenvector tests for controllability and observability to answer this question by interpreting the result of making the following entries

$$[V1, DR1] = eig(A1)$$
$$C1 * V1$$

and

$$[W1, DL1] = eig(A1')$$
$$B1' * W1$$

As a further check on your results, obtain the rank of the controllability and observability matrices using the commands "ctrb", "obsv" and "rank". Use your results to specify basis vectors for each subspace in the decomposition of the state space.

D.4.2 Weak controllability and/or observability

Enter the following system state model matrices

$$A2 = \begin{bmatrix} -1 & 5 & 7 & 9 \\ 0 & -2 & 0 & 8 \\ 0 & 0 & -3 & 6 \\ 0 & 0 & 0 & -4 \end{bmatrix} \qquad B2 = \begin{bmatrix} 1 \\ 1 \\ 0 \\ 1 \end{bmatrix}$$

$$C2 = [\,0 \quad 1 \quad .01 \quad 1\,] \qquad D2 = 0$$

and repeat the previous section for this state model. Contrast the results you obtain now with those you obtained in the previous section.

Next obtain the controllability and observability Gramians, (W_c, W_o), for the system you just entered. Do this by typing

$$Q0 = C2' * C2$$
$$QC = B2 * B2'$$
$$WO = lyap(A2', QO)$$
$$WC = lyap(A2, QC)$$

(Use "help" for information on the command "lyap".) Obtain the eigenvalues of the product of the Gramians. Record these values for future reference.

Next use the command "balreal" to obtain a balanced realization by typing

$$[A2b, B2b, C2b, g2, t2] = \text{bal}real(A2, B2, C2)$$

Determine the eigenvalues of the product of the Gramians you just entered and compare them with the entries in $g2$. Comment on any correspondences.

Next obtain a reduced order model by discarding parts of the balanced realization corresponding to elements of $g2$ that are relatively small. For example suppose $g2(3) << g2(2)$. Then since the elements of g2 are arranged in descending size we can obtain a reduced order model, $(A2r, B2r, C2r)$ by typing

$$A2r = A2b(1:2, 1:2)$$
$$B2r = B2b(1:2)$$
$$C2r = C2b(1:2)$$

Use use the command "ss2zp" to determine the poles and zeros of the transfer function for the reduced order system. Is there any obvious relation between the original system transfer function and the reduced order system transfer function?

D.4.3 Energy interpretation of the controllability and observability Gramians

Recall, from Chapter 4, that the sum of the diagonal entries in the controllability Gramian equals the energy transferred into the state over all positive time, from the input when the input is a unit impulse at time zero. We can check to see if this is the case here.

Recall from Chapter 1 that the zero state response to a unit impulse is the same as the zero input response to an initial state $x(0) = B$. Therefore we can use "initial" with the initial state set to B. As before we need to decide on the time interval and sample times for doing this. Suppose we use an interval of 10 s with samples taken every 0.1 s. Then we create a vector of sample times by typing

$$ts = 0 : 0.1 : 10$$

Notice that ts has 101 elements.

Therefore using the command "initial" with $B2$ in the position reserved for the initial state, we can obtain the resulting state by typing

$$[Y2, X2, ts] = initial(A2, B2, C2, D2, ts)$$

where X2 has four columns with each column corresponding to a different component of the state and each row corresponding to a sample time.

Recall that the energy transferred to the state is given by

$$E_c = \int_0^\infty x^T(t)x(t)\mathrm{d}t = \sum_{i=1}^{4} E_{c_i}$$

where

$$E_{c_i} = \int_0^\infty x_i^2(t)\mathrm{d}t$$

is the energy transferred to the ith component of the state. Using this fact we can compute an approximation, EAc, to E_c by entering

$$eei = X2(1:101,i)' * X2(1:101,i)$$

with $i = 1,2,3,4$ in succession and then enter

$$EAc = \frac{1}{10}\sum_{i=1}^{4} eei$$

Compare your result with the trace of the controllability Gramian by typing

$$ETc = trace(WC)$$

Recall, from Chapter 4, that the energy, E_o, which is transferred to the output from an initial state can be determined using the observability Gramian as

$$E_o = \int_0^{\infty} y^2(t)\mathrm{d}t = x^T(0)W_o x(0)$$

Now we can use the procedure just described for computing E_c to compute E_o, when the initial condition $x(0) = B1$ by using the data stored previously in $Y2$. Compare your result with the result obtained by using the observability Gramian

$$(B2)' * WO * B2$$

D.4.4 Design of reduced order models

In this part of the lab you will examine the balanced realization approach to model order reduction. The system to be reduced is a degree eight Butterworth analog filter. This system has a transfer function with constant numerator and a denominator of degree eight. The poles are located in the open left half-plane at unit distance from the origin. A list of the poles can be obtained by entering

$$[zbw, pbw, kbw] = buttap(8)$$

We can check that the poles are equidistant from the origin by examining the diagonal elements of the matrix obtained by entering

$$pbw * pbw'$$

A balanced realization (state model) for the Butterworth filter is obtained by first calculating a controller form state model and then transforming it to balanced form. Therefore enter

$$[A3, B3, C3, D3] = zp2ss(zbw, pbw, kbw)$$

and

$$[A3b, B3b, Cb3, g3, t3] = \mathrm{balreal}(A3, B3, C3)$$

Recall that the reduced order model is obtained by discarding the latter part of the state vector in balanced coordinates so that the corresponding discarded part of $g3$ is negligible in the sum of the elements in $g3$. Notice that there is a significant decrease in the size of the elements in $g3$ in going from the fifth to the sixth element. Therefore a reduced order model obtained by discarding the last three elements in the state should give a good approximation to the full order system. The state model for the reduced order approximation can therefore be obtained by entering

$$A3b5 = A3b(1:5, 1:5)$$
$$B3b5 = B3b(1:5)$$
$$C3b5 = C3b(1:5)$$

Notice that this reduced order state model approximation is also balance. This can be seen by generating the controllability and observability Gramians by typing

$$QOb5 = C3b5' * C3b5$$
$$QCb5 = B3b5 * B3b5'$$
$$Wob5 = lyap(A3b5', QOb5)$$
$$Wcb5 = lyap(A3b5, QCb5)$$

The final part of this lab concerns the performance of the reduced order model. One possible test would be to compare the step responses of the reduced and full order models.

$$[y3b, x3b] = step(A3b, B3b, C3b, D3, 1, ts)$$
$$[y3b5, x3b5] = step(A3b5, B3b5, C3b5, D3, 1, ts)$$

Note any marked differences in the plots on the screen and proceed to compare the error between the reduced order model output and the full order model output. This can be done using element by element division, i.e.,

$$rele = (y3b5 - y3b)./y3b$$
$$plot(ts, rele)$$

Repeat the foregoing by deleting more than the last three elements of the state to obtain a reduced order model and use the step response to compare the performance of this approximation with the performance of the fifth order approximation just studied.

References

[1] Anderson, B. D. O. and Moore, J. B. *Optimal Control: Linear Quadratic Methods.* Prentice–Hall, Englewood Cliffs, New Jersey, 1990.

[2] Anderson, B. D. O. and Liu, Y. "Controller reduction: Concepts and Approaches", *IEEE Trans. Automatic Control,* **TAC-34**, pp. 802–812, 1989.

[3] Bittanti, S., Laub, A. J. and Willems, J. C. (Eds.) *The Riccati Equation.* Springer–Verlag, Berlin, 1991.

[4] Blackman, P. F. *Introduction to State-variable Analysis.* The Macmillan Press, London, 1977.

[5] Brockett, R. W. *Finite Dimensional Linear Systems,* John Wiley and Sons. New York, 1970.

[6] Brogan, W. L. *Modern Control Theory.* 3rd ed., Prentice–Hall, Englewood Cliffs, N. J., 1991.

[7] Brown, R. G. *Introduction to Random Signal Analysis and Kalman Filtering.* John Wiley and Sons, Chichester, 1983.

[8] Chandrasekharan, P. C. *Robust Control of Dynamic Systems.* Academic Press, London, 1996.

[9] Chen, C. T. *Introduction to Linear Systems Theor.* Holt Rinhart and Winston, New York, 1970.

[10] Dorato, P., Abdallah, C. and Cerone, V. *Linear Quadratic Control, An Introduction.* Prentice–Hall, Englewood Cliffs, N. J., 1995.

[11] Doyle, J. C., Francis, B. A. and Tannenbaum, A. R. *Feedback Control Theory.* Maxwell Macmillan Canada, Toronto, 1992.

[12] Doyle, J. C., Glover, K., Khargonekar, P. P. and Francis, B. A. "State space solutions to standard H_2 and H_∞ control problems", *IEEE Trans. Automatic Control,* **AC-34**, pp. 831–847, 1989.

[13] Fairman, F. W., Danylchuk, G. J., Louie, J. and Zarowski, C. J. "A state-space approach to discrete-time spectral factorization", *IEEE Transactions on Circuits and Systems-II Analog and Digital Signal Processing,* **CAS-39**, pp. 161–170, 1992.

[14] Francis, B. A. *A Course in H_∞ Control Theory,* Springer–Verlag, Berlin, 1987.

[15] Furuta, K., Sano, S. and Atherton, D. *State Variable Methods in Automatic Control,* John Wiley and Sons, Chichester, 1988.

[16] Glover, K. and Doyle, J. C. "A state space approach to H_∞ optimal control", *Lecture Notes in Control and Information Science,* **135**, pp. 179–218, 1989.

[17] Golub, G. H. and Van Loan, C. F. *Matrix Computations,* 2nd ed., The Johns Hopkins University Press, Baltimore, Maryland, 1989.

[18] Green, M. and Limebeer, D. J. N. *Linear Robust Control,* Prentice–Hall, Englewood Cliffs, New Jersey, 1995.

[19] Green, M., Glover, K., Limebeer, D. J. N. and Doyle, J. "A J spectral factorization approach to H_∞ control", *SIAM Journal on Control and Optimization,* **28**, pp. 1350-1371, 1990.

[20] Green, M. "H_∞ controller synthesis by J lossless coprime factorization", *SIAM Journal on Control and Optimization*, **30**, pp. 522–547, 1992.

[21] Isidori, A. "H_∞ control via measurement feedback for affine nonlinear systems", *Int. J. Robust and Nonlinear Control,* **4**, pp. 553–574, 1994.

[22] Jonckheere, E. A. and Silverman, L. M. "A new set of invariants for linear systems-applications to reduced order compensator design", *IEEE Trans. Automatic Control*, **AC-28**, pp. 953–964, 1983.

[23] Kailath, T. *Linear Systems*, Prentice–Hall, Englewood Cliffs, N. J., 1980.

[24] Kimura, H. *Chain-Scattering Approach to H_∞ Control*, Birkhausser Boston, 1997.

[25] Kwakernaak, H. and Sivan, R. *Linear Optimal Control Systems*, Wiley Interscience, New York, N. Y., 1972.

[26] LePage, W. R. *Complex Variables and the Laplace Transform for Engineers*, Dover, 1980.

[27] Moore, B. C. "Principal component analysis in linear systems: controllability, observability, and model reduction", *IEEE Trans. Automatic Control*, **AC-26**, pp. 17–32, 1981.

[28] Missaghie, M. M. and Fairman, F. W. "Sensitivity reducing observers for optimal feedback control", *IEEE Transactions on Automatic Control*, **AC-22**, pp. 952–957, 1977.

[29] Mullis, C. T. and Roberts, R. A. "Synthesis of minimum roundoff noise in fixed point digital filters", *IEEE Trans. Circuits and Systems*, **CAS-23**, 551–562, 1976.

[30] Mustafa, D. and Glover, K. "Controller reduction by H_∞ balanced truncation", *IEEE Trans. Automatic Control*, **AC-36**, pp. 668–682, 1991.

[31] Pavel, L. and Fairman, F. W. "Robust stabilization of nonlinear plants—an L_2 approach", *Int. J. Robust and Nonlinear Control*, **6**, pp. 691–726, 1996.

[32] Pavel, L. and Fairman, F. W. "Controller reduction for nonlinear plants—an L_2 approach", *Int. J. Robust and Nonlinear Control*, **7**, pp. 475–505, 1997.

[33] Pavel, L. and Fairman, F. W. (1998). "Nonlinear H_∞ control: a J-dissipative approach", *IEEE Trans. Automatic Control*, **AC-42**, pp. 1636–1653, 1997.

[34] Pernebo, L. and Silverman, L. M. "Model reduction via balanced state space representation", *IEEE Trans. Automatic Control*, **AC-27**, pp. 382–387, 1982.

[35] Petersen, I. R. "Disturbance attenuation and H_∞ optimization: a design method based on the algebraic Riccati equation", *IEEE Trans, Automatic Control*, **AC-32**, pp. 427–429, 1987.

[36] Rudin, W. *Principles of Mathematical Analysis*, 3rd ed., McGraw-Hill Book Company, New York, 1976.

[37] Rudin, W. *Real and Complex Analysis*, 3rd ed., McGraw–Hill Book Company, New York, 1987.

[38] Scherpen, J. M. A. "H_∞ balancing for nonlinear systems", *Int. J. Robust and Nonlinear Control*, **6**, pp. 645-668, 1996.

[39] Strang, G. *Linear Algebra and its Applications*, 2nd ed., Academic Press, New York, 1980.

[40] Sveinsson, J. R. and Fairman, F. W. "Simplifying basic H_∞ problems", *Proc. Conference on Decision and Control*, **33**, 2257–2258, 1994.

[41] Tsai, M. C. and Postlethwaite, I. "On J lossless coprime factorization and H_∞ control", *Int. J. Robust and Nonlinear Control*, **1**, pp. 47–68, 1991.

[42] van der Shaft, A. J. "L_2 gain analysis of nonlinear systems and nonlinear state feedback H_∞ control", *IEEE Trans. Automatic Control*, **AC-37**, pp. 770–784, 1992.

[43] Varga, A. "A multishift Hessenberg method for pole assignment of single-input systems", *IEEE Trans. Automatic. Control*, **AC-41**, pp. 1795–1799, 1996.

[44] Vidyasagar, M. *Control System Synthesis: A Factorization Approach*, The MIT Press, Cambridge, Massachusetts, 1985.

[45] Vidyasagar, M. *Nonlinear Systems Analysis*, 2nd ed., Prentice–Hall International, Engelwood Cliffs, 1993.

[46] Young, N. *An Introduction To Hilbert Space*, Cambridge University Press, Cambridge, 1988.

[47] Zhou, K. with Doyle, J. C. and Glover, K. *Robust and Optimal Control*, Prentice–Hall, Upper Saddle River, New Jersey, 1996.

Index

Printed in the United States
54024LVS00001B/202-216